The Inorganic Heterocyclic Chemistry of Sulfur, Nitrogen and Phosphorus

The Inorganic Heterocyclic Chemistry of Sulfur, Nitrogen and Phosphorus

HENRY G. HEAL
The Queen's University of Belfast, Northern Ireland

1980

ACADEMIC PRESS

A Subsidiary of Harcourt Brace Jovanovich, Publishers

London • New York • Toronto • Sydney • San Francisco

ACADEMIC PRESS INC. (LONDON) LTD
24-28 Oval Road,
London NW1

United States Edition published by
ACADEMIC PRESS INC.
111 Fifth Avenue
New York, New York 10003

Copyright © 1980 by
ACADEMIC PRESS INC. (LONDON) LTD.

All Rights Reserved
No part of this book may be reproduced in any form by photostat, microfilm, or any other means, without written permission from the publishers

British Library Cataloguing in Publication Data

Heal, Henry G
   The inorganic heterocyclic chemistry of sulfur, nitrogen and phosphorus.
   1. Sulfur compounds   2. Nitrogen compounds
   3. Phosphorus compounds   4. Heterocyclic compounds
   I. Title.
   546'.7   QD181.S1   80-40906
   ISBN 0-12-33580-6

Printed in Great Britain by Galliard (Printers) Ltd, Great Yarmouth

Preface

Some excellent works on inorganic heterocyclics have appeared in recent years, and I must first make my excuses for writing another book on this branch of chemistry. It happened in the following way.

During the last decade, knowledge of the inorganic heterocyclics has grown so fast that the only fully comprehensive work on them is now seriously out of date; the production of an updated version would be a large undertaking and is not expected in the near future. At the same time, it has become clear that the concept of inorganic heterocyclics, embracing as it does such diverse substances as mineral silicates, carboranes, phosphazenes, and chelate complexes of metals, is, in its fullest sense, too broad to be really useful. The subject simply must be broken down into more manageable chunks, and so it usually is. Some of these, such as the borane derivatives and the silicones, are familiar to many chemists and often figure in reviews or new books. Others, such as the covalent heterocycles of arsenic or selenium, interest few people and account for only a small volume of publication. Still others are more commonly and more appropriately dealt with in other contexts, such as coordination chemistry or mineralogy. However, when all these topics have been eliminated, there remains a solid core of material which lies right outside the experience and training of most industrial and academic chemists, and which many of them now feel a need to know about. This material comprises mainly the covalent ring compounds based on various combinations of the elements nitrogen, sulfur, and phosphorus, and the related non-cyclic chemistry. No previous book has covered precisely this range of topics, and the treatments of parts of the field to be found in previous books and reviews are mostly not recent enough to include some very important and interesting new work. A compact, comprehensive and up-to-date survey is clearly needed, and this book is an attempt to provide it. The impulse to write came from two sources: from the publishers, and from my friend Dr. Henri Garcia-Fernandez who unfortunately had to withdraw from collaboration at an early stage. I am particularly grateful to Academic Press and their referees for their advice on the scope of the book, which was not easily decided upon.

The book is intended as a reference text rather than a monograph. The *Chemical Abstracts* have been scanned up to early 1979, and in addition a few unpublished results, furnished through the kindness of individuals, are mentioned. In order to keep down size and cost, I have not included all literature references but have tried to give enough of the more recent ones so that the user will quickly be led, through them and their bibliographies, to all available information on the subject. Where good reviews exist, or good review sections of research papers, these have been freely cited instead of the primary publications. I apologize to the authors of the latter for my use of this space-saving device, which most certainly was not intended as a slight but may occasionally look like one. Facts are given pretty fully when only otherwise available through a long literature search; where, however, large numbers of facts have been well compiled elsewhere, I have made a selection to illustrate classes of compounds or points of principle, and given references to the compilations.

Although myself engaged in research on sulfur–nitrogen heterocyclics, I am conscious of the small importance of my contributions compared with those of some other workers frequently named in the bibliographies, and I would not want to equate my judgment with theirs. Nevertheless, a mere recital of facts and of the uncoordinated opinions of others makes dull reading, so I have often taken the liberty of giving my own views and suggestions for future work. Some of these will no doubt seem naive to specialists in the particular fields. Inevitably, too, there will be mistakes and omissions, for which I apologize in advance. My hope for the book is simply that it will be found useful.

Writing a book strengthens one's appreciation of other books, and I want to record my debt to several which have made the task easier. These include Haiduc's *Chemistry of Inorganic Ring Systems*, Garcia-Fernandez's *Heterocycles en Chimie Minerale* and *Química Heterocíclica Inorgánica*, Armitage's *Inorganic Rings and Cages*, and Allcock's *Heteroatom Ring Systems and Polymers*. Special mention must be made of Goehring's *Ergebnisse und Probleme der Chemie der Schwefelstickstoffverbindungen*. This, published in 1957, was a pioneering work in the field, and it introduced me to a subject on which I was to research for many years. Though dated now, it still makes interesting reading, and some of the "problems" discussed in it are still worth looking at. It was continually consulted in our laboratory, where it came to be known as "the Black Book". Well thumbed and with the binding falling apart, it still stands on my shelf.

It is a pleasure to make the following acknowledgments: to my employers, the Queen's University of Belfast, for time and facilities; to the University of Southampton (where the book was begun on a study leave) for hospitality and facilities; to my friend Dr. Arthur Banister for criticism of the manuscript; to the researchers who have sent me results in advance of publication and extended hospitality on visits, including Dr. Banister, Dr. H. Garcia-Fernandez, Prof. Dr. O. Glemser, Prof. Dr. H. W. Roesky, and the late Dr. F. P. Olsen; to Mr. J. A. Beckman of the Firestone Tire and Rubber Company for technical literature; and to Academic Press for patience and encouragement.

The person who deserves the heartiest thanks of all is my wife Joan, who spent many, many hours at the typewriter, coping with my exasperating handwriting and continual amendments, and striving to bring consistency into my disorderly typography.

Henry G. Heal

May 1980

The Organization of this Book

The following notes are intended to help the reader to find particular compounds, ring systems, or subject areas in this book.

The book deals *only* with purely inorganic rings and cages, that is, containing no carbon atoms in the ring or cage skeleton. However, inorganic rings and cages with organic *ligands* are covered.

For a user not familiar with the book, the quickest way to find a particular ring system or cage is probably to look it up in the Index of Rings at the back.

For more frequent users, here is a brief explanation of how the subject matter is arranged. Sulfur–nitrogen rings come first (Chapters 2–9), followed by phosphorus–sulfur rings (Chapter 10) and then by phosphorus–nitrogen rings (Chapters 11–13). Rings containing all three elements are dealt with in appropriate chapters on *sulfur*–nitrogen rings, not under phosphorus–nitrogen rings. (S–N–P rings are too diverse to be logically treated in a chapter by themselves; the arrangement just described is at least consistent, though not ideal in all respects.) Within each broad category of rings (S–N, P–S, or P–N) saturated rings precede unsaturated ones, and among saturated S–N, P–S, or P–N rings the lower oxidation states of S or P precede the higher.

Contents

Preface

Chapter 1. Status, nomenclature, and methods of synthesis
  I. Introduction, 1
    A. Inorganic heterocycles: a modern perspective, 1
    B. The importance of inorganic heterocyclics, 3
  II. Nomenclature, 4
  III. Methods of synthesis, 8
    A. Introduction, 8
    B. Direct combination of the ring-forming elements, 9
    C. Addition polymerization, 9
    D. Heterofunctional condensations, 9
    E. Homofunctional condensations, 12
    F. Insertions, ring expansions, and ring contractions, 13
    G. Miscellaneous thermodynamically controlled processes, 14
    H. Controlled breakdown of pre-existing structures, 14
    J. Alteration of oxidation state of phosphorus or sulfur in existing rings, 14
References, 15

Chapter 2. Cyclic sulfur imides with two-coordinate sulfur, and their derivatives
  I. Introduction, scope, and history, 16
  II. Linear amides of divalent sulfur, 18
  III. Sulfur imides based on the S_8 ring, and their derivatives, 18
    A. Introduction, 18
    B. Monocyclic sulfur imides structurally related to cyclooctasulfur, 18
    C. Sulfur nitrides with coupled eight-membered, rings, 29
    D. The fused-ring sulfur nitride $S_{11}N_2$, 31
    E. The search for other fused-ring sulfur nitrides, 33
    F. Bis(heptasulfurimido) sulfoxide, $S_{15}N_2O$, 33
    G. Polymeric sulfur nitride-imides: general, 34
    H. The conformational stability of saturated S–N rings, 35

IV. Sulfur imides based on rings other than eight-membered, 36
   A. Introduction, 36
   B. Derivatives of six-membered ring imides, 36
   C. Derivatives of twelve-membered ring imides, 37
V. Cyclic sulfur hydrazides, 37
   A. Eight-membered rings, 37
   B. Six-membered rings, 38
References, 38

Chapter 3. Imides and amides of sulfur(IV) as source materials for inorganic heterocycles
I. Introduction, 41
II. Structural relationships, 41
III. Thionyl imide and its derivatives, 43
   A. Thionyl imide, 43
   B. Thionyl imide polymers and isomer, 44
   C. Organic thionylimines, 46
   D. Trimethylsilyl sulfinylamine, 46
   E. Halogeno-thionylimines and bis(thionylimino)mercury, 47
   F. Bis(thionylimino)sulfur, 48
   G. Other thionyl imide substitution products with –NSO attached to sulfur, 48
   H. The NSO^- ion, 49
IV. Sulfur diimide and the sulfodiimides, 49
   A. Sulfur diimide, 49
   B. Organo-sulfodiimides, 49
   C. Bis(trimethylsilyl) sulfur diimide and monosilylated analogs, 50
   D. Bis(trimethylstannyl) sulfur diimide, 52
   E. Bis(trifluoromethylsulfenyl) sulfur diimide, 52
   F. Halogeno-sulfodiimides, 52
V. Amidosulfurous acid and its derivatives, 52
   A. Introduction, 52
   B. Amidosulfurous acid and the ammonia–sulfur dioxide reaction, 53
   C. Imidodisulfurous acid and its salts, 54
   D. Organic derivatives of amidosulfurous acid, 55
   E. Amidosulfinyl halides, 55
VI. Thionyl diamide and its derivatives, 56
   A. Thionyl diamide, 56
   B. Imidodisulfinamide, 56
   C. Organic derivatives of thionyl diamide, 57
VII. Alkyl and fluoroalkyl sulfimides as reagents for ring synthesis, 57
   A. Introduction, 57
   B. The sulfur imide $Me_2S=NSiMe_3$, 57
   C. Bis(trifluoromethyl)sulfimide $(CF_3)_2S=NH$, 57
VIII. Mixed P–S rings containing sulfur(IV), 58
References, 58

Chapter 4. Imides and amides of sulfur(VI) as source materials for inorganic heterocycles
I. Introduction and structural relationships, 60

II. Sulfamide and its derivatives: monomeric sulfimide, 62
   A. Introduction, 62
   B. Sulfamide, 63
   C. Metal derivatives of sulfamide, 64
   D. N-Halogenosulfamides, 65
   E. Organic derivatives of sulfamide and of monomeric sulfimide, 65
III. Cyclic and linear polymers of sulfimide, 67
   A. Introduction, 67
   B. The preparation of cyclic and linear sulfimide polymers, 67
   C. Properties and derivatives of the cyclic sulfimides, 70
   D. Properties of the linear polymeric sulfimides, 71
   E. Four-membered sulfimide rings, 74
IV. Mixed six-membered rings containing the sulfimide group, 74
   A. Introduction, 74
   B. The compound $N_3S_3O_4H$, 74
   C. Preparation and properties of $N_3P_2SCl_4O_2Me$ isomers, 75
   D. The cyclic anion $O_2S[NP(NH_2)_2]_2N^-$, 75
V. Mixed four-membered rings containing the sulfimide group, 76
   A. Introduction, 76
   B. The one-ring compound $(MeN)_2SO_2PF_3$, 76
   C. Spiro compounds, 76
VI. Further imido derivatives of sulfur(VI), used or potentially usable in heterocyclic synthesis, 77
   A. The S,S-dialkylsulfodiimides, 77
   B. The mercurial $Hg[NS(O)F_2]_2$, 79
References, 79

Chapter 5. Sulfanuric halides and related compounds

I. Introduction: structural relationships, 81
II. The sulfanuric halides, 83
   A. Preparation of the sulfanuric chlorides, 83
   B. Physical properties and structure of sulfanuric chlorides, 84
   C. The cyclic trimeric sulfanuric fluorides, 85
   D. The cyclic sulfanuric fluoride tetramer, 86
   E. High polymeric sulfanuric fluorides, 86
   F. Mixed sulfanuric chlorofluorides, 87
   G. Reactions and derivatives of the sulfanuric halides, 87
   H. The sulfanuric anion $(N_3S_3O_4F_2)^-$, 89
   J. Isomerism of sulfanuric derivatives, 89
III. Mixed sulfanuric rings wth other sulfur groups, 92.
   A. Introduction, 92
   B. The anion $N_3S_3O_3F_2^-$ and compounds $N_3S_3O_2XF_2$ (X = F, Cl), 92
   C. Preparation and properties of $[NS(O)Cl][NSCl]_2$ and $[NS(O)Cl]_2[NSCl]$, 93
   D. Preparation and properties of $[NS(O)Cl][NSO_2][NS]$, 93
IV. Sulfanuric rings containing a saturated segment, 94
V. Mixed sulfanuric–phosphazene rings, 94
   A. Introduction, 94
   B. Preparation and properties of $[NS(O)Cl][NPCl_2]_2$ and $[NS(O)Cl]_2[NPCl_2]$, 94

C. Preparation and properties of [NS(O)F]$_2$[NPCl$_2$], 95
D. Reactions and derivatives of the mixed sulfanuric–phosphazene halides, 95
E. The eight-membered ring [NPCl$_2$]$_3$[NS(O)Cl], 96

References, 96

Chapter 6. Formally unsaturated sulfur nitrides and sulfur nitride ions
I. Introduction and plan of treatment, 98
II. Thiazyl polymers, (SN)$_n$, and thiazyl monomer, 99
   A. Tetrasulfur tetranitride, 99
   B. Disulfur dinitride, S$_2$N$_2$, 111
   C. Thiazyl polymer, (SN)$_x$, 112
   D. Thiazyl monomer, SN, 114
III. Other unsaturated sulfur–nitrogen ring molecules and ions, 114
   A. Introduction, 114
   B. Tetrasulfur dinitride, S$_4$N$_2$, 115
   C. The thiotrithiazyl ion, S$_4$N$_3{}^+$, and its salts, 117
   D. The cyclopentathiazenium cation, S$_5$N$_5{}^+$, and salts, 120
   E. The S$_6$N$_4{}^{2+}$ cation and salts, 121
   F. The cations S$_3$N$_2{}^+$ and S$_4$N$_4{}^{2+}$, 122
   G. The cation S$_3$N$_2{}^{2+}$, 123
   H. The anion S$_3$N$_3{}^-$, 123
IV. Sulfur–nitrogen cages based on tetrasulfur tetranitride, 123
   A. Introduction, 123
   B. The anion S$_4$N$_5{}^-$, 124
   C. The cation S$_4$N$_5{}^+$, 124
   D. The sulfur nitride S$_5$N$_6$, 125
V. Unsaturated non-cyclic ions, 125
   A. The thiazenium ion, SN$^+$, 125
   B. The perthionitrate anion, NS$_4{}^-$, 125

References, 126

Chapter 7. Unsaturated cyclic sulfur nitride S-halides and their S-derivatives
I. Introduction and history, 130
II. Thiazyl halides (NSX)$_n$, (NS)$_n$X$_2$, and (NS)$_3$X, and their substitution products, 131
   A. Thiazyl halides: general, 131
   B. Cyclo-(NSF)$_4$ and cyclo-(NSCl)$_4$; cyclo-(NS)$_4$F$_2$ and cyclo-(NS)$_4$Cl$_2$, 131
   C. Cyclo-(NSF)$_3$ and cyclo-(NSCl)$_3$, 133
   D. The compound (NS)$_3$Br$_2$, 136
   E. The compounds (NS)$_3$X (X = Cl, Br, I), 136
   F. Monomeric thiazyl halides, 136
   G. Substitution products of the cyclic thiazyl halide oligomers, 137
III. The sulfur nitride halides S$_3$N$_2$X$_2$, 139
   A. Introduction and history, 139
   B. Preparation and properties of [S$_3$N$_3$Cl]$^+$Cl$^-$, 139
   C. Other compounds S$_3$N$_2$X$_2$ (X = halogen) of unknown structure, 140
IV. Covalent compounds containing the five-membered S$_3$N$_2$ ring, 141
   A. Introduction, 141
   B. Preparation of S$_3$N$_2$ derivatives, 141

C. Properties of S_3N_2 derivatives, 142
D. Structures of S_3N_2 derivatives, 143
V. Sulfur–nitrogen halides of unknown structure, 143
References, 143

Chapter 8. Unsaturated cyclic sulfur nitride S-oxides and S-oxide ions
I. Introduction and history, 145
II. Compounds with eight-membered S–N rings, 146
  A. $S_4N_4O_4$, 146
  B. $S_4N_4O_2$, 146
III. Compounds with six-membered S–N rings, 148
  A. The ion $S_3N_3O_4^-$, 148
  B. The compound $S_3N_3O_2Cl$, 148
IV. Compounds with five-membered S–N rings: the oxide S_3N_2O, 148
V. A sulfur nitride oxide ion based on the S_4N_4 cage: $S_4N_5O^-$, 150
VI. Sulfur nitride oxides of unknown structure, 150
References, 150

Chapter 9. Bonding and electron-counting in S–N heterocycles
I. Introduction, 152
II. S–N bond parameters in general, 153
  A. Length–force constant correlation, 153
  B. Length–order correlation, 153
  C. Energy–order correlation, 154
III. Bond angles and lengths in formally saturated S–N rings, 156
  A. Configurations at nitrogen, 156
  B. Configurations at sulfur, 157
IV. Bond lengths and angles in formally unsaturated S–N rings, 157
  A. Configurations at nitrogen, 157
  B. Configurations at sulfur, 159
V. Synoptic treatments of bonding in unsaturated S–N heterocycles, 159
  A. Introduction, 159
  B. Aromaticity in near-planar rings, 159
  C. The cage or cluster viewpoint, 161
  D. Transannular bonding in general, 164
  E. Conclusion, 164
References, 165

Chapter 10. Phosphorus-sulfur rings and cages
I. Introduction and history, 166
II. Small-molecule phosphorus sulfides and derivatives, 167
  A. Preparation, characterization, and structure, 167
  B. Reactions, 172
  C. Uses of the phosphorus sulfides, 175
  D. The oxosulfide $P_4S_6O_4$, 175
  E. Cage and ring compounds from halogenation of phosphorus sulfides, 176
III. Organo-substituted P–S rings, 178
  A. Six-membered P_3S_3 rings, 178
  B. Five-membered P_3S_2 rings, 178

C. Five-membered P_4S rings, 179
D. Four-membered P_2S_2 rings, 180
IV. Phosphorus sulfide high polymers, 181
References, 182

Chapter 11. Saturated phosphorus–nitrogen heterocycles: the cyclo- and closo-phosphazanes
I. Introduction, 184
  A. Scope, status, and treatment, 184
  B. Conditions for stability of dimeric cyclophosphazanes, 186
II. Saturated phosphorus(III)–nitrogen heterocycles, 187
  A. Introduction, 187
  B. Reactions of phosphorus trihalides with amines and ammonia, 188
  C. Cyclodiphosph(III)azanes: preparation, structure, and properties, 189
  D. Cyclotri- and cyclotetra-phosph(III)azanes, 192
  E. Cage phosph(III)azanes: preparation structure and properties, 192
  F. Heterocyclic hydrazido derivatives of phosphorus(III), 195
  G. Phosph(III)azane polymers, 196
III. Saturated phosphorus(V)–nitrogen heterocycles, 197
  A. Introduction, 197
  B. Reactions of phosphorus(V) halides with amines and ammonia, 197
  C. Cyclodiphosph(V)azanes: preparation, structure, and properties, 199
  D. Cyclophosph(V)azanes: trimers and higher oligomers, 205
  E. Spirophosph(V)azanes, 207
  F. Cage phosph(V)azanes, 208
  G. Heterocyclic hydrazido derivatives of phosphorus(V), 209
  H. Homologues of the cyclodiphosph(V)azanes with sulfonyl groups as ring components, 210
IV. The cage phosphazane-phosphazene anion $P_{12}N_{14}S_{12}^{6-}$, 210
References, 211

Chapter 12. The phosphazenes
I. General introduction, 214
II. Literature, 215
III. History, 216
IV. General properties of the cyclic phosphazenes, 217
V. Primary preparations of the cyclic phosphazenes, 217
  A. Introduction, 217
  B. Syntheses from ammonium halides and phosphorus(V) compounds, 218
  C. Synthesis from aminochlorophosphoranes, 221
  D. Oxidative phosphazene syntheses from phosphorus(III) halides, 221
  E. Other primary phosphazene syntheses, 221
VI. Phosphazene polymers other than simple rings, 221
  A. Phosphazene high polymers, 221
  B. Linear phosphazene oligomers, 223
  C. Fused-ring, spiro, and coupled-ring phosphazene compounds, 224
VII. Reactions and derivatives of the cyclic phosphazenes, 226
  A. Isomerism among substitution products of the cyclophosphazenes, 226
  B. Complexes with Lewis bases, 227

C. Nucleophilic substitution reactions of the phosphazenes: introduction, 228
D. Halogen and pseudohalogen exchange, 228
E. Nucleophilic substitution by –OH, –OR, –SR, 229
F. Nucleophilic substitution by amino groups, 231
G. Nucleophilic substitution by carbanions, 232
H. Addition of phosphazenes to Lewis acids, including the proton, 234
J. Catalytic arylation of phosphazenes, 237
K. Reduction of phosphazenes, 237
VIII. Molecular structure and its rationalization, 237
IX. Electronic structure and the question of aromaticity, 242
References, 246

Chapter 13. Polymeric phosphorus nitrides and related compounds
I. Introduction, 250
II. History, 251
III. The phosphorus nitrides and phospham, 251
A. The gas-phase molecule PN, 251
B. Amorphous phosphorus nitrides, including "phosphorus mononitride", $(PN)_x$, 252
C. Triphosphorus pentanitride, 253
D. Phosphams, $(PN_2H)_x$, 254
IV. Polymeric phosphorus oxonitride and thionitride, 255
A. Phosphorus oxonitride, $(OPN)_x$, 255
B. Phosphorus thionitride, $(SPN)_x$, 256
V. Desirable lines of work in this area, 256
References, 257

Appendix, 259
Index, 260

1
Status, nomenclature, and methods of synthesis

I. Introduction

A. Inorganic heterocycles: a modern perspective

Many knowledgeable and experienced chemists are hardly aware of the existence of a subject "inorganic heterocycles". Texts on it rarely meet the eye, even on the shelves of large libraries, and there must be few undergraduate university courses in which it plays more than an insignificant part. To some chemists with well ordered, conventional schemes of ideas, it seems awkward to classify and embarrassing to think about; like fluorocarbon chemistry, it does not fall neatly into any recognized major branch of chemistry, and so tends to be ignored by the practitioners of all. This came home to me in 1965, when my contribution at the Prague IUPAC Polymer Symposium on some derivatives of cyclic sulfur imides was classified in the organic section, though none of the compounds mentioned in it contained any carbon! Inorganic heterocycles have made less of a theoretical splash than borane chemistry or coordination chemistry, and they have not, until recently, formed the subject of any research fashion or bandwagon based upon possible commercial or military applications. Yet, scarcely noticed by the chemical world at large, they have grown into a substantial branch of knowledge with thousands of publications, which now clamors for attention, and which more and more chemists feel a need to learn about.

The first inorganic heterocycle, $(NPCl_2)_3$, was described by Liebig and Wöhler in 1834, and the second, S_4N_4, in 1835 by Gregory; but decades were to pass before Butlerov (1860) first recognized the importance of molecular structures in chemistry, and it was over a century before the structures of these two compounds were actually determined. Inorganic heterocyclic chemistry developed slowly during the nineteenth century. Notable periods were the 1880s (Demarçay's work on sulfur–nitrogen compounds) and the 1890s (Stokes's work on the phosphazenes). Kekulé's theory of the cyclic structure of benzene (1865) must have set inorganic researchers thinking about rings;

ring formulas were suggested, for example, by Stokes for the phosphazenes (1896) and by Hantzsch for sulfimide trimer (1901). Schenck, writing in Abegg's *Handbuch der anorganischen Chemie* (1907), thought that Stokes's ring formulas had "no claim to special probability". Though consistent with elemental compositions and molecular weights, they could not at that date be so reliably established as that of benzene, which was supported by the chemical evidence of isomeric substitution products. However, by 1930 the ring formulas of Stokes and Hantzsch were generally accepted. The early decades of the present century otherwise saw little progress on inorganic heterocycles; this was a period dominated by complexes and the Werner coordination theory. Our subject really came into its own shortly before the Second World War, when X-ray and electron diffraction began to be widely applied to structure determination. As more and more molecular structures were solved over the following decades, it became vastly easier to interpret reactions and plan syntheses in the field of inorganic heterocycles. After the war, too, there were great improvements in laboratory methods which revolutionized the practical study of these often difficult substances: inert-atmosphere and high-vacuum techniques became standard, recording infrared and n.m.r. spectrometers appeared, and chromatography was greatly improved. New reagents such as metal alkyls became readily available; indeed, specialist firms such as Alfa Inorganics began to supply a wide range of useful organometallics and other research intermediates. Thus new reactions can be tried out after much less preliminary work than formerly, and their products characterized with a speed, completeness, and precision not dreamed of a few years ago. The pace of discovery has consequently quickened, though most branches of inorganic heterocyclic chemistry still demand special skills and experience which are confined to a few laboratories.

The first book to deal largely with inorganic heterocycles seems to have been Goehring's little classic *Ergebnisse und Probleme der Chemie der Schwefelstickstoffverbindungen* (1957), still interesting reading. Of the several texts that have appeared subsequently, it seems right to mention specially Haiduc's comprehensive two-volume *Chemistry of Inorganic Ring Systems* (1970). Since 1970 the subject has advanced so fast that Haiduc's intended second edition was by 1978 proving difficult as a publishing venture. Meanwhile, the First International Conference on Inorganic Ring Systems was held at Besançon in 1975, and the second, in which 88 papers were presented, at Göttingen in 1978. Future international conferences are expected at least triennially.

Research on inorganic heterocycles is going on in many countries, including most of the larger industrially advanced ones, and key contributions have come from many places. As shown by the chapter bibliographies in this book, and by the lists of participants at conferences, German chemists are playing a major part in the development of the subject.

B. The importance of inorganic heterocyclics

Industrially the cyclosiloxanes have been important since the 1940s as intermediates in the manufacture of silicone products, and the cage compound tetraphosphorus trisulfide has been used in the manufacture of matches since 1898. For several decades there has been a continuous trickle of patents connected with various inorganic heterocyclics, but an actual new application has only very recently arisen, with the marketing of a phosphazene elastomer by the Firestone Tire and Rubber Company. Very recently, too, there has been intense technical interest in the metallic polymer $(SN)_x$. Both of these unusual substances are made from heterocyclic precursors. It remains true, however, that interest in inorganic heterocycles still centers largely on their fundamental chemistry.

Why are chemists attracted to the compounds discussed in this book as subjects for fundamental research? Everyone who works in the field has his own reasons for liking it, but it does seem reasonable to single out the following explanations for its appeal.

One of the major groups of compounds to be described here is the chlorophosphazenes and their derivatives. These constitute (along with the cyclosiloxanes) the best examples known of inorganic homologous series, with many members (in principle an unlimited number) capable of being isolated and individually studied. This is the most organic-like department of inorganic heterocyclic chemistry. The P–N ring skeletons are very stable and persist through many substitution reactions, like carbon skeletons. The course of these reactions can be interpreted by reasoning broadly similar to that used by organic chemists, but the problems are more complicated and more challenging than those of carbon or siloxane chemistry because of the more flexible stereochemistry and bonding possibilities of the phosphorus atom. There is an intriguing suspicion of aromaticity in these P–N rings. In total contrast are the S–N rings. Some of these are unsaturated in a formal sense, but their chemistry has little in common with that of the carbon–carbon double bond. They do not form homologous series, and the reasons why some unsaturated S–N rings exist, and others (apparently) do not, are decidedly mysterious and tantalizing to the theoretician. Compared with the cyclophosphazenes, these S–N rings are quite labile and can be transformed into each other under mild conditions in unexpected and puzzling reactions. Indeed, the reaction chemistry of these substances deserves to rank with that of the boranes for novelty and interest. A third topic which is full of interest and potential for development is saturated cage compounds, found among the phosphazanes (P–N cages) and phosphorus sulfides (P–S cages). The serious study of these is only just starting, but it is already clear that the isomeric transformations of the cages, and their formation and dismemberment, open up fascinating possibilities for experimental and theoretical study. These are some of the highlights of the chemistry in this book; there are many other interesting small fields and by means of numbered structural formulas.

II. Nomenclature

At the time of writing, the nomenclature of inorganic heterocycles is in an unsatisfactory state. There is no universally accepted system. The current publications of the International Union of Pure and Applied Chemistry[8,9] on nomenclature scarcely recognize the existence of inorganic heterocyclics as a distinct class of substances, and give no direct guidance on their naming, though the subject is under urgent consideration by IUPAC's Inorganic Nomenclature Commission (see Appendix). Nomenclature has, however, given rise to fewer difficulties in the writing of this book than might have been expected, since it has been possible to name many compounds by well known semi-systematic or trivial names, and to refer to even the more difficult cases by means of numbered structural formulas.

A chemist working in this field is likely to encounter problems of nomenclature in the following two ways.

(1) Finding references to a particular compound or ring in *Chemical Abstracts* (CA). This can usually be done without prior knowledge of the complicated conventions of nomenclature used in CA. One simply looks up the ring in the Index of Ring Systems published in each semi-annual volume and in the quinquennial and decennial indexes. Although it is only claimed to cover organic rings, the Index of Ring Systems includes the great majority of inorganic rings too. Here, against the formula of the ring, will be found the name under which it is indexed in CA. Unfortunately this does not always work. For example, $S_4N_3^+$, the cyclic structure of which was proved in 1962, has never appeared in the Index of Ring Systems. In such cases the name can be hunted via the Formula Index, in which the compound will certainly be listed. This latter method can only be used when the ligands or counter-ions associated with the ring are known, but in practice this is rarely a problem.

(2) Naming a new compound of known structure, usually for the purpose of publication. Should the compound be based on a previously known ring, procedure (1) may be followed and the compound named by analogy. If a new ring is involved, it must be named from first principles. This can be a fairly tricky exercise, to which a few paragraphs are now devoted.

A few years ago Haiduc devised a system, which has much to recommend it, for use in his book,[5] but it has not formally been adopted as standard. In the absence of formal standards, the nearest thing to an internationally recognized norm is the system used in *Chemical Abstracts*. If one wishes to work in this field, to publish, or to use CA, some familiarity with this system is desirable.

II. NOMENCLATURE

The CA system has grown up by the adoption of a succession of expedients, each of which seemed reasonable at the time and all of which accorded with such IUPAC rules as were then current. The net result has most of the complexities of organic nomenclature plus a few special difficulties of its own. A fairly complete account of it can be found in ref. 3. Since IUPAC is expected soon to publish a better system (see Appendix), only a summary of CA nomenclature principles will be given here, accompanied by examples (Fig. 1.1) which illustrate the main features as they apply to the compounds described in this book.

A few of our compounds, such as tetrasulfur tetranitride, S_4N_4, are indexed in CA under binary names (Fig. 1.1A). CA uses the standard IUPAC sequence given in Rule 2.161 of ref. 8, which leads to the names phosphorus nitrides, phosphorus sulfides, and nitrogen sulfides. The first two of these accord with common practice, but nearly all chemists call binary sulfur–nitrogen compounds sulfur nitrides, not nitrogen sulfides, because nitrogen is the more electronegative partner. However, in the indexes of CA they are called nitrogen sulfides.

Ring nomenclature is far more often used in CA for inorganic heterocycles than is binary nomenclature, and of course it is usually essential. Rings with more than ten atoms, and smaller rings if bridged, are given the names of corresponding carbocyclic rings, with prefixes and locant numbers added to show the nature and positions of the "hetero" atoms, that is, *all* the atoms in the purely inorganic rings under discussion. The names of inorganic heterocycles with ten or fewer ring atoms are based on Hantzsch–Widman stems according to organic practice, i.e. ring size and degree of unsaturation are denoted by particular terminations (e.g. -ole, -epine, -ocine, etc.).[3,4,9] In this system, organic heterocycles often have short names because the carbon atoms are not mentioned in the name, but for an inorganic heterocycle all the atoms of the ring have to be listed, resulting in longer names. For rings of all sizes, the locant number 1 is assigned to an atom which comes highest in the "table of precedence"[3] or to a bridgehead. Substituents are named according to IUPAC practice.[8,9] Sometimes sulfur or phosphorus atoms with different valencies are present in a ring. Under the CA system, one or more of these is distinguished by giving its oxidation number. Many chemists, however, especially German ones, prefer ligancy numbers such as λ^3, λ^5, etc. (ligancy = number of bonds to the atom).

It will already be apparent that the CA system is cumbersome and difficult to apply. It also has the following defects. Simple structures (examples **E** and **J** in Fig. 1.1) often have very long names. The names of ligand atoms sometimes precede and sometimes (example **N**) follow those of the ring atoms, and it is difficult to see at a glance which atoms comprise the ring. Some names (example **K**) seem to imply a relationship to hydrocarbons which is wholly unreal.

(A) 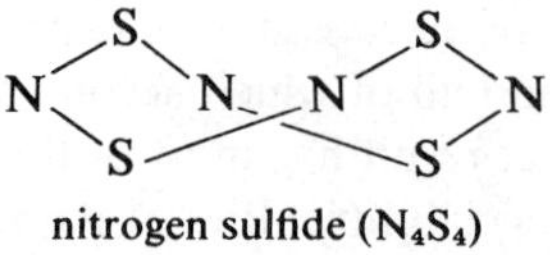 nitrogen sulfide (N₄S₄)

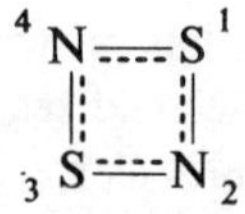

(B) 1,3,2,4-dithiadiazete

(C) 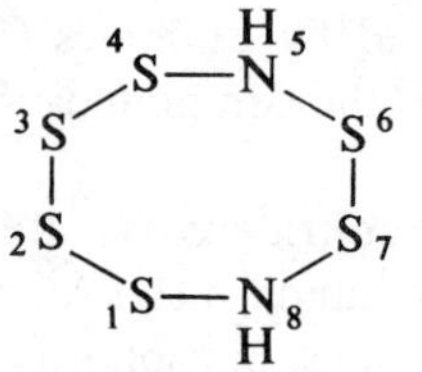 1,2,3,4,6,7,5,8-hexathiadiazocine

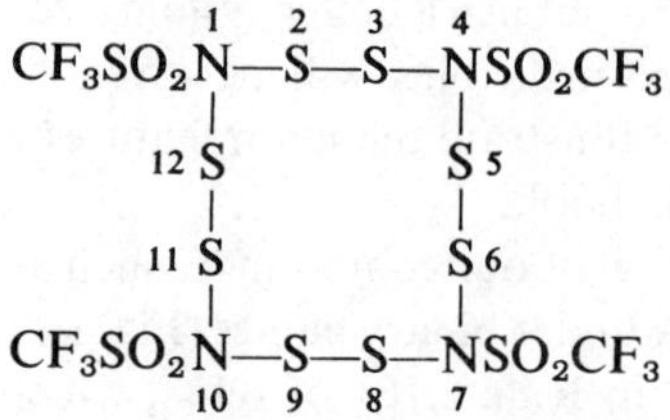

(D) 1,4,7,10-tetrakis(N-trifluoromethylsulfonyl)-1,4,7,10-tetraaza-2,3,5,6,8,9,11,12-octathiacyclododecane

(E) 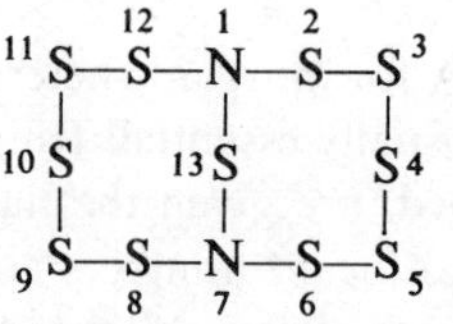 1,7-diaza-2,3,4,5,6,8,9,10,11,12,13-undecathiabicyclo[5.5.1]tridecane

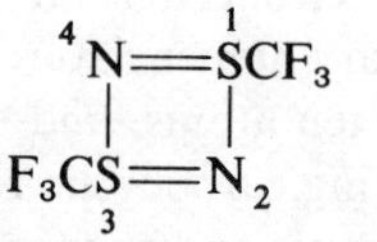

(F) 1,3-bis(trifluoromethyl)-1H,3H-1,3,2,4-dithiadiazete

(G) 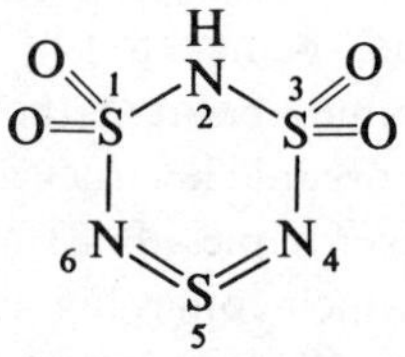 2H-1,3,5,2,4,6-trithiatriazine-1,1,3,3-tetroxide

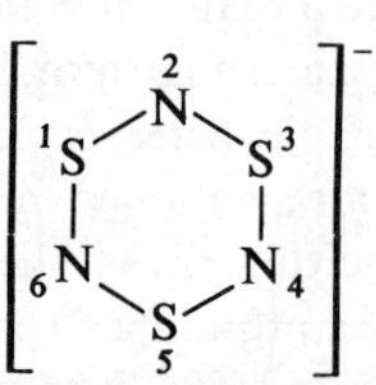

(H) 1,3,5,2,4,6-trithia(5-S^IV)triazine ion (1−)

(I) 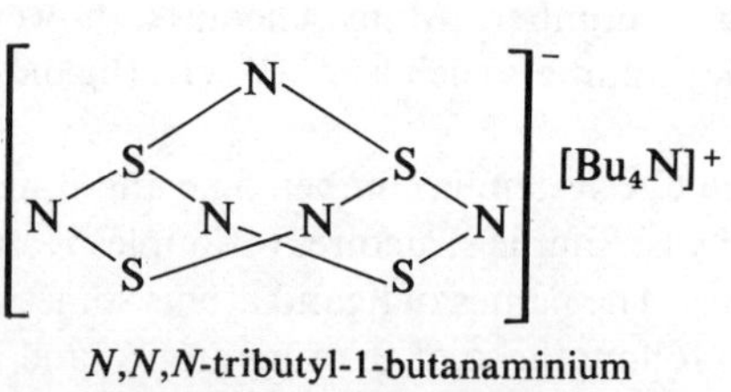 N,N,N-tributyl-1-butanaminium di-μ-thioxodi-μ₃-thioxopentanitrate (1−)

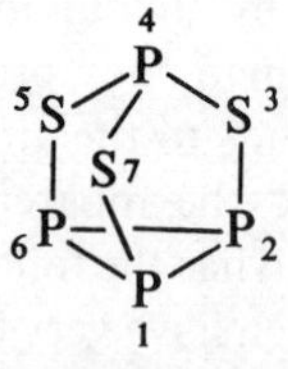

(J) 3,5,7-trithia-1,2,4,6-tetraphosphatricyclo[2.2.1.0²,⁶]heptane

(K) 3,6-diiodo-2,5,7-trithia-1,3,4,6-tetraphosphabicyclo[2.2.1]heptane

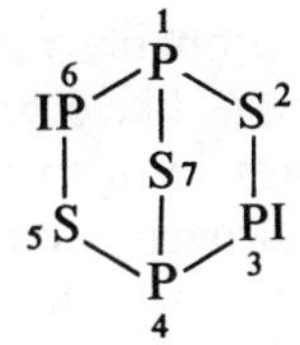

(L) 1,3-di-*tert*-butyl-2,4-dichloro-1,3,2,4-diazadiphosphetidine

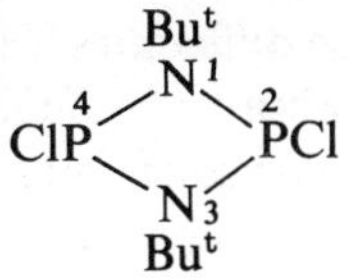

(M) 2,2,4,4,6,6,8,8,10,10-decachloro-2,2,4,4,6,6,8,8,10,10-decahydro-1,3,5,7,9,2,4,6,8,10-pentaazapentaphosphecine

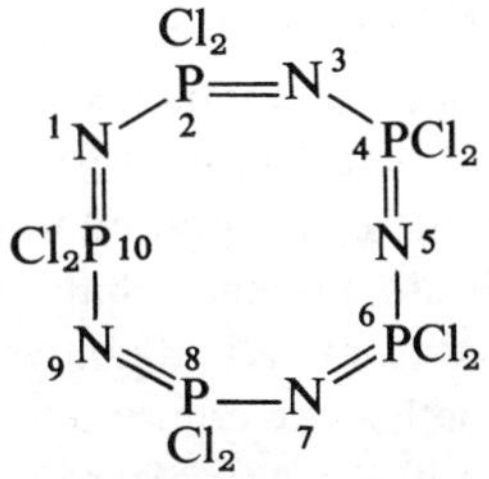

(N) 4,6-dichloro-1,3,5,7,9,10-hexamethyl-2,8-dithia-1,3,5,7,9,10-hexaaza-4,6-diphospha(4,6-P^V)dispiro[3.1.3.1]decane 2,2,8,8-tetraoxide

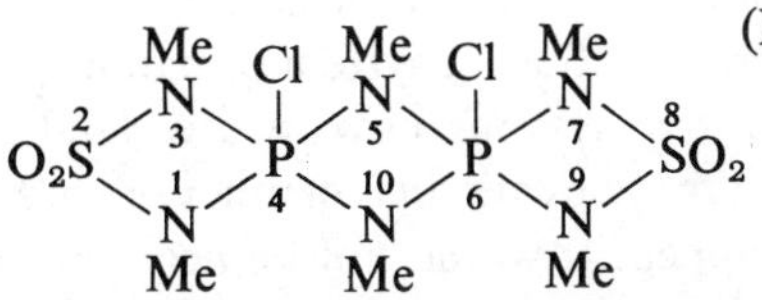

Figure 1.1. *Chemical Abstracts* nomenclature of inorganic heterocycles. All names are those under which the compounds are indexed in recent volumes. (A) Binary nomenclature. (B) Hantzsch–Widman name for unsaturated ring. (C) Hantzsch–Widman name for unsaturated ring; most chemists would regard this ring as saturated, but it is unsaturated under the Hantzsch–Widman definition —see article 146 of ref. 3. (D) Name for saturated ring larger than ten atoms, based on corresponding cycloalkane. (E) von Baeyer bridged-ring nomenclature (IUPAC standard). (F) Hantzsch–Widman name with "indicated hydrogen" (see ref. 3). (G) Hantzsch–Widman name for unsaturated ring, with prefix *H*- to indicate one "saturated atom", and oxygen ligands in suffix. (H) Based on Hantzsch–Widman name of hypothetical parent acid. (I) IUPAC coordination nomenclature, treating the ion as a cluster of five nitrogen atoms bridged by sulfurs. (J) and (K) von Baeyer bridged-ring nomenclature. (L) Hantzsch–Widman name for saturated ring. (M) Hantzsch–Widman name for unsaturated ring, with "decahydro" added because this molecular skeleton "is not treated normally in substitutive nomenclature" (ref. 3, article 158). (N) Standard CA nomenclature for spiro system, with "abnormal valencies" (ref. 3, article 158) of phosphorus atoms indicated; in the alternative version of this name showing ligancies instead of oxidation states, the part 4,6-diphospha(4,6-P^V)- becomes 4λ^5,6λ^5-diphospha-.

The prefixes *H*- or hydro- sometimes figure in the names of compounds containing no hydrogen, and oxo- in those of compounds containing no oxygen (examples **F, I,** and **M**). In example **H**, one sulfur atom is singled out as being in the (IV) oxidation state, though in fact all three sulfurs are equivalent; this has happened because the ion is conceptually derived from an acid with an unsymmetrical structure which is not known and probably never will be. Some of these difficulties arise because the CA system is, like organic nomenclature, substitutive, though it is often needless masochism to think of the ligands on inorganic heterocycles as replacing hydrogen atoms in hypothetical parent substances. In example **M**, for instance, the expression 2,2,4,4,6,6,8,8,10,10-decahydro- could be omitted with no loss of information.

Reduced to basics, the problem of naming inorganic heterocycles is not nearly as difficult as this historical accretion of expedients and prejudices would seem to suggest. It is only necessary to identify and list the ring atoms and bridgeheads, and to list any ligands, using locant numbers as needed in both cases. This information completely defines the compound for indexing purposes. Such a system of naming would be additive, as opposed to the substitutive systems now in use. Under it, the absurd pretence that inorganic heterocycles are, in a manner of speaking, derived from cycloalkanes could be dropped. It would not be necessary to show whether the compound is saturated or unsaturated, or where the double bonds, if any, are (often impossible to decide in the case of inorganic heterocycles). Moreover, once the positions of ligands are specified by locant numbers, it is not necessary to give oxidation numbers or ligancy numbers of ring atoms. Given an additively specified structure, the reader can fill in double bonds and oxidation numbers as he sees fit. If the substance is an ion, the charge *must* be stated in its name, but once this is done, special terminations such as -onium, -enium, and -ate become superfluous.

An improved nomenclature would require one further feature: concise names for cage compounds based on simple molecular polyhedra, which can have extraordinarily complicated names when treated as bridged ring systems (see Chapter 11, Sec. II.E, for an illustration). This could be achieved, for example, by means of short prefixes used analogously to cyclo-, e.g. dio- (for a tetrahedron, derived from the word diamond), cubo-, icoso-, etc.

III. Methods of synthesis

A. Introduction

This section is intended as an overview, not an exhaustive treatment, of the diverse approaches to the synthesis of inorganic S–N, P–S, and P–N heterocycles. The aims are: to show the generality of some widely used methods; to mention some important new reagents; and to help the user to find synthetic routes to hitherto unknown heterocycles. The general principles of inorganic ring synthesis have been reviewed by Nöth[10] and by Wannagat and collaborators[14] but with emphasis on boron and silicon compounds.

B. Direct combination of the ring-forming elements

The main application of this method is to the cage molecules of the phosphorus sulfides (Chapter 11). It is not a practical route to rings containing nitrogen, though PN (Chapter 13, Sec. III.A) can be made by direct combination under drastic conditions, and subsequently polymerizes to solid phosphorus nitrides which may contain rings.

C. Addition polymerization

Rings made up of regularly alternating components (alternation heterocycles) can occasionally be made by the addition-polymerization of suitable monomers, e.g.

$$\text{HNSO (l)} \xrightarrow[-70°]{\times 4} (\text{HNSO})_4 \text{ (s)} \quad \text{(Chapter 3, Sec. III.B)}$$

$$\text{NSF (g)} \xrightarrow[20°]{\times 3} (\text{NSF})_3 \text{ (s)} \quad \text{(Chapter 7, Sec. II.F)}$$

Such reactions are seldom of preparative value for inorganic heterocycles, though unstable monomers which polymerize additively *in situ*, and are not isolated, may arise in some of the condensation reactions dealt with in the following paragraphs.

D. Heterofunctional condensations

In this method, different ring components couple together by reactions between unlike functional groups in which small molecules (or occasionally ions) are eliminated. This is the most important and most flexible way of making inorganic rings, being suitable for homocycles, for alternation heterocycles, and for heterocycles with irregular atom sequences.

In principle, all reactions of this kind are condensation-polymerizations capable of yielding linear polymers as well as rings of more than one size, and linear polymers do often constitute a substantial or even predominant part of the products. This point can be illustrated by the well known basic reaction of silicone chemistry:

$$\text{Me}_2\text{SiCl}_2 + \text{H}_2\text{O (l)} \xrightarrow[20°]{} \frac{1}{x} [\text{Me}_2\text{SiO}]_x + 2\text{HCl}$$

the products of which include, in comparable amounts, rings $[\text{Me}_2\text{SiO}]_{3,4,5}$ and linear polymers based on the repeating unit $[\text{Me}_2\text{SiO}]$.

The conditions of condensation reactions may be chosen to favor either kinetic or thermodynamic control of the products, though such a choice has often not consciously been made. An example of the deliberate and successful use of kinetic control is the synthesis by Becke-Goehring and Jenne (Chapter 2, Sec. IV.B) of derivatives of the 6-membered saturated 1,4-S_4N_2 ring, which is thermodynamically unstable with respect to 8-membered rings. The overall reaction is

$$2S_2Cl_2 + 4EtNH_2 \xrightarrow[20°]{\text{ether}} S_4(NEt)_2 + 2EtNH_3{}^+Cl^- \downarrow$$

The first step is probably

$$S_2Cl_2 + EtNH_2 \rightarrow ClS_2NHEt + HCl$$

The Ruggli–Ziegler dilution technique employed entails the continuous slow mixing of the reagents at exact stoichiometric equivalence so that no excess of either is ever present. In these circumstances the intermediate ClS_2NHEt is more likely to meet another molecule of its own kind than either S_2Cl_2 or $EtNH_2$, so favoring the desired cyclization sequence

$$2ClS_2NHEt \rightarrow ClS_2N(Et)S_2NHEt + HCl$$
$$\downarrow$$
$$S_4(NEt)_2 + HCl$$

against competing reactions. In some other reported ring-forming condensations, kinetic control undoubtedly operates, but in more complicated and obscure ways. For example, in the standard synthesis of phosphazene rings (Chapter 12, Sec. V.B) from PCl_5 and NH_4Cl, the temperature is too low to allow thermodynamic equilibration between rings of different sizes, and their relative proportions depend on dilution in a manner suggesting kinetic control.

In contrast to these examples, many syntheses of inorganic heterocycles have been carried out under conditions of thermodynamic product control; that is, the proportions of different products are determined by an equilibrium set up at some stage in the synthesis. One example is the reactions of PCl_5 with primary amines (Chapter 11, Sec. I.B) which afford monomers $RN{=}PCl_3$, cyclic dimers $(RNPCl_3)_2$, or occasionally both, depending on their relative stability. The reaction of ammonia with S_2Cl_2 gives a mixture of cyclic sulfur imides (Chapter 2, Sec. III.B) the composition of which seems to depend, at least in part, on a series of equilibria involving the intermediate $NS_4{}^-$.

As this last example illustrates, the mechanisms of many ring-forming condensation reactions are far from simple or obvious, and their course cannot readily be foreseen by blackboard exercises.

For many years, ammonia or organic amines provided the nitrogen component in syntheses of inorganic nitrogen heterocycles by condensation. Recently, however, there have been important developments. One of these is the use of organosilicon– or organotin–nitrogen compounds. This idea seems to have sprung from the work of Burg and Kuljan (1950),[12] who showed that although trisilylamine is only weakly basic, the reaction

$$(H_3Si)_3N + BCl_3 \rightarrow (H_3Si)_2NBCl_2 + H_3SiCl$$

goes readily at $-78°$. In the 1960s, the principle of splitting the Si–N bond by reaction with non-metal halides began to be exploited for the preparation of inorganic heterocycles, for example in the reaction

$$(Me_3Si)_2NMe + O_2S\begin{smallmatrix}\nearrow N=PCl_3 \\ \searrow N=PCl_3\end{smallmatrix} \rightarrow O_2S\begin{smallmatrix}\nearrow N=PCl_2\searrow \\ \searrow N=PCl_2\nearrow\end{smallmatrix}NMe + 2Me_3SiCl$$

(Chapter 4, Sec. IV.C). The publications of, for example, H. W. Roesky and collaborators since 1970 afford many instances of the use of silyl– and stannyl–nitrogen compounds for ring synthesis and for the replacement of exocyclic ligands (see bibliographies of Chapters 3, 4, 6, 7, 8). The advantages are as follows. First, these reagents are sometimes more stable and easily handled than the parent compounds; for example, $(Me_3SiN)_2S$ occurs whereas $(HN)_2S$ is unknown, and Me_3SiNSO is far more stable than HNSO. Secondly, they react readily and smoothly with a wide range of halogen compounds, including relatively weak acceptors such as PCl_3, yet as nucleophiles are milder and less destructive than ammonia and amines. Thirdly, the unwanted product is not a hydrogen halide, as from condensations with conventional amines, but an organosilicon or organotin halide R_3SiX or R_3SnX, which in contrast does not react further, and which lends itself to separation from the wanted products by distillation *in vacuo* or filtration. In all these reactions a silicon–nitrogen or tin–nitrogen bond is replaced by a stronger silicon–halogen bond, this replacement providing the driving force (e.g. Si–N 335 kJ, Si–Cl 391 kJ, Si–F 383 kJ). Bis(trimethylsilyl)amine, $(Me_3Si)_2NH$, also called hexamethyldisilazane, is sold commercially and serves as a source of other useful trimethylsilyl–nitrogen compounds,[7] including tris(trimethylsilyl)amine. A desired intermediate containing a reactive Me_3Si- group is typically made from a starting material with more than one such group using the principle just mentioned, e.g.

$$(Me_3Si)_3N + SOCl_2 \rightarrow Me_3SiNSO + 2Me_3SiCl$$

(Chapter 3, Sec. III.D), or from a metal derivative of bis(trimethylsilyl)amine and an appropriate halide, e.g. the preparation of $(Me_3SiN)_2S$ (Chapter 3, Sec. IV.C).

The relevant tin chemistry may be found in refs. 6 and 11.

A heterofunctional condensation using $MeN(SiMe_3)_2$ has been employed to introduce a nitrogen bridge across a phosphorus–nitrogen ring (Chapter 11, Sec. II.F).

Imino hydrogen atoms are as a rule decidedly acidic, so it is often possible to prepare mercury, silver, or alkali-metal derivatives of compounds containing an imino function. These derivatives undergo condensation reactions with covalent halides, eliminating metal halide. Such condensations have been employed for the synthesis of inorganic silicon heterocycles.[14] In the areas of chemistry covered by this book, their main application has been to the replacement of ligands and the coupling of rings, e.g.

$$S_7NLi + MeI \rightarrow LiI + S_7NMe$$
$$2S_7NHgPh + SOBr_2 \rightarrow 2PhHgBr + (S_7N)_2SO$$

(Tables 2.3 and 2.4), but there seems no reason why they could not be exploited in ring-closure reactions.

Organotin sulfides have been used in the synthesis of phosphorus–sulfur cages, the principle again being a heterofunctional condensation with elimination of organotin halide:

$$\alpha\text{- or }\beta\text{-}P_4S_3I_2 + (Me_3Sn)_2S \rightarrow \alpha\text{- or }\beta\text{-}P_4S_4 + 2Me_3SnI$$

(Chapter 10, Sec. II.E).

The technique of heterofunctional condensation was originally used with simple starting molecules, each of which supplied one atom of the ring. An important modern trend, however, is to begin the ring synthesis from bifunctional molecules already containing several of the atoms required. New starting materials suitable for this approach are reported from time to time. Examples are: $(Me_3SiN)_2S$, mentioned above, which provides –N=S=N– groups; $HN(SO_2Cl)_2$, first described in 1962 and since used in two heterocyclic syntheses (Chapter 4, Secs. III.B.4 and IV.B); and the S,S-dialkyl sulfodiimides $R_2S(NH)_2$ and their N,N'-dibromo derivatives $R_2S(NBr)_2$ (Fig. 4.3). Condensations employing $R_2S(NBr)_2$ have a special feature: "positive bromine" in the starting material becomes "negative bromine", Br^-, in the product, while P(III) or S(IV) in the partner reagent is oxidized to P(V) or S(VI) in the product (Fig. 4.3). The same principle is used in oxidative phosphazene syntheses (Chapter 12, Sec. V.D).

E. Homofunctional condensations

In this method, ring components (which have almost invariably been of the same kind, though in principle they need not be) couple together by the elimination of small molecules between like functional groups. As in the previous method, linear polymers result besides rings. The principle is well known from silicone chemistry, e.g. in the second reaction of the sequence

$$Ph_2SiCl_2 \xrightarrow{H_2O} Ph_2Si(OH)_2 \xrightarrow{-H_2O} (Ph_2SiO)_{3,4}$$

and has been employed to a limited extent for the synthesis of heterocycles of sulfur, nitrogen, and phosphorus. Thus sulfamide, $SO_2(NH_2)_2$, when heated above 140°, condenses with itself, eliminating ammonia and forming the cyclic trimer $(SO_2NH)_3$ and linear polymers (Chapter 4, Sec. III.B). The phosphorus(v) oxoamides and thioamides behave similarly (Chapter 11, Secs. III.B and C).

The homofunctional condensation of a substance with itself can naturally only produce heterocycles of the alternant type.

F. Insertions, ring expansions, and ring contractions

It is convenient to group under this head some interesting syntheses of S–N and P–S rings which set their chemistry apart from that of carbon rings.

Unsaturated, "electron-rich" S–N rings (Chapters 6, 7, and 8) are rather unstable thermally, so it is not surprising that atoms or ring segments can readily be inserted into them or removed from them, near or slightly above room temperature. Here are examples.

$$S_4N_4 + \tfrac{1}{3}(NSCl)_3 + FeCl_3 \rightarrow [S_5N_5]^+[FeCl_4]^-$$

(ring expansion; Chapter 6, Sec. III.D)

$$3S_4N_4 + 2S_2Cl_2 \rightarrow 4[S_4N_3]^+Cl^-$$

(ring contraction; Chapter 6, Sec. III.C)

$$S_4N_4 + NaN_3 \rightarrow Na^+[S_4N_5]^- + N_2$$

(ring bridging by N^-; Chapter 6, Sec. III.G)

$$(S_2N_2SnMe_2)_2 + 2SOF_2 \rightarrow 2S_3N_2O + 2Me_2SnF_2$$

(replacement of tin in the S_2N_2Sn ring by sulfur; Chapter 8, Sec. IV). Knowledge of these processes is empirical, and it is not safe to assume the most obvious mechanisms; for example, there is evidence that the third reaction goes via $S_3N_3^-$.[2]

The chemistry of P–S rings affords more straightforward examples, for instance

$$(CF_3P)_4 + \tfrac{1}{8}S_8 \rightarrow (CF_3)_4P_4S$$

(insertion of a sulfur atom into a P_4 ring; Chapter 10, Sec. III.C). The temperatures required suggest that thermal opening of the ring of the starting material is an essential step.

G. Miscellaneous thermodynamically controlled processes

Certain transaminations between non-cyclic molecules give rise to rings, e.g.

$$(Me_2N)_3P \begin{array}{c} \xrightarrow{PhNH_2} (PhNPNHPh)_2 + Me_2NH \\ \xrightarrow{(MeHN)_2} P_2(NMe)_6 + Me_2NH \end{array}$$

(Chapter 11, Secs. II.C and F).

The thermal rearrangement of a phosphazane ring to a cubic cage has been described:

$$2(MeNPF_3)_2 \rightarrow (MeNPF_3)_4$$

(Chapter 11, Sec. III.F).

The potential of both types of reaction is probably great but remains to be fully evaluated.

H. Controlled breakdown of pre-existing structures

The stepwise degradation of complicated inorganic molecules to simpler ones has been studied mainly for the purpose of analyzing mechanisms (Chapter 4, Sec. III.C), but also offers some possibilities, as yet little exploited, for the synthesis of rings from cages or polymers. The cage molecule S_4N_4, for example, is transformed into compounds containing 6-membered S_3N_3 rings by reaction with chlorine or triphenylphosphine (Chapter 6, Sec. II.A). The cage phosphorus sulfides P_4S_3 and P_4S_7 give P–S rings on reaction with iodine or bromine respectively (Chapter 10, Sec. II.E). The Wade–Williams theory of atom clusters (Chapter 9, Sec. V.C) holds some promise as a way of rationalizing reactions of this kind and so encouraging further study.

J. Alteration of oxidation state of phosphorus or sulfur in existing rings

The oxidation state of sulfur or phosphorus atoms in a ring or cage can often be altered without breaking down the structure. Thus $(NSCl)_3$ can be oxidized to $[NS(O)Cl]_3$ (Chapter 5, Sec. II.A), S_4N_4 oxidized to $S_4N_4^{2+}$ (Chapter 6, Sec. III.F), and P(III)–N rings and cages oxidized to the corresponding P(V) compounds (Chapter 11, Secs. II.C, E, and F). An interesting recent oxidation reagent is trifluoroperoxoacetic acid, which adds oxygen ligands at low temperatures to divalent sulfur atoms in S_7NH (Chapter 2, Sec. III.B.4) as well as to cyclo-S_6, -S_7, -S_8, and -S_{10}.[13]

References

1. Appel, R. and Eichenhofer, K.-W., *Chem. Ber.* **104**, 3859 (1971).
2. Bojes, J., Chivers, T., Drummond, I. and MacLean, G., *Inorg. Chem.* **17**, 3668 (1978)
3. *Chemical Abstracts, Cumulative Index Guide*, 76–85 (1972–6), Appendix IV, "Selection of Index Names for Chemical Substances"
4. Fletcher, J. H., Dermer, O. C. and Fox, R. B., *Adv. Chem. Ser.* no. 126, "Nomenclature of Organic Compounds: Principles and Practice", American Chemical Society, Washington (1974)
5. Haiduc, I., *The Chemistry of Inorganic Ring Systems*, Wiley-Interscience, London (1970)
6. Jones, K. and Lappert, M. F., in *Organotin Compounds*, ed. A. K. Sawyer, vol. 3, Marcel Dekker, New York (1971)
7. Kruger, C. R. and Niederprum, H., *Inorg. Synth.* **8**, 15 (1966)
8. International Union of Pure and Applied Chemistry, *Nomenclature of Inorganic Chemistry*, second edition, Butterworths, London (1971)
9. International Union of Pure and Applied Chemistry, *Nomenclature of Organic Chemistry*, Butterworths, London (1971)
10. Nöth, H., Plenary lecture at the Second International Symposium on Inorganic Ring Systems, Göttingen, 1978. (Abstracts published by Akademie der Wissenschaften, Göttingen)
11. Pommier, J.-C. and Pereyre, M., "Organotin Compounds: New Chemistry and Applications", *Adv. Chem. Ser.* no. 157, ed. J. J. Zuckerman, American Chemical Society, Washington (1976)
12. Scherer, O. J., *Organometal. Chem. Rev.* **A3**, 281 (1968)
13. Steudel, R. and Steidel, J., *Angew. Chem. Int. Ed. Engl.* **17**, 134 (1978)
14. Wannagat, U., Schlingmann, M. and Autzen, H., *Z. Naturforsch.* B **31b**, 621 (1976)

2
Cyclic sulfur imides with two-coordinate sulfur, and their derivatives

I. Introduction, scope, and history

This chapter deals with a well defined group of cyclic sulfur–nitrogen compounds, the formulas of which can be written without double bonds and which are therefore often described as saturated. Most of them contain 8-membered rings with the same puckered or crown shape as the well known S_8 molecule, and in principle derived from it by replacement of up to four sulfur atoms by 3-coordinate nitrogen. The third valencies of the nitrogen atoms can serve as the means of attaching a wide variety of ligands. If these are hydrogen atoms, we have the cyclic sulfur imides (Fig. 2.1). The imides are usually regarded as parent compounds, and many derivatives, both organic and inorganic, can be made by replacing their hydrogen atoms with other groups (Tables 2.3 to 2.5). This area of chemistry is rather like traditional organic substitution chemistry, since the 8-membered S–N rings are stable enough to survive treatment with the milder substitution reagents. In an interesting type of substitution, S–N rings are coupled or fused to each other through their nitrogen atoms, giving rise to new sulfur nitrides, a sulfur nitride-oxide, and sulfur–nitrogen polymers. In theory at least, one can see here the prospect of an extensive chemistry of saturated sulfur–nitrogen frameworks.[30,33]

The history of this subject begins with the preparation by H. Wölbling in 1908 of the imide $S_4(NH)_4$.[18] This showed the relationship of $S_4(NH)_4$ to S_4N_4, but gave no hint that it was one of a series of imides structurally related to S_8. The next important event was the isolation in 1923, by Macbeth and Graham,[18] of impure heptasulfur imide, S_7NH, the second member of this series to be found; in 1951 it was better characterized by Goehring, Herb, and Koch. A growing belief that these compounds were essentially S_8 with sulfur atoms replaced by NH groups was confirmed by Lund and Svendsen's (1957) crystallographic determination of the structure of $S_4(NH)_4$, and soon afterwards by Weiss's crystallographic studies of S_7NH and $1,5-S_6(NH)_2$. Strange to say, chromatography, the ideal method of purifying the imides and their derivatives, was not applied to the imides until 1959; it soon resulted in the

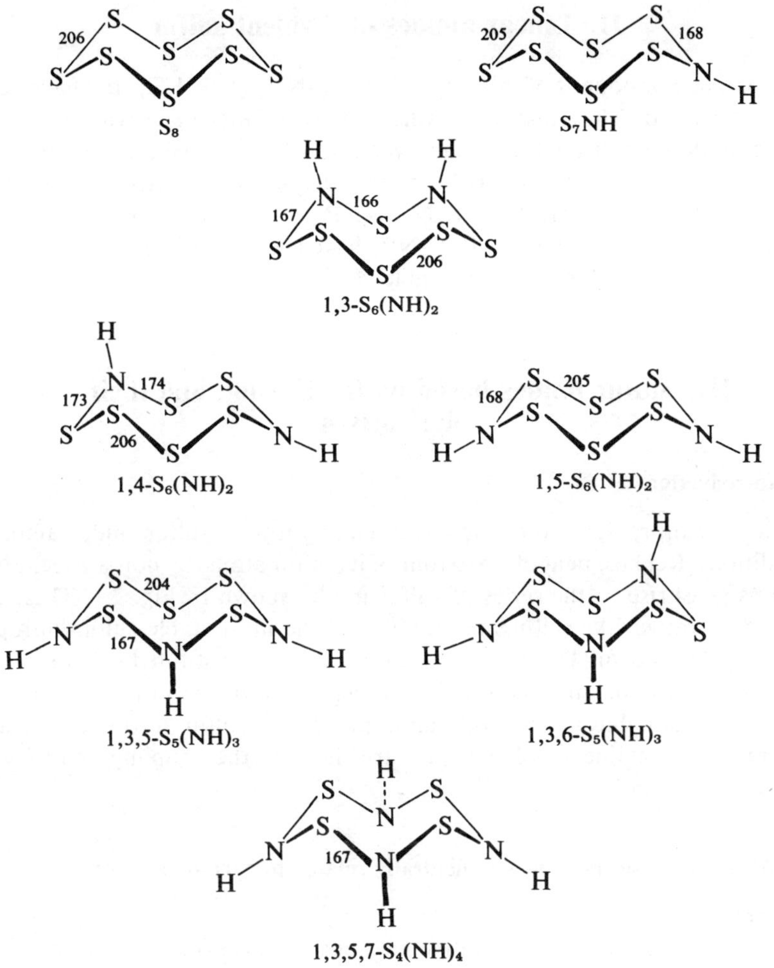

Figure 2.1. Molecular structures of cyclooctasulfur and the sulfur imides $S_{8-n}(NH)_n$; bond lengths are in pm.[48,49,58,63]

discovery of all the missing members of the series of imides shown in Fig. 2.1. In the last decade the subject has developed along several lines, notably structure determination and synthesis. The chief areas of ignorance apparent at the moment are mechanisms, on which little work has ever been done in this field, and rings other than 8-membered, which occur in rather few compounds.

None of the cyclic sulfur imides is now sold commercially, but heptasulfur imide, in particular, is not difficult to prepare in experimental quantities, and is safe and can easily be stored for long periods.

II. Linear amides of divalent sulfur

The parent compounds of this class, $H_2N-S_x-NH_2$ ($x = 1, 2$, etc.) have never been described. They may form when chlorosulfanes react with ammonia, but if so, the complicated secondary reactions which immediately follow make them undetectable (Section III.B below; Chapter 6, Sec. II.A). However, a number of organo derivatives $R_2N-S_x-NR_2$, bis(dialkylamino)sulfanes, are known.[28] Many sulfenamides $R-S-NR'_2$ have been described, including some with $R' = H$, which seem fairly stable.[35]

III. Sulfur imides based on the S_8 ring, and their derivatives

A. Introduction

Cyclooctasulfur, S_8, is the stable molecular form of sulfur under ambient conditions. Replacement of up to four of its sulfur atoms by non-adjacent NH groups gives rise to the series of sulfur imides shown in Fig. 2.1. These are physically rather like sulfur and are among the most stable sulfur–nitrogen heterocycles known. Their hydrogen atoms can be substituted in a variety of ways without damaging the sulfur–nitrogen rings. Among the many derivatives so obtainable are several sulfur nitrides, a sulfur nitride-oxide, and several series of linear polymers, all arising from the coupling or fusion of 8-membered rings.

B. Monocyclic sulfur imides structurally related to cyclooctasulfur

1. Preparation

As Table 2.1 shows, all preparative methods except (iv) give mixtures of the imides in which S_7NH predominates, but varying considerably in composition with regard to the other imides. A method should be chosen according to the imide or imides mainly required. A brief account and assessment of the methods follows.

(i) In the standard preparation due to Becke-Goehring, Jenne, and Fluck (1958), S_2Cl_2 is added slowly to a concentrated solution of ammonia in dimethylformamide at $-5°$ to $-10°$.[27] For work-up the product is diluted with aqueous HCl, precipitating a doughy mass of mixed imides and sulfur, which is best dealt with by chromatography on silica gel. This is a good route to S_7NH. It is the best method for 1,4-$S_6(NH)_2$, which follows S_7NH off the column and is well separated from it and the other diimides. With patience, it also yields worthwhile and roughly equal amounts of the 1,3 and 1,5 diimides,

Table 2.1. Yields* of cyclic sulfur imides from various preparative methods.
All yields are reported as grams of sulfur imide from an amount of starting material containing 100 g sulfur

Preparative reaction	Solvent†	S₇NH	1,3-S₆(NH)₂	1,4-S₆(NH)₂	1,5-S₆(NH)₂	1,3,5-S₅(NH)₃	1,3,6-S₅(NH)₃	S₄(NH)₄	Total yield of imides (% on S)	Method in Sec. III.B.1	Ref.
S₂Cl₂ + NH₃	HMPA	34	2.0	2.9	2.0	—	—	—	38	(i)	7
	DMF	18 (av.)	1.9	4.0	1.9	0.1	0.3 (av.)	—	24	(i)	27
	water/CS₂	17 (av.)	0.8	3.0	0.9	—	—	—	20	(i)	1
	CCl₄	10	—	—	—	—	—	—	10	(i)	3
S$_x$Cl₂ + NH₃ (x = 2–7)	DMF	12–32	—	—	—	—	—	—	11–30	(i)	11
S₈ + NaN₃	HMPA	39	0.15 total			—	—	—	37	(ii)	7
S₄N₄ + N₂H₄	CCl₄	63	12	1.6	2.6	6.5	—	trace	78	(iii)	13, 15
S₄N₃Cl + N₂H₄	CCl₄	62	8.8	3.4	3.4	1.1	0.5	—	72	(iii)	13, 56
S₄N₄ + SnCl₂	benzene/methanol	probably very small						86	59 (av.)	(iv)	18, 20

* Some judgment has had to be used in coping with internal inconsistencies in the published information.
† DMF = dimethylformamide; HMPA = hexamethylphosphoramide.

but these overlap in the chromatographic separation and are best purified by combining chromatography with fractional crystallization from CS_2.[27]

The main drawbacks of this method are the time and attention required, and the fairly high cost of the solvent, dimethylformamide, needed in large quantities.

Some variants have been tested. Yields in hexamethylphosphoramide (Table 2.1) are even better but this solvent is prohibitively expensive. Alternatively, a solution of S_2Cl_2 in CS_2 can be added to cooled concentrated aqueous ammonia; yields are nearly as good as with dimethylformamide, but the method has only been tried on a small scale.[1] In the non-polar solvent CCl_4, substantial but smaller yields[3] have been reported (cf. Chapter 6, Sec. II.A.1). Replacement of S_2Cl_2 by higher chlorosulfanes S_nCl_2 offers no important advantage.[11]

(ii) The following method is simple and can be strongly recommended for gram quantities of S_7NH.[7] A solution of S_8 and NaN_3 (about 1:4 mol ratio) in hexamethylphosphoramide is stirred at room temperature for 3 days, during which nitrogen is evolved. The mixture is worked up as for (i), giving an excellent yield of S_7NH but only small yields of the diimides.

Chivers and coworkers have offered an interesting, though incomplete, explanation[7] of what happens in methods (i) and (ii). They postulate a set of equilibria in basic solvents:

$$S_7N^- \rightleftharpoons NS_4^- + \tfrac{3}{8}S_8 \rightleftharpoons \tfrac{1}{2}S_6N_2^{2-} + \tfrac{1}{2}S_8$$

involving the dark blue ion NS_4^- (Chapter 6, Sec. V.B) the color of which is strikingly evident in the reaction mixtures. Addition of aqueous HCl is thought to convert the various imide anions present into the imides themselves, which precipitate. In support, it is noted that a tetrahydrofuran solution of $[Bu_4N]^+[NS_4]^-$ gives with aqueous HCl a mixture of imides like that from method (i); moreover, S_7NH and the 1,4 diimide quickly rearrange in hexamethylphosphoramide solution to give much NS_4^-. But an explanation is still required for the different product compositions from methods (i) and (ii).

(iii) Tetrasulfur tetranitride (Chapter 6, Sec. II.A) or thiotrithiazyl chloride (Chapter 6, Sec. III.C) is reduced in CCl_4 by means of hydrazine adsorbed on silica gel.[13,15] At 46°, this requires about 8–10 hours for the first compound and 50–60 hours for the second. The method gives good yields of S_7NH (Table 2.1) but is not as convenient for this imide as (i) or (ii), since S_4N_4 and $[S_4N_3]^+Cl^-$ are not available commercially and must be made first. The peculiar virtue of the present method is its high yields of the 1,3 diimide and the 1,3,5 triimide, for both of which it is the method of choice. With relatively little 1,5 diimide present, chromatographic purification of the 1,3 isomer becomes much easier than in method (i). The explosive nature of S_4N_4 is not a real drawback since good results can be obtained with the crude mixture of

S_4N_4 and sulfur from the standard S_4N_4 synthesis (Chapter 6, Sec. II.A.1) which is safe.[15]

(iv) Reduction of S_4N_4 in benzene at 80° with methanolic $SnCl_2 \cdot 2H_2O$ is the only satisfactory route to $S_4(NH)_4$,[18,20] which is formed fairly pure and in high yield. This imide probably results in small amounts from the other methods given above, but has not been properly looked for. It is inconvenient to chromatograph because of its low solubility in most organic solvents.

The 1,3,6 triimide is only formed in small amounts, whatever the method of preparation used (Table 2.1), and no good route to it is known.

2. General properties and manipulation

These imides all form odorless fairly air-stable crystals. Pure S_7NH has a barely perceptible yellow tinge, but the other imides are colorless. Their toxicity is unknown, but no physiological activity has been reported. $S_4(NH)_4$ is said to explode mildly when struck hard, but no accidents are known to have occurred with it; the others are not explosive. Purified imides, and more especially impure mixtures of them, should be stored in the freezer compartment of a refrigerator to avoid slow decomposition. The 1,4 diimide darkens to orange in a few hours when exposed to both light and air, and should be kept in the dark.

Some properties of the imides are summarized in Table 2.2. The following points deserve comment. Solubilities in CS_2 decline, and solubilities in polar solvents such as methanol increase, with increasing numbers of NH groups.[27] The solubilities of the diimides in CS_2 are in the inverse order of their dipole moments, and similarly for the triimides. For S_7NH (alone) there exist solubility data for other solvents over a range of temperature near 25°.[24] Most of the

Table 2.2. Properties of the cyclic sulfur imides.[27,28]

Compound		M.p. (°C) (28)	Solubility in CS_2 at 18° (molal) (28)	R_f on silica gel t.l.c.* (27)	Dipole moment (debyes)† (28)
S_7NH		113.5	0.57	0.58	1.28
$S_6(NH)_2$	1,3	130 dec.	0.038	0.22	2.28
	1,4	133	0.15	0.34	1.23
	1,5	155 dec.	0.023	0.24	1.74
$S_5(NH)_3$	1,3,5	128 dec.	0.008	0.09	2.8
	1,3,6	133 dec.	0.038	0.16	1.0
$S_4(NH)_4$		145	probably 0.005	very low	—

* These values apply to plates not activated by heating, on which S_8 has $R_f = 1$. Activation lowers all R_f values but does not affect the order of elution.
† In CS_2 solution at 25°.

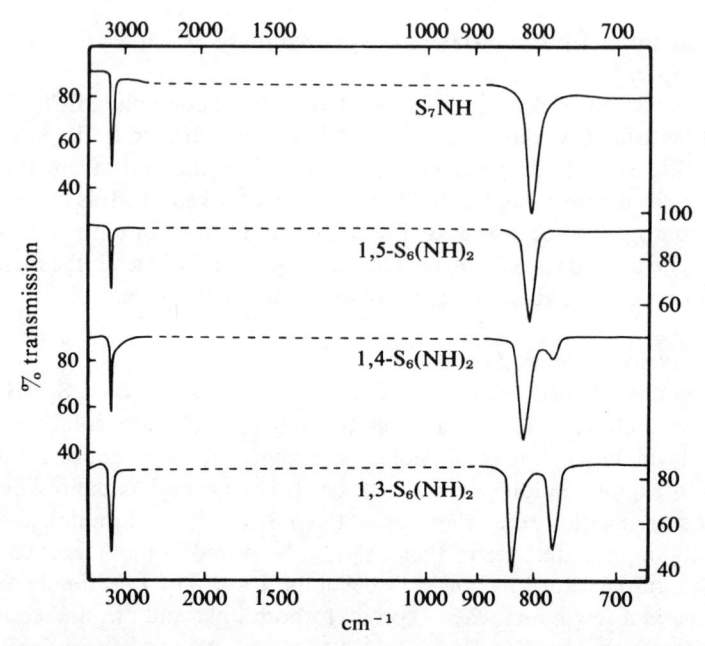

Figure 2.2. Infrared absorption spectra of heptasulfur imide and the isomeric hexasulfur diimides in the N–H and S–N stretching regions;[28] CS_2 solutions, 0.2 mm cell; the broken regions contain a rather weak NH bending mode at about 1280 cm^{-1} and several strong solvent absorptions.

imides can conveniently be recrystallized from CS_2 (which must be freshly distilled), or from benzene or toluene. Individual imides are most easily identified, and mixtures analyzed both qualitatively and quantitatively, by means of the infrared spectra of their CS_2 solutions, which change in a characteristic way with the number and location of the NH groups (Fig. 2.2). Melting points are affected by decomposition and are not very useful for identification. Thin-layer chromatograms on silica gel (developed with CS_2 and visualized by spraying with 0.1M $AgNO_3$) give a quick indication of the composition of mixtures of the imides, and show up any sulfur present, which will not readily be evident from the infrared spectrum.

3. Molecular structures and parameters

Full X-ray crystal-structure determinations have been performed on all the imides except the 1,3,6 triimide; data and references appear in Fig. 2.1. The well established correlation between dipole moments (Table 2.2) and structure leaves no doubt that the 1,3,6 isomer has indeed the structure attributed to it in Fig. 2.1. All the molecules have the same kind of puckered crown configuration as cyclooctasulfur. The 1,4 diimide is chiral, and its crystals contain equal numbers of molecules of the two enantiomers. Complete vibrational analyses have been performed on S_7NH and $S_4(NH)_4$.[58] The meaning of the molecular parameters for bonding is discussed in Chapter 9, Sec. III.

4. Reactions

The reactions of the sulfur imides will be discussed in the following order: thermal decomposition; oxidation; reduction; reactions involving acidic behavior of the imides; reactions involving their basic behavior. Many derivatives have been prepared by reactions of the last two categories.

A purified sulfur imide, if stored for a few weeks at room temperature, is often found to develop a small degree of contamination with S_8 and with the other imides; this decomposition may proceed via S_4N^- produced by traces of basic impurities (Section 1 above). S_7NH and the three diimides decompose in the molten state below 160° within minutes.[55] The only sulfur–nitrogen compound produced in important quantities is S_4N_4, yields of which range from 20 to 30% of theoretical. The decomposition is probably initiated by the opening of S–S bonds, which is known from the behavior of S_8 to begin in this temperature region.

Few oxidation reactions of the sulfur imides have been adequately studied.[18] $S_4(NH)_4$ reacts with ozone, or with air on heating, to give what is believed to be a thionyl imide tetramer $(OSNH)_4$ (Chapter 3, Sec. III.B).[12] Trifluoroperoxoacetic acid adds an oxygen ligand to one sulfur atom of S_7NH, giving a very unstable compound S_7NHO.[59] The red solution of S_7NH in concentrated sulfuric acid has been attributed[40] to the oxidation product $[S_7N-NS_7]^+$ but more likely arises from $[S_3N_2]^+$.[38] $S_4(NH)_4$ is converted by chlorine or sulfuryl chloride into $(NSCl)_3$ (Chapter 7, Sec. II.C) and by bromine or N-bromosuccinimide into $[S_4N_3]^+Br^-$ (Chapter 6, Sec. III.C).[2] S_7NH is oxidized by $(NSCl)_3$ in benzene at 80°, in presence of pyridine, to S_4N_4 and S_4N_2 (Chapter 6, Secs. II.A and III.B).[21]

The only reduction reactions proper reported for the sulfur imides are with hydrogen iodide;[18,28] in anhydrous formic acid iodine is produced according to the equations

$$S_4(NH)_4 + 8HI \rightarrow 4I_2 + 4S + 4NH_3$$
$$S_7NH + 2HI \rightarrow I_2 + 7S + NH_3$$

Most chemical studies on the sulfur imides have concerned their behavior as acids or bases, in reactions in which the sulfur–nitrogen ring is preserved. Such reactions have been the main source of the derivatives listed in Tables 2.3, 2.4, and 2.5. They will now be reviewed.

The sulfur imides are weak Brønsted acids. Chivers and Drummond[10] have estimated pK_a (aqueous) for S_7NH at about 5, similar to that for carboxylic acids. This value, based on conductivities, is unconvincing; probably the true value is in the region 9–12,[33] similar to that for succinimide (9.66). As either figure would suggest, the imides are readily deprotonated by strong bases.[10,47,62] A yellow solution of S_7N^- results when S_7NH in tetrahydrofuran is stirred with powdered potassium hydroxide at $-62°$ for a few minutes; the filtered solution is stable for a day below $-62°$, but at $0°$ quickly turns blue by decomposition to NS_4^- (Section 1 above). The mono- and di-anions of 1,4- and 1,5-$S_6(NH)_2$ can be made from the imides and ethyllithium in tetrahydrofuran, but the anions of the 1,3 isomer seem to be unstable.[62] Deprotonation of $S_4(NH)_4$ with KH gives $S_3N_3^-$ (Chapter 6, Sec. III.H).[8] The imide anions are strong nucleophiles. By reaction with appropriate halides, S_7N^- gives derivatives of S_7NH in which the hydrogen atom is replaced by the following groups: alkyl, formate ester, Me_3Si, $Ph_2P(S)$ (Tables 2.3 and 2.4). The sulfur imides undergo several condensation reactions for which the presence of a base is required, suggesting that the imide anions are first formed. In presence of pyridine, S_7NH with S_xCl_2 ($x = 1, 2, 3, 5$) gives $(S_7N)_2S_x$ (Section C below), and with $SOCl_2$ it gives $(S_7N)_2SO$ (Section F below). The diimides with S_2Cl_2 give linear polymers of which the lowest oligomers have been isolated (Section G below). 1,3-$S_6(NH)_2$ condenses with S_5Cl_2 in presence of pyridine to give the fused-ring nitride $S_{11}N_2$ (Section D below). Several of the imides have been acetylated and benzoylated in presence of pyridine (Table 2.4), and the hydroxymethyl derivatives of S_7NH and $S_4(NH)_4$ result from the imide and formaldehyde in presence of sodium hydroxide (Table 2.4).

The sulfur imides do not function as Brønsted bases under ordinary conditions; in some of the standard "superacid" media they would probably be oxidized. However, several reactions are known which should probably be attributed to their reactivity as Lewis bases. $S_4(NH)_4$ forms 1:1 adducts with aluminum chloride and bromide[2] and with silver perchlorate,[46] but other donor reactions of the imides are accompanied by elimination or rearrangement and give rise to sulfur imides substituted on nitrogen (Table 2.3). Thus S_7NH reacts readily with BCl_3 and BBr_3:

$$S_7NH + BCl_3 \xrightarrow[0°]{CS_2} S_7NBCl_2 + HCl$$

but not with the weaker acceptor BF_3.[26,28] A similar reaction occurs with diborane:[41]

$$S_7NH + \tfrac{1}{2}B_2H_6 \xrightarrow[0°]{\text{ether or THF}} S_7NBH_2 + H_2$$

The planar or near-planar hybridization of the nitrogen atoms in the imides implies (Chapter 9, Sec. III.A) that their lone pairs are not readily available in the unperturbed molecules, but models suggest that reorganization to sp^3 (tetrahedral) hybridization is not sterically difficult, and it may take place on the approach of a strong electrophile such as BCl_3.

Table 2.3. Covalent boron, silicon, phosphorus, and metal substitution products of cyclic sulfur imides

Compound	Description	Preparation	Ref.
$S_4(NAg)_4$	red-brown explosive solid	$AgNO_3 + S_4(NH)_4$	18, 20, 65
$(CuNS)_x$	dark brown solid	$CuCl + S_4(NH)_4$	18, 20, 65
$Cu_2Cl_2H_2(NS)_4$	yellow solid	$CuCl_2 + S_4(NH)_4$	18, 20, 65
$(S_7N)_2Hg_2$	light-yellow solid	$S_7NH + Hg_2(NO_3)_2$	18, 20, 65
$(S_7N)_2Hg$	nearly white solid, dec. 20°	$S_7NH + (HgOAc)_2$	51
S_7NHgMe	buff solid, dec. 80°	$S_7NH + MeHgOAc$	51
S_7NHgPh	buff solid, dec. 80°	$S_7NH + PhHgOAc$	51
$(S_6N_2Hg)_x$, 1,3-, 1,4-, and 1,5-isomers	buff-yellow solids, dec. 0–20°	appropriate $S_6(NH)_2$ isomer + $Hg(OAc)_2$	51
$1,4\text{-}S_6(NHgMe)_2$	buff solid, dec. 30°	$1,4\text{-}S_6(NH)_2 +$ MeHgOAc	51
$S_6(NHgPh)_2$, 1,3-, 1,4-, and 1,5-isomers	buff solids, dec. 41–68°	appropriate $S_6(NH)_2$ isomer + PhHgOAc	51
$(HgNS)_x$	yellow solid	$S_4(NH)_4 + Hg_2(NO_3)_2$	18, 20, 65
$Hg_5(NS)_8$	greenish crystals, expl. 160°	$S_4(NH)_4 + Hg(OAc)_2$	18, 20, 65
$Hg(NS)_2$	greenish solid, dec. 140°	$S_4(NH)_4 +$ excess $Hg(OAc)_2$	18, 20, 65
$S_4(NHgPh)_4$	(impure) beige powder	$S_4(NH)_4 + PhHgOAc$	51
S_7NBH_2	crystalline ether adducts	$S_7NH + B_2H_6$	41
S_7NBCl_2	colorless, m.p. 44°	$S_7NH + BCl_3$	26, 28, 41
S_7NBBr_2	red, viscous, impure	$S_7NH + BBr_3$	26, 28, 41
$(S_7N)_2BHal$ and $(S_7N)_3B$	not fully characterized	$S_7NHgPh + BHal_3$ (Hal = Cl, Br)	52
S_7NSiMe_3	m.p. 34°	$(Me_3Si)_2NH + S_7NH$	6
		$S_7N^- + Me_3SiCl$	42
S_7NSnMe_3	orange crystals, dec. 50°	$S_7NHgPh + Me_3SnCl$	52
S_7NPbMe_3	impure yellow solid	$S_7NHgPh + Me_3PbCl$	52
S_7NPbEt_3	impure yellow solid	$S_7NHgPh + Et_3PbCl$	52
$S_7NP(S)Ph_2$	colorless, m.p. 123°	$S_7NHgPh + Ph_2PCl$	52
		$S_7N^- + Ph_2PCl$	36
$1,4\text{-}S_6[NP(S)Ph_2]_2$	colorless, mp. 94°	$1,4\text{-}S_6(NHgPh)_2 + Ph_2PCl$	52

Table 2.4. Organic derivatives of the cyclic sulfur imides.

Compound	Description	Preparation	Ref.
Alkyl derivatives			
S_7NR			
Me	pale greenish-yellow, m.p. 23°	$MeNH_2aq + S_xCl_2$ ($x = 5,7$)	22
		$S_7N^- + MeI$	47
Et	yellow oil	S_7N^- + alkyl iodide	47
Pr^i	yellow oil		47
allyl	yellow oil		47
Bz	yellow, m.p. 45°		47
1,3-$S_6(NMe)_2$	colorless, m.p. 30°	$MeNH_2aq + SCl_2$	22
1,4-$S_6(NH)(NMe)$	solid	$1,4$-$S_6N_2H^- + MeI$	62
1,4-$S_6(NMe)_2$	colorless, m.p. 25°	$1,4$-$S_6N_2^{2-} + 2MeI$	62
		$MeNH_2aq + S_2Cl_2$	22
1,5-$S_6(NH)(NMe)$	n.m.r. and i.r. given	$1,5$-$S_6N_2H^- + MeI$	62
1,5-$S_6(NMe)_2$	colorless, m.p. 84°	$1,5$-$S_6N_2^{2-} + 2MeI$	62
		$MeNH_2aq + S_3Cl_2$	22
1,3,5-$S_5(NMe)_3$	colorless, m.p. 45°	$MeNH_2aq + SCl_2$	22
1,3,6-$S_5(NR)_3$			
Me	colorless, mp. 55°	$MeNH_2aq + SCl_2$	22
Bz	n.m.r. and i.r. given	$BzNH_2 + SCl_2$	54
β-PhEt		β-$PhEtNH_2 + SCl_2$	54
$S_4(NR)_4$			
Me	colorless, m.p. 126°	$MeNH_2 + SCl_2$	60
Et	colorless, m.p. 143°	$EtNH_2 + SCl_2$	54
		$S_4(NAg)_4 + EtI$	20
Bz	colorless, m.p. 150°	$BzNH_2 + SCl_2$	54
β-PhEt	colorless, m.p. 89°	β-$PhEtNH_2 + SCl_2$	54
Acyl derivatives			
S_7NCOR			
Me	m.p. 104°	$S_7NH + MeCOCl + K_2CO_3$	13, 44
CH_2Cl	m.p. 110°		
Ph	m.p. 138°		
Bz	m.p. 110°		
$CH_3(CH_2)_7$	liquid	S_7NH + carboxylic acid in presence of dicyclohexylcarbodiimide or ethoxyacetylene	28
$CH_3(CH_2)_{17}$	m.p. 85°		28
PhCH=CH	m.p. 146°		28
o-ClC_6H_4	m.p. 92°		28
p-$NO_2C_6H_4$	m.p. 136°		28
$S_7NCO.CONS_7$	colorless, m.p. 174°	S_7NHgPh + oxalyl chloride	52
$S_7NCO(CH_2)_2$-$CONS_7$	colorless, m.p. 195°	S_7NH + succinyl chloride + py	25
1,3-$S_6(NCOR)_2$			
Me	colorless, m.p. 130°	1,3-$S_6(NH)_2$ + acyl chloride + py	25
Bz	unstable		25

III. EIGHT-RING IMIDES

Table 2.4. (*continued*)

Compound	Description	Preparation	Ref.
Acyl derivatives			
1,4-S_6(NCOR)$_2$			
Me	colorless, m.p. 77°	1,4-S_6(NH)$_2$ + acyl	25
Bz	colorless, m.p. 156°	chloride + py	25
1,5-S_6(NCOR)$_2$			
Me	colorless, m.p. 158°	1,5-S_6(NH)$_2$ + acyl	25
Bz	colorless, m.p. 183°	chloride + py	25
S_4(NCOR)$_4$			
Me	m.p. 87.5°	S_4(NH)$_4$ + MeCOCl + K_2CO_3	2
PhNH	cream needles m.p. 224°	S_4(NH)$_4$ + PhNCO	28
Formate esters			
S_7NCO$_2$R			
Et	yellow liquid	S_7N$^-$ + ClCO$_2$R	42
Bz	colorless, m.p. 99°		42
Hydroxymethyl derivatives and their esters			
S_7NCH$_2$OH	yellow needles m.p. 111°	S_7NH + HCHO + NaOH	28
1,3,5-S_5(NCH$_2$OH)$_3$	colorless needles m.p. 148°	1,3,5-S_5(NH)$_3$ + HCHO + NaOH	45
S_4(NCH$_2$OH)$_4$	white needles m.p. 173°	S_4(NH)$_4$ + HCHO + NaOH	28
S_4(NCH$_2$OCOR)$_4$			
Me	m.p. 88°	S_4(NCH$_2$OH)$_4$ + acyl chloride	2
p-NO$_2$C$_6$H$_4$	creamy plates m.p. 222°		2
Other derivatives			
S_7NCH$_2$N(CH$_2$)$_5$	m.p. 60°	S_7NH + N-hydroxymethyl-piperidine	6
S_7NSNR$_2$			
Me	m.p. 55°	S_7NH + (R$_2$N)$_2$S	9
Et	yellowish m.p. 30°		
Pr^i	crystals m.p. 66°		
C$_6$H$_{11}$	m.p. 98°		

5. Metal derivatives

These consist of ionic derivatives of the "a" class alkali metal ions, and derivatives, probably covalent, of the "b" class post-transition metal ions.

The anions of the sulfur imides have been mentioned in Section 4. Air-sensitive solid sodium salts have been made from S_7NH and S_4(NH)$_4$ with sodium triphenylmethyl.[28] Their constitutions are uncertain. The reaction of

$S_4(NH)_4$ with potassium hydride gives mainly $K^+[S_3N_3]^-$ and $K^+[NS_4]^-$ (Chapter 6, Secs. III.H and V.B), not potassium salts of the imide itself[8] (cf. ref. 20).

The most stable and best-characterized derivatives of the imides with posttransition metals are the mercury compounds shown in Table 2.3 which are all readily precipitated by metathesis of the imides with mercury(II) acetate, organomercury(II) acetates, or mercury(I) nitrate in polar organic solvents.[51] Most of the simple mercury compounds decompose fairly quickly at room temperature or even below; decomposition of $(S_7N)_2Hg$ is an excellent source of S_4N_2 (Chapter 6, Sec. III.B). The phenylmercury compounds are rather more stable (Table 2.3), and make useful intermediates, capable of reacting with various compounds containing labile halogen to give derivatives such as $(S_7N)_2SO$ and $S_7NP(S)Ph_2$ (Table 2.3).[52] All these mercury compounds are buff to yellow to green powders, the monomeric ones being somewhat soluble in CS_2. From their reactions[52] it is evident that those derived from S_7NH and the diimides, at least, contain the intact S–N rings of the parent imides. Copper derivatives (Table 2.3) and very unstable silver and thallium derivatives can also be made from the imides,[20,28,52,65] but have not been adequately investigated.

6. Organic derivatives

Many alkyl derivatives of the sulfur imides have been described (Table 2.4). They are fairly stable, colorless or almost colorless, low-melting solids, easily recrystallized from hexane. Although they can be made in some cases from the imides themselves by conversion into their anions (Section 4 above) and subsequent treatment with an alkyl halide, a better route is reaction of a chlorosulfane with the appropriate alkylamine. Somewhat surprisingly, aqueous methylamine works well in this preparation,[22,23] and an appropriate choice of chlorosulfane among the two commercially available and two others easily prepared will give a product mixture rich in any desired methylated sulfur

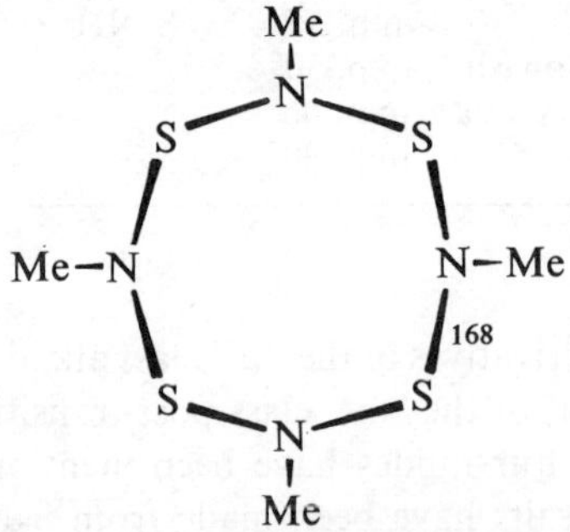

Figure 2.3. Molecular structure of $S_4(NMe)_4$;[39] bond length in pm.

imide except $S_4(NMe)_4$.[22] The structure of $S_4(NMe)_4$ is shown in Fig. 2.3 and discussed in Chapter 9, Sec. III.A.[39] It preserves the puckered ring of $S_4(NH)_4$ (Fig. 2.1). Consistently with the near-planar configuration about nitrogen, this compound is a very weak base.[39] S_7NMe similarly has been shown not to combine with BCl_3.[23] The chromatographic R_f values of all the methyl imides of this group, and dipole moments of some, have been reported.[22]

The acyls and formate esters, the hydroxymethyl derivatives and their esters, and the R_2NS derivatives are mostly colorless or pale yellow covalent compounds (Table 2.4), and are obtainable by standard methods indicated in the table.

No derivative of a sulfur imide with a chiral organic group has yet been reported. Such derivatives might provide a way of resolving the enantiomers of 1,4-$S_6(NH)_2$, which could in turn throw light on the conformational stability of the 1,4-S_6N_2 ring (see Section H below).

C. Sulfur nitrides with coupled eight-membered rings

Four sulfur nitrides with formula **I** have been prepared. The preparative method was first worked out by Becke-Goehring, Jenne, and Rekalic[4] for

```
    S—S—S        S—S—S
    |   |        |   |
    S   N—Sₓ—N   S
    |   |        |   |
    S—S—S        S—S—S
```

(I)

$S_{15}N_2$ (**I**, $x = 1$) and $S_{16}N_2$ ($x = 2$), and later extended to $S_{17}N_2$ ($x = 3$) and $S_{19}N_2$ ($x = 5$) by Shahid.[14] The appropriate chlorosulfane S_xCl_2, dissolved in CS_2, is added gradually to a stirred solution of S_7NH and the stoichiometric amount of dry pyridine in CS_2, at room temperature:

$$2S_7NH + S_xCl_2 + 2py \rightarrow (S_7N)_2S_x + 2pyH^+Cl^- \downarrow$$

The chief requirement for success is a good, freshly purified, sample of the chlorosulfane. Failing this, the resulting mixture of $(S_7N)_2S_x$ compounds, containing possibly also S_8, will be almost impossible to separate. SCl_2, especially, is exceedingly unstable and must be used immediately after purification (refs. 22, 56, and 60 will be found helpful). It is strongly recommended to distil chlorosulfanes in an all-glass rotary film evaporator, at 0.01 Torr, near or below room temperature, rather than in the usual primitive contraption with an air leak.

$S_{15}N_2$, $S_{16}N_2$, and $S_{17}N_2$ are yellow crystalline solids; $S_{19}N_2$ was obtained as a red viscous liquid which would not crystallize but which was shown chromatographically to be one substance.[31] All decompose slowly on long storage at room temperature. They are moderately soluble in CS_2 ($S_{17}N_2$, 0.0346 molal at 20°)[56] and almost insoluble in other indifferent solvents. A good method of recrystallization is to add a little CCl_4 to a CS_2 solution of the nitride, and leave the solution to evaporate slowly in a loosely covered beaker. These nitrides are not readily characterized by elemental analysis, but their i.r. spectra in CS_2 solution are distinctive (Fig. 2.4). Table 2.5 gives their melting points and dipole moments. Crystal-structure determinations[17] on $S_{16}N_2$ and $S_{17}N_2$ confirm the formula **I** and show a planar arrangement of each nitrogen atom with its three sulfur neighbors (discussed in Chapter 9, Sec. III.A).

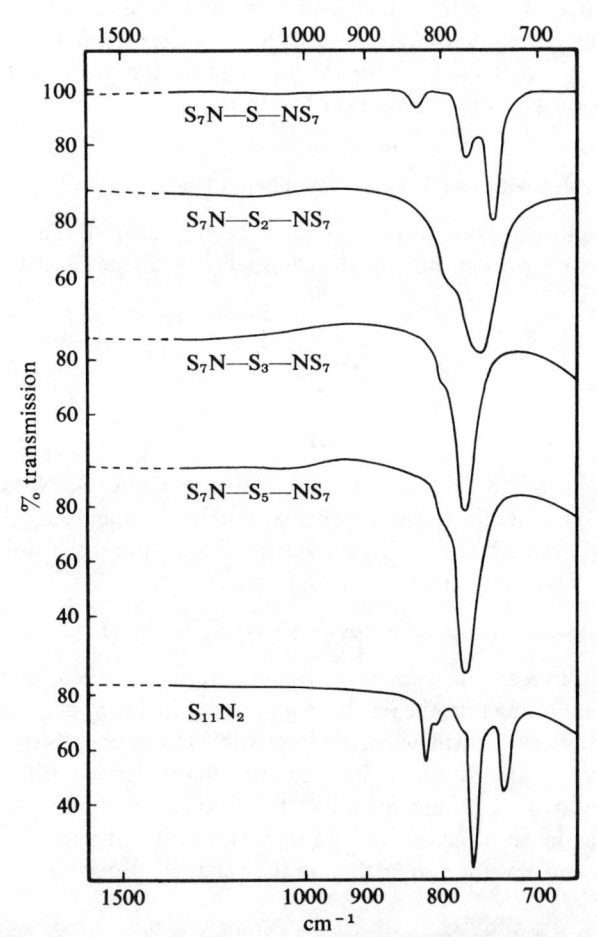

Figure 2.4. Infrared absorption spectra of the coupled-ring sulfur nitrides $S_7N-S_x-NS_7$ ($x = 1, 2, 3, 5$) and the fused-ring nitride $S_{11}N_2$.[31] CS_2 solutions, 2–4%; 0.2 mm cell; all strong absorptions in the sodium chloride region are shown.

III. EIGHT-RING IMIDES   31

Table 2.5. Properties of fused- and coupled-ring sulfur nitrides.[31]

	Fused-ring		Coupled-ring		
	$S_{11}N_2$	$S_{15}N_2$	$S_{16}N_2$	$S_{17}N_2$	$S_{19}N_2$
M.p. (°C)	150 dec.	137	122	97	oil
Dipole moment (debyes)	0.46	0.79	1.46	0.88	

These nitrides have seldom been made and their chemistry is scarcely known. When heated for a few minutes just above their melting points, $S_{15}N_2$, $S_{16}N_2$, and $S_{17}N_2$ decompose[31,55] giving some S_4N_4. This is produced in largest yield from $S_{15}N_2$, which alone of the nitrides contains in its molecule the S–N–S–N sequence present in S_4N_4. $S_{16}N_2$ yields S_7NH when treated with piperidine in dimethylformamide.[4]

The first member just described of the homologous series $S_7N-S_x-NS_7$, with $x = 0$, would be particularly interesting. Unsuccessful attempts have been made to prepare it by the action of iodine on $(S_7N)_2Hg$ and on S_7NHgPh. Reaction occurred at once in CS_2 solution at 0°, with quantitative formation of HgI_2 or PhHgI respectively, but the CS_2-soluble products appeared from their i.r. spectra and chromatographic behavior to consist mainly of mixed compounds of the series $S_7N-S_x-NS_7$, with x averaging about 6 in the case of S_7NHgPh. Probably S_7N-NS_7 would be unstable, with a strong tendency to eliminate N_2.[52]

D. The fused-ring sulfur nitride $S_{11}N_2$

This compound, with formula **II**, was first prepared by Heal and Shahid in 1969, after trials with molecular models had shown that two rings with the conformation of S_8 could probably be fused together at the 1,3 positions without much strain.[31] It is prepared by a condensation of the same type as in

```
S—S—N—S—S          S—S—NH
|    |    |         |    |
S    S    S         S    S
|    |    |         |    |
S—S—N—S—S          S—S—NH
      (II)              (III)
```

Section C above, and carried out under similar conditions, but between S_5Cl_2 and 1,3-hexasulfur diimide (**III**) in a 1:1 mol ratio.[31] The primary yield of $S_{11}N_2$ is small, and large amounts of polymers are also formed, as would be expected from a coupling reaction between two bifunctional molecules. However, the polymers seem to be thermodynamically unstable, and break down when refluxed with CS_2 to give a total yield of $S_{11}N_2$ up to 38%. The use of Ruggli–Ziegler dilution conditions maximizes the primary yield of $S_{11}N_2$, but does not in practice offer any advantage over ordinary mixing technique followed by decomposition of the polymers. Gel-permeation chromatography on polystyrene is essential for the isolation of $S_{11}N_2$ in a pure state from the product mixture. Since the starting material **III** is tedious to make pure (Section B.1 above), it is worth knowing that the crude mixture of **III** with its 1,5 isomer also gives good results in the preparation of $S_{11}N_2$.

The properties of this interesting compound have been fairly fully reported.[32] There are two crystalline forms, low-temperature α (unsymmetrical octahedra), and β (thin platelets) stable above 21°. At this temperature both forms have solubility 2.8 g per 100 g CS_2; the compound is almost insoluble in other indifferent organic solvents. X-ray crystallography has shown that the two forms differ only in the mode of packing, both being made up of molecules as shown in Fig. 2.5,[17] with rings in the crown conformation of S_8 and planar coordination of the nitrogen atoms (Chapter 9, Sec. III.A). The crystals have a very pale amber color, and the electronic absorption spectrum in hexane solution closely resembles that of S_8.[32] The i.r. spectrum in CS_2 solution is distinctive (Fig. 2.4) and provides the best means of identification; complete vibrational spectra have been reported.[32] From differential thermal analysis curves for its decomposition in molten sulfur between 140° and 190°, ΔH_f° of $S_{11}N_2$ has been estimated to be 282 kJ mol^{-1}.[32] With certain assumptions an S–N bond-energy term of 248 kJ mol^{-1} can be deduced, about 20 kJ mol^{-1} higher than the single-bond energy calculated by Pauling's method.[32] Neither this small disparity nor the bond length is convincing evidence for pπ–dπ bonding (Chapter 9, Sec. III.A).

The lone pairs of planar-hybridized nitrogen atoms cannot be localized in sp^3 orbitals. It is no surprise, therefore, that $S_{11}N_2$ has a very small dipole moment, 0.46 D, and no basic character, being insoluble in and unaffected by

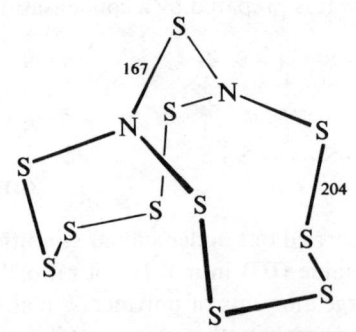

Figure 2.5. Molecular structure of the fused-ring sulfur nitride $S_{11}N_2$;[32] bond lengths are in pm.

cold concentrated sulfuric and acetic acids, and refusing to combine with BCl_3 in CS_2.[32]

E. The search for other fused-ring sulfur nitrides

Molecular models suggest that structures **IV**, **V**, and **VI** would also be fairly free from strain; moreover, cyclo-S_6 (compare **IV**) and cyclo-S_{10} (compare **V**) are well characterized, though rather unstable, molecular forms of sulfur.

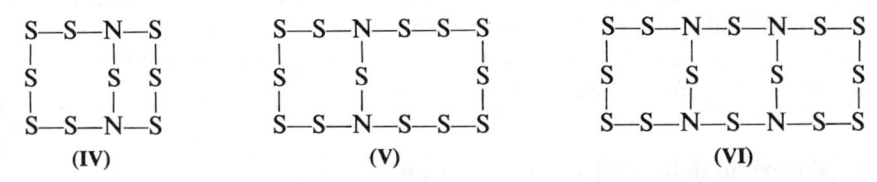

```
S—S—N—S          S—S—N—S—S—S              S—S—N—S—N—S—S
|   |  |          |   |     |              |   |     |     |
S   S  S          S   S     S              S   S     S     S
|   |  |          |   |     |              |   |     |     |
S—S—N—S          S—S—N—S—S—S              S—S—N—S—N—S—S
   (IV)              (V)                         (VI)
```

However, attempts to condense 1,3-$S_6(NH)_2$ with S_3Cl_2 to give **IV**, and with S_7Cl_2 to give **V**, were unsuccessful, despite the use of the Ruggli–Ziegler dilution technique.[31] In both cases the only fused-ring sulfur nitride found in the products was $S_{11}N_2$ (**II**). The problem seems to be that **IV** and **V** must both be thermodynamically unstable with respect to $S_{11}N_2$, while even under the best conditions of kinetic control the primary yields of **IV** and **V** are very small.

Structure **VI** has not been made, but with its three 8-membered rings might well be thermodynamically very stable, and preparable with good planning. Old-fashioned "lasso chemistry" would suggest the condensation of SCl_2 with 1,3-$S_6(NH)_2$ as a possible route to this compound, but for kinetic reasons the formation of linear polymers would probably proceed much faster than the closure of the middle ring of **VI**. Perhaps the best strategy would be deliberately to make a mixture of linear polymers from SCl_2 and 1,3-$S_6(NH)_2$ and then decompose them gently, e.g. by refluxing with CS_2, in the hope that they would eliminate **VI** in a thermodynamically controlled process. A compound **VI** might be difficult to recognize and work up, since it would probably be only sparingly soluble in CS_2 and almost insoluble in all other indifferent solvents.

F. Bis(heptasulfurimido) sulfoxide, $S_{15}N_2O$

This sulfur–nitrogen compound **VII** was prepared independently by Steudel and Rose[57] and by Heal and Ramsay.[52] It is another derivative of heptasulfur

```
S—S—S   O   S—S—S
|   |   ‖   |   |
S   N—S—N   S
|   |   |   |
S—S—S   S—S—S
       (VII)
```

imide and is structurally related to the nitride $S_{15}N_2$ (**I**, $x = 1$). It is formed in up to 50% yield, from $SOBr_2$ and S_7NHgPh,[52] or from $SOCl_2$ and S_7NH or S_7NHgPh in presence of pyridine,[52,57] or from $S_{15}N_2$ (Section C above) and trifluoroperoxoacetic acid.[59] It is conveniently purified by gel-permeation chromatography on polystyrene.[50,52]

$(S_7N)_2SO$ crystallizes from CS_2 in colorless needles which melt with decomposition at 125–127° (slow heating) or 136° (fast heating). It is fairly stable in moist air.[57] The structure has not been determined crystallographically but the vibrational spectra,[50,52,57] so far as they have been assigned, seem consistent with the presumed structure **VII**. There is a strong S=O stretch at 1 16 cm^{-1}, in the usual region for thionyl compounds.

G. Polymeric sulfur nitride-imides: general

The formation of sulfur nitrides $S_7N-S_x-NS_7$ from S_7NH and S_xCl_2 has been described in Section C. When the bifunctional sulfur diimide isomers 1,3-, 1,4-, and 1,5-$S_6(NH)_2$ are used instead of S_7NH in reactions of this kind, linear polymers are produced. This is the general situation. In a few special cases where ring-closure is sterically possible, cyclic oligomers may also be expected; thus $S_{11}N_2$ (Section D above) can be regarded as the first ($n = 1$) of a series of cyclic oligomers with the general formula **VIII** formed by condensation of 1,3-$S_6(NH)_2$ with S_5Cl_2.

$$\begin{bmatrix} N-S-N-S_5 \\ | \quad\quad\quad | \\ S \quad\quad\quad S \\ | \quad\quad\quad | \\ S-S-S \end{bmatrix}_n$$

(**VIII**)

The linear oligomers formed from all three sulfur diimides with S_2Cl_2 have been investigated.[29] The formation of small-ring oligomers is geometrically impossible with this chlorosulfane, and, by the use of a 2:1 ratio of imide to chlorosulfane, the linear polymers produced in appreciable amounts were limited to the first few members of each of the homologous series **IX**, **X**, and **XI**. In the simplest case, that of the 1,5 diimide, compounds **XI** with $n = 1, 2, 3$, and 4 were isolated by preparative-scale thin-layer chromatography on silica gel, as small yellow crystals. With the other diimides, however, complications arose, as follows.

Molecular models show that the first oligomer from the 1,3 diimide (**IX**, $n = 1$) can exist in two topologically non-equivalent forms (loosely describable as *cis–trans* isomers). These were partly separated by chromatography. One has a single N–H stretching band in its i.r. spectrum and the other a

(IX) from 1,3-S$_6$(NH)$_2$

(X) from 1,4-S$_6$(NH)$_2$

(XI) from 1,5-S$_6$(NH)$_2$

close doublet, attributed to intramolecular hydrogen-bonding. Both isomers changed in a few hours at 20° in CS$_2$ solution into an equilibrium mixture containing about 73% of the hydrogen-bonded isomer. The 1,4 diimide is chiral, and its first coupling product (**X**, $n = 1$) should exist as diastereoisomers with rings of the same or opposite chiralities respectively. These were in fact successfully separated and shown by chromatography to epimerize in a few hours at room temperature; their i.r. spectra are indistinguishable. Of course, increasingly complicated possibilities of isomerism arise for the higher oligomers of **IX** and **X**.[30,31]

H. The conformational stability of saturated S–N rings

For any 8-membered saturated S–N ring, or for S$_8$ itself, it is possible to imagine the ring being turned "inside out" without bond breakage. In all but two of the rings shown in Fig. 2.1, the result would be a ring indistinguishable from the original one; with the chiral rings of the 1,4 diimide and the 1,3,6 triimide, however, the product would be the enantiomeric ring of opposite chirality. Does this happen, and how could it be studied?

A straightforward way, not yet tested, would be to resolve the enantiomers of the 1,4 diimide or 1,3,6 triimide and study their racemization. It would be necessary to show that any racemization observed did not proceed by way of ring opening; this would be established by the absence of other products such as plastic sulfur, S$_8$, or other cyclic imides.

The experiments described above in Section G on the compounds **IX** ($n = 1$) and **X** ($n = 1$) throw light on this problem. The "*cis–trans*" isomerization of the first, and the epimerization of the second, both took place at rates which (qualitatively) appeared not to depend on concentration, implying first-order intramolecular processes.[29] In neither case were any by-products produced, such as would almost inevitably arise if bonds were broken. Moreover, the half-reaction times, about 3 hours in each case at 20°, imply activation energies of approximately 96 kJ mol^{-1}, too low if a bond breakage were the rate-determining step. It seems reasonable to speculate that in both cases the mechanism is an inversion of one of the rings, and that we have here in principle a method for studying such inversions.

IV. Sulfur imides based on rings other than eight-membered

A. Introduction

Crystalline sulfur allotropes are now known containing cyclic molecules with six, seven, nine, ten, eleven, twelve, eighteen, and twenty atoms,[43] besides the common S_8. Although all these unusual forms are thermodynamically unstable under ordinary conditions with respect to S_8, some of them, notably S_{12} and S_{18}, are pretty stable kinetically. The question arises whether sulfur imides could exist in which sulfur atoms of these rings are replaced by –NH–.

No saturated rings other than the 8-membered ones of Fig. 2.1 have ever been detected in the products from the standard preparations of sulfur imides described above in Section III.B. However, it must be remembered that strongly basic solvents or reagents are used in most of these preparations, and that sulfur imide rings are reversibly converted by strong bases into NS_4^- (Section III.B above); hence thermodynamic equilibration would be promoted in the reaction mixtures and rings having other than eight members would tend to be destroyed. A minor crystalline product from the hydrazine reduction of S_4N_4 (Section III.B.1 above) was originally identified as $S_{11}NH$, with the ring structure of S_{12},[16] but reinterpretation of the X-ray diffraction data has shown its lattice to be composed of a mixture of S_8 molecules and molecules of 8-membered cyclic diimides;[61] chromatography of its solution in CS_2 confirms this. An early report of the reduction of S_4N_2 to a 6-membered imide $S_4(NH)_2$ seems to have been discounted by later work.[31]

A few alkyl-substituted sulfur imides with 6-membered rings have been prepared in the absence of strong bases, as follows.

B. Derivatives of six-membered ring imides

When S_2Cl_2 reacts with primary amines under Ruggli–Ziegler dilution conditions

$$2S_2Cl_2 + 6RNH_2 \xrightarrow[0°]{\text{diethyl ether}} S_4(NR)_2 + 4[RNH_2]^+Cl^-$$

compounds are formed, in 25–85% yield, which are thought to have the constitution **XII** on the evidence of analysis and molecular weight,[5,23] confirmed in the methyl case by i.r.[23] The dilution technique ensures the absence of excess base (amine) at all stages and maximizes the probability of ring

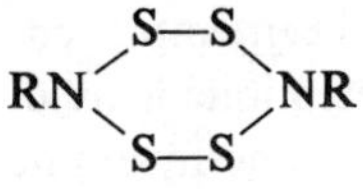

$$\text{RN} \underset{S-S}{\overset{S-S}{\diamond}} \text{NR}$$

(XII)

(R = Me, Et, n-dodecyl, cyclohexyl, Bz, PhCH$_2$CH$_2$)

closure. Without it, yields of the rings are very small and large quantities of polymers are formed instead.[5]

Compounds **XII** form colorless low-melting crystals soluble in the usual organic solvents and recrystallizable from them. The methyl,[23] ethyl, and dodecyl compounds[5] decompose slowly but the others are stable at room temperature. No crystallographic or chemical studies have been reported.

C. Derivatives of twelve-membered ring imides

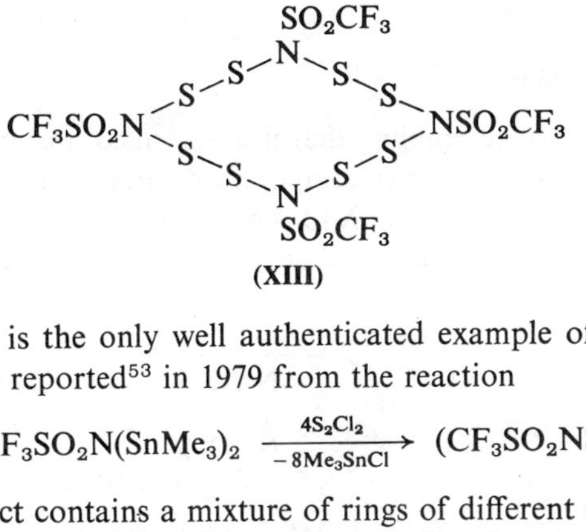

$$\text{CF}_3\text{SO}_2\text{N} \underset{S-S-N-S-S}{\overset{S-S-N-S-S}{\diamond}} \text{NSO}_2\text{CF}_3$$
(with SO$_2$CF$_3$ groups on the two ring N atoms)

(XIII)

Compound **XIII** is the only well authenticated example of a 12-membered S–N ring. It was reported[53] in 1979 from the reaction

$$4\text{CF}_3\text{SO}_2\text{N}(\text{SnMe}_3)_2 \xrightarrow[-8\text{Me}_3\text{SnCl}]{4\text{S}_2\text{Cl}_2} (\text{CF}_3\text{SO}_2\text{NS}_2)_4$$

The crude product contains a mixture of rings of different sizes. Compound **XIII** remains behind after subliming off the smaller rings *in vacuo* up to 140°, and can be recrystallized from methylene chloride. It melts with decomposition at 213°.

The sulfur–nitrogen ring in **XIII** does not correspond in shape to the S$_{12}$ ring[43] but nevertheless has unstrained dihedral angles at sulfur.

V. Cyclic sulfur hydrazides

A. Eight-membered rings

The reactions of hydrazine with chlorosulfanes do not seem to have been studied, but hydrazine is so easily oxidized, e.g. by sulfur itself, that this would be an unpromising approach to sulfur–nitrogen rings containing N–N bonds. The hydrazine dicarboxylate EtO$_2$CN(H)–(H)NCO$_2$Et is, in contrast, very resistant to oxidation, and has been employed in the synthesis of compounds **XIV** and **XV**.[37] The dilution technique is used, very much as in Section IV.B above, but with the addition of trimethylamine as a hydrogen chloride acceptor. With S_2Cl_2, **XIV** and **XV** are both produced, together with polymer; S_6Cl_2 affords **XV**. These cyclic hydrazides separate from hexane as colorless crystals, **XIV** melting at 110° and **XV** at 80°. Their crystal structures have not been determined, but the presence of N–N bonds is shown by the mass spectra and by the formation of the hydrazine dicarboxylate when either compound is desulfurized with Raney nickel.[37]

$$\begin{array}{cc} \text{(XIV)} & \text{(XV)} \end{array}$$

B. Six-membered rings

A preparative method resembling that just described, but using SCl_2 instead of S_2Cl_2, affords **XVI**, the ethyl compound melting at 150° and the t-butyl compound at 170°. The structure **XVI** is supported by osmometric molecular weight determinations, i.r., and the formation of hydrazine on treatment with trifluoroacetic acid.[64]

(XVI)

References

1. Alford, J. R., Bigg, D. C. H. and Heal, H. G., *J. Inorg. Nucl. Chem.* **29**, 1538 (1967)
2. Banister, A. J. and Younger, D., *J. Inorg Nucl. Chem.* **32**, 3763 (1970)
3. Becke-Goehring, M., Jenne, H. and Fluck, E., *Chem. Ber.* **91**, 1947 (1958)
4. Becke-Goehring, M., Jenne, H. and Rekalic, V., *Chem. Ber.* **92**, 855, 1237 (1959)
5. Becke-Goehring, M. and Jenne, H., *Chem. Ber.* **92**, 1149 (1959)
6. Becke-Goehring, M., *Angew. Chem.* **73**, 589 (1961)
7. Bojes, J. and Chivers, T., *J. Chem. Soc. Dalton Trans.* 1715 (1975)
8. Bojes, J., Chivers, T., Drummond, I. and MacLean, G., *Inorg. Chem.* **17**, 3668 (1978)
9. Bruce, R. B., Gillespie, R. J. and Slim, D. R., *Can. J. Chem.* **56**, 2927 (1978)
10. Chivers, T. and Drummond, I., *Inorg. Chem.* **13**, 1222 (1974)

REFERENCES

11. Fehér, F., Kreutz, R. and Minz, Fr.-R., *Z. Naturforsch. B* **31b**, 918 (1965)
12. Fluck, E. and Boeing, H., *Chem. Ztg. Chem. App.* **94**, 331 (1970)
13. Garcia-Fernandez, H. *Bull. Soc. Chim. France* 3647 (1967)
14. Garcia-Fernandez, H., Heal, H. G. and Shahid, M. S., *Compt. rend. C* **272**, 60 (1971)
15. Garcia-Fernandez, H., *Bull. Soc. Chim. France* 1210 (1973)
16. Garcia-Fernandez, H., Heal, H. G. and Teste de Sagey, G., *Compt. rend. C* **278**, 517 (1974)
17. Garcia-Fernandez, H., Heal, H. G. and Teste de Sagey, G., *Compt. rend. C* **282**, 241 (1976)
18. *Gmelins Handbuch der Anorganischen Chemie*, vol. 9, part B, section 3, "Schwefel", Verlag Chemie, Weinheim (1963)
19. Goehring, M. and Hohenschutz, H., *Naturwiss.* **40**, 291 (1953)
20. Goehring, M., *Ergebnisse und Probleme der Chemie der Schwefelstickstoffverbindungen*, Akademie-Verlag, Berlin (1957)
21. Golloch, A. and Kuss, M., *Z. Naturforsch. B.* **29b**, 320 (1974)
22. Gordon, W. I. and Heal, H. G., *J. Inorg. Nucl. Chem.* **32**, 1863 (1970)
23. Gordon, W. I., Ph.D. thesis, Queen's University of Belfast (1971)
24. Hamada, S., Kudo, Y. and Kawano, M., *Bull. Chem. Soc. Japan*, **48**, 2963 (1975)
25. Heal, H. G. and Kane, J., *J. Chem. Eng. Data* **10**, 386 (1965)
26. Heal, H. G., *J. Chem. Soc.* 4442 (1962)
27. Heal, H. G. and Kane, J., *Inorg. Synth.* **11**, 184 (1968)
28. Heal, H. G., in *Inorganic Sulphur Chemistry*, ed. G. Nickless, Elsevier, Amsterdam (1968)
29. Heal, H. G. and Kane, J., *J. Polymer Sci. C* 3491 (1968)
30. Heal, H. G., *Int. J. Sulfur Chem. C* **6**, 27 (1971)
31. Heal, H. G., *Adv. Inorg. Chem. Radiochem.* **15**, 375 (1972)
32. Heal, H. G., Shahid, M. S. and Garcia-Fernandez, H., *J. Inorg. Nucl. Chem.* **35**, 1693 (1973)
33. Heal, H. G., paper presented at the First International Symposium on Inorganic Heterocyclics, Besançon, France, 1975
34. Hecht, H.-J., Reinhardt, R., Steudel, R. and Bradaczek, H., *Z. Anorg. Allg. Chem.* **426**, 43 (1976)
35. Houben-Weyl, *Methoden der Organischen Chemie*, vol. IX, Thieme-Verlag, Stuttgart (1955)
36. Kanamueller, J. M., *J. Inorg. Nucl. Chem.* **36**, 3855 (1974)
37. Lingmann, H. and Linke, K.-H., *Angew. Chem. Int. Ed. Engl.* **9**, 956 (1970)
38. Lipp, S. A., Chang, J. J. and Jolly, W. L., *Inorg. Chem.* **9**, 1970 (1970); cf. Gillespie, R. J. et al., *Can. J. Chem.* **53**, 3147 (1975)
39. Macdonald, A. L. and Trotter, J., *Can. J. Chem.* **51**, 2504 (1973)
40. Machmer, P., *Z. Naturforsch. B* **24b** 1056 (1969)
41. Mendelsohn, M. H. and Jolly, W. L., *Inorg. Chem.* **11**, 1944 (1972)
42. Mendelsohn, M. H. and Jolly, W. L., *J. Inorg. Nucl. Chem.* **35**, 95 (1973)
43. Meyer, B., *Adv. Inorg. Chem. Radiochem.* **18**, 287 (1976)
44. Molina, E. Vegas, Martinez-Ripoll, M. and Garcia-Blanco, S., second meeting of the first International Symposium on Inorganic Heteroatom Ring Systems, Madrid, 1977, p. 328 of Proceedings (publ. by C.N.R.S., Bellevue, 92190 Meudon, France)
45. Molina, E. Vegas, Martinez-Ripoll, M. and Garcia-Fernandez, H., *ibid.*, p. 329

46. Nabi, S. N., *J. Chem. Soc. Dalton Trans.* 1152 (1977)
47. Olsen, B. A. and Olsen, F. P., *Inorg. Chem.* **8**, 1736 (1969)
48. Postma, H. J. van Bolhuis, F. and Vos, A., *Acta Cryst.* **B27**, 2480 (1971)
49. Postma, H. J., van Bolhuis, F. and Vos, A., *Acta Cryst.* **B29**, 915 (1973)
50. Ramsay, R. J., Ph.D. thesis, Queen's University of Belfast (1974)
51. Ramsay, R. J., Heal, H. G. and Garcia-Fernandez, H., *J. Chem. Soc. Dalton Trans.* 234 (1976)
52. Ramsay, R. J., Heal, H. G. and Garcia-Fernandez, H., *J. Chem. Soc. Dalton Trans.* 237 (1976)
53. Roesky, H. W., Diehl, M., Krebs, B. and Hein, M., *Z. Naturforsch.* B **34b**, 814 (1979)
54. Ross, L. A., Roscoe, J. S. and Pace, A., Jr., *J. Chem. Eng. Data* **8**, 611 (1963)
55. Shahid, M. S., Heal, H. G. and Garcia-Fernandez, H., *J. Inorg. Nucl. Chem.* **33**, 4364 (1971)
56. Shahid, M. S., Ph.D. thesis, Queen's University of Belfast (1971)
57. Steudel, R. and Rose, F., *Z. Naturforsch.* B **30b**, 810 (1975)
58. Steudel, R., *J. Phys. Chem.* **81**, 343 (1977)
59. Steudel, R. and Rose, F., *Z. Naturforsch.* B **33b**, 122 (1978)
60. Stone, B. D. and Nielsen, M. L., *J. Am. Chem. Soc.* **81**, 3580 (1959)
61. Teste de Sagey, G., paper presented at the First International Symposium on Inorganic Heterocyclics, Besançon, France, 1975
62. Tingle, E. M. and Olsen, F. P., *Inorg. Chem.* **8**, 1741 (1969)
63. Van de Grampel, J. C. and Vos, A., *Acta Cryst.* **B25**, 611 (1969)
64. Weinstein, B. and Chang, H.-H., *J. Heterocycl. Chem.* **11**, 99 (1974)
65. Weiss, J. *Fortschr. Chem. Forsch.* **5**, 635 (1966)

3
Imides and amides of sulfur(IV) as source materials for inorganic heterocycles

I. Introduction

The imides and amides of sulfur(IV), in their monomeric forms, are not themselves cyclic compounds, but a number of inorganic heterocyclics are derived from them. There is, namely, at least one cyclic oligomer of thionyl imide; also, certain compounds which contain the –N=S=O, –N=S=N–, or =S=NR groups, and so belong to the class of sulfur(IV) imides, are now being used as intermediates for the synthesis of inorganic heterocyclics. The present chapter covers first these topics, and also contiguous areas of non-cyclic chemistry, which are included partly for their intrinsic interest and possibilities, and partly in order to place the central theme in context. It concludes with a note on some S–N–P rings containing sulfur(IV).

II. Structural relationships

The compounds to be described are classified here essentially according to the "nitrogen system" of E. C. Franklin (1935), who regarded the imide group, =NH, as the "nitrogen equivalent" of =O, and the amide group, –NH$_2$, as "equivalent" to –OH. This is the usual basis for nomenclature and systematization, but it is unsatisfactory on two counts. First, some important parent compounds in the scheme, such as sulfurous acid, have only a hypothetical existence. Secondly, no account is taken of donor–acceptor compounds; thus sulfamic acid or amidosulfuric acid, really sulfur trioxide–ammonia, H$_3$N→SO$_3$, is usually named as if it were the hypothetical isomer H$_2$NSO$_3$H. However, in such a confusing field, even an inadequate signpost is better than none, and no alternative is in fact available.

For the purposes of inorganic heterocyclic chemistry, some imides and amides of sulfur(IV) can be regarded under Franklin's system as structurally related to sulfur dioxide (Fig. 3.1) or sulfurous acid (Fig. 3.2). These diagrams also show certain oligomeric compounds, namely, the cyclic tetramer of thionyl imide mentioned above (Fig. 3.1) and two condensation dimers (Fig.

3.2) derived respectively from amidosulfurous acid and thionyl diamide. In principle, one might expect several series of polymers with the repeating unit –S(O)–NH–, but apart from those just mentioned, only one, formulated as HO–(SONH)$_4$–S(O)OH, has been reported.[16]

The above compounds will be treated in the sequence of Figs. 3.1 and 3.2, starting on the left in each case. The sulfimides R$_2$SNR' fall outside the scope of Figs. 3.1 and 3.2, but can be viewed under Franklin's system as derivatives of the sulfoxides R$_2$SO. They are treated next. The mixed P–N–S rings dealt with last have no analogs under Franklin's viewpoint.

```
                        O=S=O    sulfur dioxide
                          |
         ┌────────────────┴────────────────┐
      HN=S=O  thionyl imide            HN=S=NH  sulfur diimide
         |                                 |
    polymers                          derivatives
```

(HNSO)$_x$ (RN)$_2$S   organo-sulfodiimides
(including probable (Me$_3$SiN)$_2$S ⎫ bis(trimethyl-silyl) and
 cyclic tetramer)   (Me$_3$SnN)$_2$S ⎭   -stannyl) sulfur diimides
                     (CF$_3$SN)$_2$S   bis(trifluoromethylsulfenyl) sulfur diimide
                     (BrN)$_2$S ⎫
                     (IN)$_2$S   ⎬ halogeno-sulfur diimides
                     HNSNCl   ⎭

derivatives

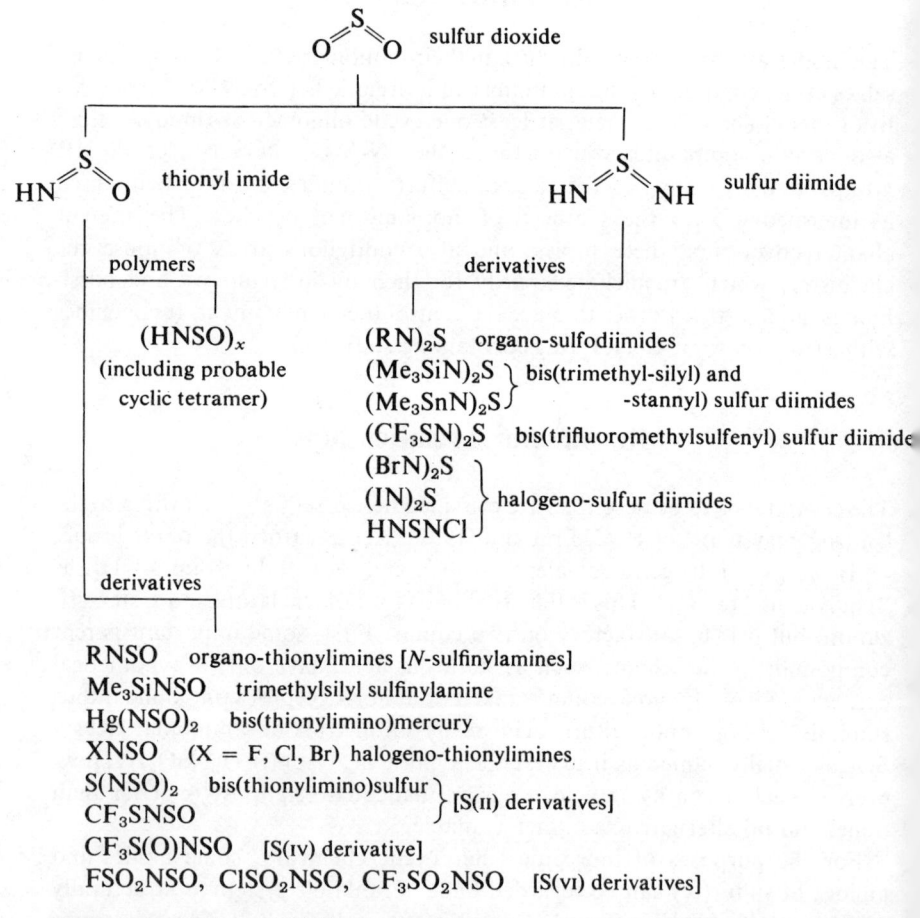

RNSO   organo-thionylimines [N-sulfinylamines]
Me$_3$SiNSO   trimethylsilyl sulfinylamine
Hg(NSO)$_2$ bis(thionylimino)mercury
XNSO   (X = F, Cl, Br) halogeno-thionylimines
S(NSO)$_2$ bis(thionylimino)sulfur ⎫
CF$_3$SNSO   ⎬ [S(II) derivatives]
CF$_3$S(O)NSO   [S(IV) derivative]
FSO$_2$NSO, ClSO$_2$NSO, CF$_3$SO$_2$NSO   [S(VI) derivatives]

Figure 3.1. Sulfur(IV) imides and imide derivatives structurally related to sulfur dioxide; the parent compound sulfur diimide has not been proved to exist, though all the derivatives shown have been made.

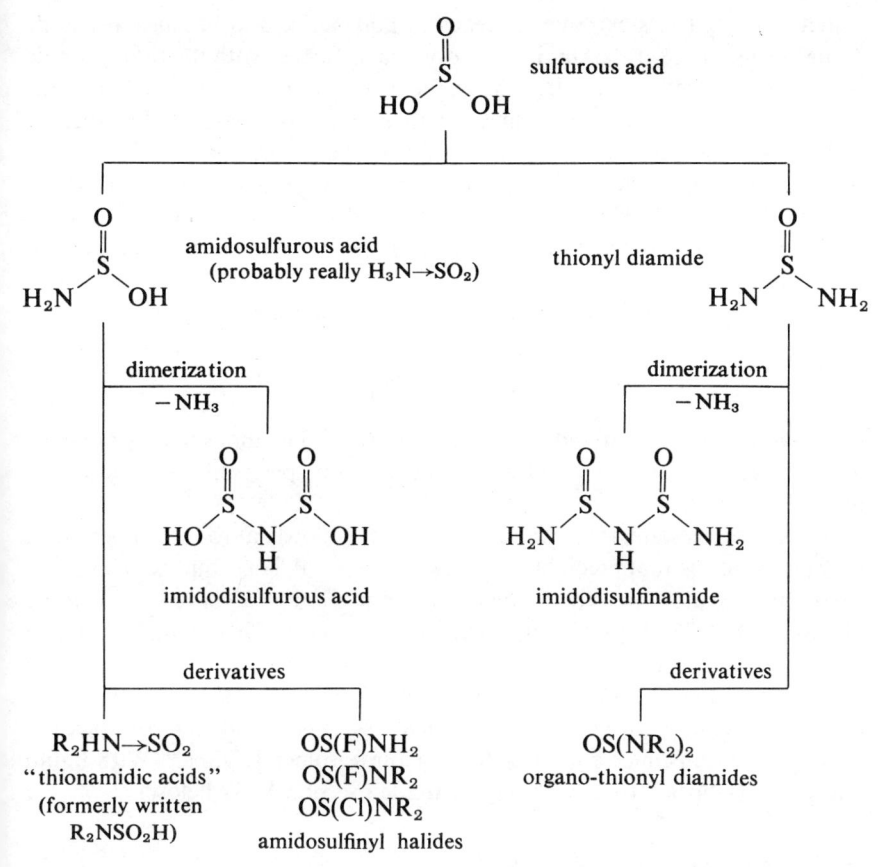

Figure 3.2. Sulfur(IV) amides and amide derivatives structurally related to sulfurous acid; sulfurous acid and imidodisulfurous acid are known only as salts.

III. Thionyl imide and its derivatives

A. Thionyl imide

This compound, discovered by P. W. Schenk in 1942, is best prepared by his method, the reaction of thionyl chloride with ammonia:[1,22,26,39]

$$SOCl_2 + 3NH_3 \rightarrow HNSO + 2NH_4Cl$$

Schenk was guided by the work of Michaelis and Storbeck, who about 1893 had prepared organo-thionylimines, RNSO, similarly from primary amines and $SOCl_2$. The preparation of the parent compound, however, was far more difficult owing to its extreme tendency to polymerize and because any additional ammonia over that in the equation reacts further with the thionyl imide. Schenk succeeded by mixing the gaseous reactants at low pressures and with exactly the correct stoichiometry. Subsequent investigators, while successfully repeating this preparation, have found that a little SO_2 is also formed and can be kept below 5% by working at low pressures (4 Torr $SOCl_2$, 12 Torr NH_3).[7,8,22] Thionyl imide also results, less efficiently, from the gas-phase hydrolysis of thiazyl fluoride NSF,[22] or the action of SO_2 on ammonia.[19]

Gaseous monomeric thionyl imide has the planar *cis* structure

$$H-N=S=O$$

as shown by its infrared and microwave spectra.[22] The molecular parameters are accurately known. The ultraviolet absorption spectrum has been determined.[1]

In the gaseous state, thionyl imide decomposes slowly at room temperature; decomposition is inappreciable in a few hours at 20 Torr, but extensive in a week. In a vacuum line, the imide can be condensed to a white solid melting at about $-85°$. The liquid begins to polymerize above $-70°$, so little is known of its physical properties as a monomer.[39]

The instability of thionyl imide has made its reactions as a monomer difficult to study, but photolysis in an argon matrix has been shown to produce the *trans* geometric isomer and also the structural isomer HOSN.[45] With liquid ammonia it appears to give thionyl diamide (Section VI.A below).[39]

B. Thionyl imide polymers and isomer

These substances present a complicated picture. Most of the facts, and their usual interpretations, are shown in Fig. 3.3, but it must be emphasized that the interpretations are speculative in several respects. The main points are as follows.

(1) There is a group of three metastable yellow solid polymers (enclosed in a box in Fig. 3.3) which may well be identical but have not been proved to be. One of these, obtained by air oxidation of heated $S_4(NH)_4$,[11] has been fully characterized by analysis, molecular weight, and i.r., and has almost certainly the cyclic tetrameric formula shown. Another, formed when liquid monomeric thionyl imide stands at $-70°$ to $-60°$, is also tetrameric.[26,39] The third, resulting from the action of liquid HCl on imidodisulfinamide (Section VI.B below)[14] is, like the other two, soluble in organic solvents and therefore probably also of low molecular weight.

(2) All three of the compounds just mentioned change on standing at room

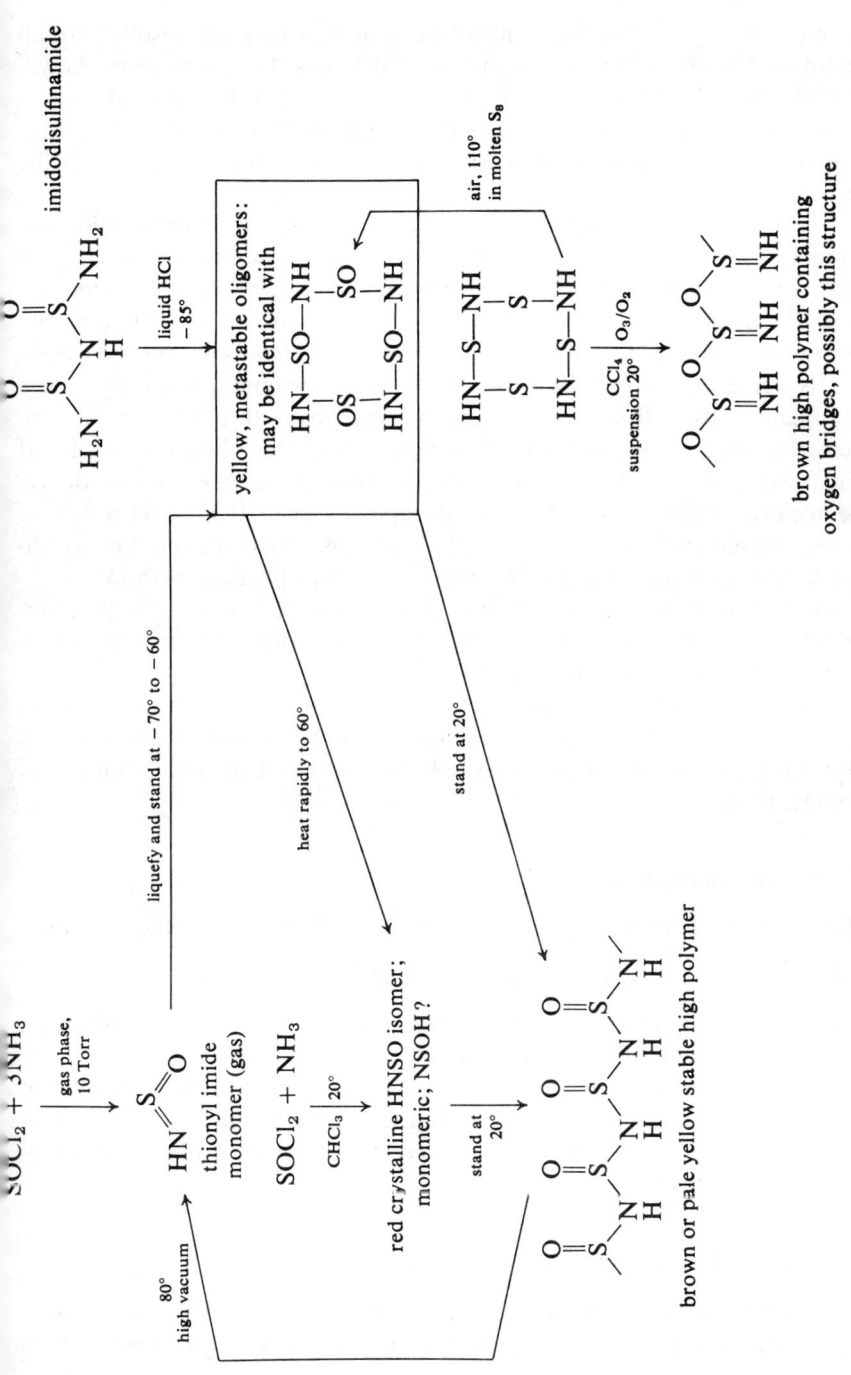

Figure 3.3. The chemistry of thionyl imide, its isomer, and its polymers.

temperature to a stable pale yellow to brown high polymer, insoluble in all indifferent solvents. On i.r. evidence its structure is thought to be as shown, and this is supported by the formation in 40% yield of thionyl imide monomer on heating the polymer *in vacuo* at 80°. This polymer is reversibly hygroscopic. It is a semiconductor in which the current appears to be carried by protons.[26,39]

(3) A red crystalline substance said to be isomeric with thionyl imide and monomeric in nitrobenzene is reported from the reaction of thionyl chloride with ammonia in chloroform at room temperature.[5,13] This compound has not recently been investigated and its status is obscure. Its very low volatility throws doubt on the formulas originally suggested, H–S–N=O and H–O–S=N, and suggests that it may be a salt or a mixture containing a salt as a main component. Its visible-u.v. spectrum in methanol, with a peak at 360 nm, resembles that of ethyl nitrosyl mercaptan,[14] which might be taken to support the formula H–S–N=O. A peak at 360 nm occurs in the spectrum of the recently discovered cyclic $S_3N_3^-$ ion (Chapter 6, Sec. III.H), and a salt of $S_3N_3^-$ would analyze with the same S/N ratio as HSNO. It would be worthwhile to reinvestigate the red "isomer" by modern physical methods.

(4) Ozone oxidation of $S_4(NH)_4$ gives a second brown high polymer, clearly distinguished from the other by its i.r. spectrum, and thought to contain oxygen bridges as opposed to terminal oxygen atoms.[12]

The remaining problems posed by this group of compounds could almost certainly be solved by application of modern physical methods of structure determination to the yellow metastable polymers and the red "isomer" of thionyl imide.

C. Organic thionylimines

The preparation of *N*-sulfinylaniline (phenyl thionylimine) by the reaction

$$SOCl_2 + PhNH_2 \rightarrow PhNSO + 2HCl$$

has been mentioned. This is a general reaction of primary amines; it is usually carried out in benzene or ether solution. The organo-thionylimines (*N*-sulfinylamines) are mostly fairly stable liquids, some of which can be distilled at atmospheric pressure. They are easily hydrolyzed. Unlike the parent compound, they have no tendency to polymerize. They have found applications in organic synthesis.[9,23]

D. Trimethylsilyl sulfinylamine

This compound, which seems to have possibilities as an intermediate in synthesis [compare bis(trimethylsilyl) sulfur diimide below] was first prepared in

1966 in a somewhat analogous way to the organo-thionylimines by the reaction[40]

$$(Me_3Si)_3N + SOCl_2 \xrightarrow[70°]{AlCl_3} Me_3SiNSO + 2Me_3SiCl$$

It is a colorless liquid boiling at 109°. It is the source of two sulfur–nitrogen heterocycles (Fig. 3.4; Chapter 4, Sec. IV.B; Chapter 7, Sec. IV; and this chapter, ref. 7) and of the interesting compounds next to be described.

E. Halogeno-thionylimines and bis(thionylimino)mercury

These compounds were obtained in 1969 by Verbeek and Sundermeyer[46] using the reaction scheme of Fig. 3.4. The most stable of the halogeno compounds is the chloride Cl–N=S=O. It reacts vigorously with mercury and

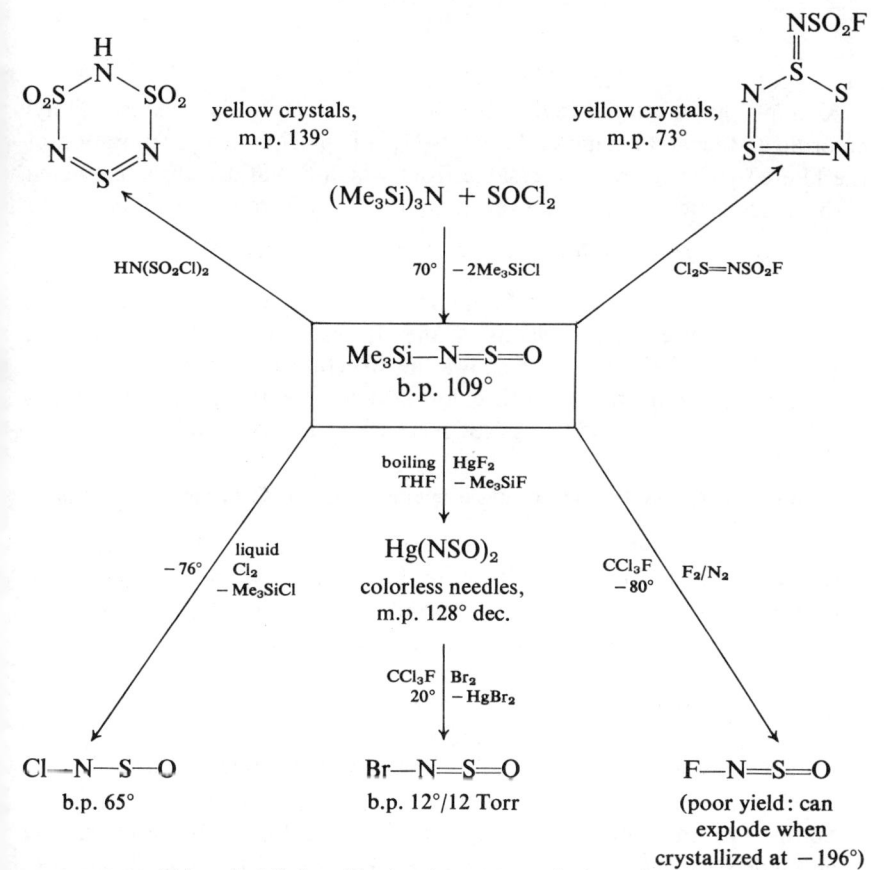

Figure 3.4. Trimethylsilyl sulfinylamine as an intermediate in heterocyclic synthesis (upper, Section III.D of this chapter), and as a source of further potential intermediates (lower, Section III.E of this chapter).

explosively with water. The halogeno compounds resemble thionyl imide itself in having the *cis* configuration, but differ from it in being non-planar.[10,29] The mercury and halogeno compounds have possibilities as sources of NSO radicals or fragments for the synthesis of inorganic heterocyclics.

F. Bis(thionylimino)sulfur

This compound, with the formula $S_3N_2O_2$ and formerly called thiodithiazyl dioxide, was discovered by Goehring and Heinke in 1951. The structure was thought to be cyclic until Weiss, in an X-ray diffraction study reported in 1961, showed the compound to be a thionylimino derivative of sulfur(II):[47]

$$S(=N-S-N=S)$$
$$\parallel \quad \quad \quad \parallel$$
$$O \quad \quad \quad \quad O$$

A simple preparation is to dissolve tetrasulfur tetranitride in excess thionyl chloride and leave to stand 48 hours. S_4N_3Cl is precipitated; evaporation of the filtered solution leaves a residue from which $S(NSO)_2$ can be extracted with benzene. Better yields, however, are obtained from the reaction[3]

$$2Me_3SiNSO + SCl_2 \rightarrow 2Me_3SiCl + S(NSO)_2$$

$S(NSO)_2$ forms yellow crystals melting at 100° and recrystallizable from benzene, or better trimethylchlorosilane. It reacts with PCl_5 to give the heterocycle $SP_2N_3Cl_5$ (Fig. 3.5 shows the structure of this heterocycle and an alternative route to it). With chlorine it affords the cyclic compounds $(NSOCl)(NSCl)_2$ and $(NSOCl)_2(NSCl)$ (Chapter 5, Sec. III.B).

G. Other thionyl imide substitution products with –NSO attached to sulfur

Besides the compound $S(NSO)_2$ just described, there exist other compounds with the –NSO group attached to sulfur, and containing that element in the oxidation states +2, +4, and +6 respectively. They are all prepared analogously to thionyl imide itself and its organo derivatives, by reaction of a thionyl halide with a suitable amine.

Sulfur(+2): the compound CF_3SNSO, a colorless liquid, comes from CF_3SNH_2 and SOF_2.[18] It is little known.

Sulfur(+4): the compound $CF_3S(O)NSO$, a yellow liquid, b.p. 48°/15 Torr, arises from $CF_3SN(SiMe_3)_2$ and $SOCl_2$.[33] It too has received little study.

Sulfur(+6): the compounds FSO_2NSO[32] and CF_3SO_2NSO,[31] liquids, have been made by reactions of $SOCl_2$ with FSO_2NH_2 and $CF_3SO_2NH_2$ respectively. The trifluoromethyl compound is inert to water, whereas the fluoro compound hydrolyzes explosively. Both have been used in the synthesis of the rare 4-membered S–N ring (Chapter 4, Sec. III.E), and the fluoro compound also for the preparation of cyclic sulfur nitride-oxides (Chapter 8, Sec. II.B). The chloro compound $ClSO_2NSO$ is known.[38]

H. The NSO⁻ ion

The evidence for this ion and/or its isomers is confusing and needs re-evaluation with the help of new physical-structural data.

Red to violet salts believed at the time to contain NSO^- as anion were prepared from the alleged red isomer of thionyl imide (Section B above) by treatment with organosodium or organolithium compounds or with silver nitrate,[5,14] but difficulty was experienced in getting reproducible analyses. It seems possible that the recently discovered ions $S_4N_5^-$ or $S_3N_3^-$ (Chapter 6, Secs. III.G and H) may have been somehow involved in these results. The red-violet solution of the alleged NaNSO soon decomposed, precipitating sulfur and forming in solution HSO_3^-, $S_2O_3^{2-}$, and $S_3O_6^{2-}$.

IV. Sulfur diimide and the sulfodiimides

A. Sulfur diimide

This compound, the parent of a series of well characterized organo-sulfodiimides $(RN)_2S$, is not known, though a chloro derivative, HN=S=NCl, has been reported (Section E below).

Sulfur diimide might be expected to result from ammonolysis of sulfur(IV) halides. The reaction between sulfur tetrafluoride and ammonia, however, leads to other products including S_4N_4. With $N_3S_3Cl_3$, also a sulfur(IV) compound, liquid ammonia gives a red solution. This was once suspected to contain sulfur diimide, but recent work has shown the presence of large amounts of $S_4N_5^-$ (Chapter 6, Sec. IV.B).

B. Organo-sulfodiimides

The first of these compounds was reported in 1957. They result from several reactions.[23] For example, they can be made from primary aromatic amines by reaction with SF_4:

$$2ArNH_2 + SF_4 \rightarrow ArN=S=NAr + 4HF$$

or by heating the amine with elemental sulfur and mercury acetamide.[6]

The dialkyl compounds are bright yellow liquids decomposing at or slightly above room temperature. The diaryl compounds are more stable orange to red liquids or low-melting red crystals. Geometric isomers of the dimethyl and di-t-butyl compounds have been discovered in n.m.r. studies at low temperatures.[17]

Compounds of this class are attracting interest in connection with organic synthesis.[23] They can act as ligands in transition-metal complexes.[27]

C. Bis(trimethylsilyl) sulfur diimide and monosilylated analogs

Bis(trimethylsilyl) sulfur diimide has recently become very important for the synthesis of new sulfur–nitrogen ring systems.[34] It was first prepared by Wannagat and Kuckertz in 1962, by the reaction

$$(Me_3Si)_2NM + SCl_2 \rightarrow MCl + Me_3Si-N=S=N-SiMe_3$$

(M = alkali metal). Better yields result when $SOCl_2$ is used instead of SCl_2. The compound is a yellow oil boiling at 73–74°/24 Torr.

In typical synthetic applications of this compound, it is treated with a covalent fluoride or chloride, and Me_3SiF or Me_3SiCl is eliminated. The driving force of these reactions seems to be the weakness of Si–N bonds relative to the Si–F, Si–Cl, or Si–O bonds formed. Even the reaction

$$(Me_3SiN)_2S + SO_2(NCO)_2 \rightarrow Me_3SiO(CN_3S_2O_2) + Me_3SiNCO$$

will go, though only one of the two silicon atoms changes its partner atom.

Figure 3.5 shows how seven unusual heterocyclics have been obtained by the reaction of various covalent halides and oxides with N,N'-bis(trimethylsilyl) sulfur diimide.[34] Five of them have been prepared only in this way. Usually the diimide acts as a source of –N=S=N– groups which become incorporated in the heterocyclic ring. The following reaction sequence shows the stepwise use of this principle via a silylated intermediate:[25]

$$2(Me_3SiN)_2S + SCl_2 \rightarrow Me_3SiNSN-S-NSNSiMe_3 + 2Me_3SiCl$$
(orange-yellow needles, m.p 69°)
$$Me_3SiNSN-S-NSNSiMe_3 + SCl_2 \rightarrow S_4N_4 + 2Me_3SiCl$$

The relatively little-known monosilylated sulfur diimides arise from the reaction[2]

$$RNSO + LiN(SiMe_3)_2 \rightarrow RN=S=NSiMe_3 + \frac{1}{n}[LiO(SiMe_3)_3]_n$$

They have been used for the preparation of long S–N chains, as in the following example:[2]

$$2Bu^tN=S=NSiMe_3 + SCl_2 \rightarrow (Bu^tNSN)_2S + 2Me_3SiCl$$

(compare Chapter 6, Sec. II.A.4).

IV. SULFUR DIIMIDE

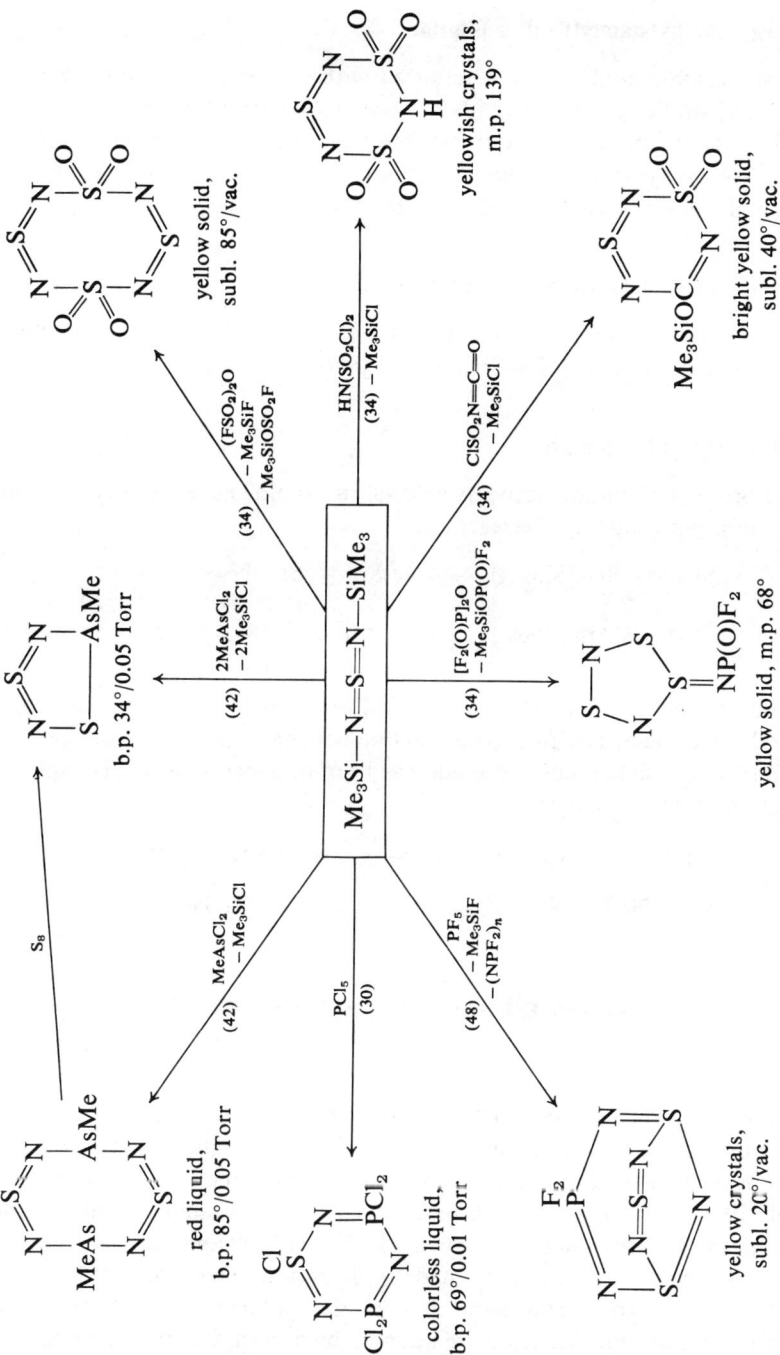

Figure 3.5. Synthesis of inorganic heterocycles using bis(trimethylsilyl) sulfur diimide; references are the numbers in parentheses beside the arrows; see also refs. 24 and 41.

D. Bis(trimethylstannyl) sulfur diimide

This solid compound has been reported without details to result from the action of $(Me_3Sn)_3N$ on S_4N_4.[35] It has been used with $MeSiCl_3$ to synthesize a yellow crystalline compound thought to have the structure $MeSi(-NSN-)_3$-SiMe.[36] As a synthetic reagent it may be expected to behave like its silyl analog just described; cf. ref. 37 and literature there cited.

E. Bis(trifluoromethylsulfenyl) sulfur diimide

This compound, with the formula $(CF_3SN)_2S$, is a bright red liquid boiling at 142°, obtained from $CF_3SN=SF_2$ and CF_3SNH_2.[18]

F. Halogeno-sulfodiimides

The dibromo and diiodo derivatives of sulfur diimide have recently been prepared in a pure state by the reactions[44]

$$SF_4 + 2BrN(SiMe_3)_2 \rightarrow 4Me_3SiF + Br-N=S=N-Br$$
(yellow crystals, m.p. 0.5°)
$$SF_4 + 2IN(SiMe_3)_2 \rightarrow 4Me_3SiF + I-N=S=N-I$$
(orange crystals, m.p. 106°)

carried out in 1,1,2,2-tetrafluoro-1,2-dichloroethane at 0°. Both compounds explode on impact. The analogous reaction with chlorine did not succeed, but the monochlorinated sulfur diimide has been obtained as a highly explosive gas by the reaction sequence [25]

$$(Me_3SiN)_2S + Cl_2 \xrightarrow[-70°]{Freon\ 114} MeSiNSNCl + R_3SiCl$$
$$Me_3SiNSNCl + HCl \xrightarrow[-78°]{} H-N=S=N-Cl$$

V. Amidosulfurous acid and its derivatives

A. Introduction

Amidosulfurous acid, $OS(OH)NH_2$, would be isomeric with the 1:1 adduct of sulfur dioxide and ammonia, O_2SNH_3. In principle the two structures are interconvertible by the migration of a proton. Primary amine–SO_2 adducts would likewise be isomeric with organic derivatives of amidosulfurous acid. A number of compounds are known which clearly belong to this group. In most cases no attempt has been made to determine their structures. Usually there would be three reasonable possibilities: adduct (= zwitterion), un-ionized acid, or ion. The structure adopted by a given compound might well vary with its physical state, and in solution could depend on the nature of the solvent.

In the early decades of this century, the idea of donor–acceptor complexes was seldom discussed by organic chemists. Consequently the "thionamidic acids" then known were formulated as acids, $OS(OH)NR_2$, without structural evidence and without regard to the possibility that they might be adducts. This was the viewpoint even in *Houben-Weyl* published in 1958.[9] Similarly, all the early thinking about the reaction products of ammonia with sulfur dioxide was in terms of acid rather than adduct formulas.

In 1937 Jander suggested yet another kind of formulation for the 1:1 compound between ammonia and SO_2, namely as a salt, $[(H_3N)_2SO]^{2+}SO_3^{2-}$. This was designed to fit his own theories of acids and bases in non-aqueous solvents. No evidence for this structure was offered and there seems no reason to take it seriously. It is really not at all likely that the nitrogen atoms in thionyl diamide, $OS(NH_2)_2$, which because of inductive effects should be only weakly basic, would add *two* protons to give Jander's cation.

In recent years the adduct formulation has often been employed for these compounds. This is probably justified, but reliable generalizations will not be possible until we have more physical-structural evidence.

B. Amidosulfurous acid and the ammonia–sulfur dioxide reaction

The existence of a 1:1 compound between ammonia and sulfur dioxide, perhaps amidosulfurous acid, is well established, but only preliminary indications have been obtained as to its structure.

The fact that ammonia and sulfur dioxide combine, with precipitation of white or yellow solids, has been known for over a century and a half, but the nature of the products has baffled one generation of chemists after another. In 1911 Fritz Ephraim thought it "astonishing" that this reaction had not been fully elucidated. He would have been more astonished to learn that it would still not be properly understood in 1974. Since then definite progress has been made, but the story is not yet complete. The main difficulty has been in characterizing the solid products.

Dry ammonia and sulfur dioxide gases do not react at room temperature, but at lower temperatures a yellow-white 1:1 solid adduct, with a characteristic i.r. spectrum different from that of the components, condenses out from an equimolar mixture of the gases, with which it appears to be in equilibrium:[21]

$$NH_3 + SO_2 \rightleftharpoons NH_3 \cdot SO_2$$

The i.r. spectrum has been tentatively interpreted as fitting a structure like that of the sulfite ion with one oxygen atom replaced by NH_3 (C_s symmetry), but this is provisional owing to the lack of Raman data.[21] There is i.r. evidence for the formation of one and possibly two adducts $(NH_3)_2SO_2$ in the presence of excess ammonia.[21] With the help of this information a better interpretation might be possible of the data of Scott and Lamb[43] who explained quantitatively the vapour-pressure–temperature curves of the solids formed below 0° at various ratios of the gases by supposing that only two compounds are present, $SO_2 \cdot NH_3$ and $SO_2 \cdot 2NH_3$, and that they form a continuous series of solid solutions. From the absence of an O–H stretch in the i.r. spectrum of $NH_3 \cdot SO_2$, and other spectral features, it no longer seems possible to accept the view of earlier workers that the 1:1 adduct is amidosulfurous acid and the 2:1 adduct its ammonium salt; adduct formulations are probably correct.

When SO_2 and ammonia react at higher temperatures the products are complicated. Goehring and Kaloumenos obtained S_4N_2 by vacuum distillation of the raw product of reactions at 50°, and S_4N_4 from hydrolysis of this product.[13,20]

C. Imidodisulfurous acid and its salts

Imidodisulfurous (= imidodisulfinic) acid, if it existed, would have the formula HOS(O)–NH–S(O)OH (see Fig. 3.2), and can be derived in principle by the condensation of two molecules of amidosulfurous acid; i.e. it has the same relationship to amidosulfurous acid as imidodisulfinamide has to thionyl diamide. The acid is unknown but the literature contains reports of five compounds which might be its salts.[16] None has had its structure determined, and there has been no recent work on them. Our limited information comes from the following sources.

Ephraim and Piotrowski (1911) argued that the red color of the products from sulfur dioxide and excess ammonia at high temperatures arises from a compound with the same empirical formula, $SO_2 \cdot 2NH_3$, as the white compound described recently by Scott and Lamb (Section B above), but with the actual constitution $[NH_4^+]_3[N(SO_2)_2]^{3-}$, i.e. that of a triammonium salt of imidodisulfurous acid. An unstable red silver salt analyzing fairly well for $AgN(SO_2Ag)_2$ was obtained by metathesis from the red products of both the NH_3–SO_2 reaction and the NH_3–$SOCl_2$ reaction. The cause of the yellow, orange, or red colors produced in the SO_2–NH_3 reaction under various conditions is obscure, but in view of the later isolation of S_4N_4 (orange) and S_4N_2 (red) from the products, Ephraim and Piotrowski's explanation must be in doubt.

Divers and Ogawa (1901) obtained a colorless crystalline compound analyzing well as the diammonium salt of imidodisulfurous acid, $[NH_4^+]_2[HN(SO_2)_2]^{2-}$, from the $SO_2 \cdot 2NH_3$ adduct, after allowing the latter to decompose at 35° with loss of some ammonia. A potassium salt $K^+{}_2[HN(SO_2)_2]^{2-}$ and a barium salt $Ba^{2+}[NH_4{}^+]_2[HN(SO_2)_2{}^{2-}]_2$ were prepared from it by metathesis.

D. Organic derivatives of amidosulfurous acid

Zipp has recently found, by the study of vapor-pressure–composition diagrams, that a wide range of primary aromatic amines reversibly form 1:1 complexes, and *only* 1:1 complexes, with SO_2.[50] The vibrational spectra of these complexes show N–S bonding. If they were amidosulfurous acids OS(OH)NArH, one would expect evidence in the phase diagram of 1:2 SO_2–amine "complexes", i.e. amine salts of the amidosulfurous acids, but there is none. It seems therefore that the pure solid 1:1 complexes are simple donor–acceptor adducts $ArH_2N{\rightarrow}SO_2$. Similar adducts are formed by secondary and tertiary aromatic amines. The behavior of aliphatic primary and secondary amines and sulfur dioxide has not been adequately investigated. 1:1 complexes (originally reported as amidosulfurous acids) are formed in ether, but methylamine and ethylamine in the gas phase yield the thionylimines RNSO.[9,19]

Although the existence of organo-amidosulfurous acids with the structure OS(OH)NRR' is doubtful, the esters OS(OR")NRR' certainly exist. Examples were prepared in 1958 by Zinner as colorless oils, distillable under reduced pressure, by the reaction[49]

$$2RR'NH + ClS(O)OR'' \rightarrow [RR'NH_2]^+Cl^- + OS(OR'')NRR'$$

E. Amidosulfinyl halides

Amidosulfinyl fluoride, $OS(F)NH_2$, was reported in 1956 by Goehring and Voigt to arise from the reaction[15]

$$OSF_2 + 2NH_3 \xrightarrow[-78°]{CCl_2F_2} OS(F)NH_2 + NH_4F$$

It was obtained partly as a volatile ether-soluble gas, and partly as an involatile ether-insoluble yellow solid which may have been the polymer

$$\left[\begin{matrix} F \\ | \\ S\!\!-\!\!NH \\ | \\ OH \end{matrix}\right]_x$$

Neither form was isolated in a pure state. However, by the use of piperidine or diethylamine in place of ammonia, and working in ether at low temperatures, one may obtain organo-aminosulfinyl fluorides $OS(F)NC_5H_{10}$ and

OS(F)N(C$_2$H$_5$)$_2$ as relatively stable yellow liquids distillable under reduced pressure.[15] The chloride OS(Cl)NH$_2$ is not known, but organo-chlorides resembling the corresponding fluorides have been made, for example, by the reaction [9]

$$R_2NH \cdot SO_2 + SOCl_2 \rightarrow OS(Cl)NR_2 + HCl + SO_2$$

VI. Thionyl diamide and its derivatives

A. Thionyl diamide

The only information on this compound comes from an experiment by Schenk (1964), who condensed ammonia on to monomeric thionyl imide and pumped off the excess at $-70°$. He found that the imide took up just 1 mol of ammonia and deduced that the reaction

$$OSNH + NH_3 \rightarrow OS(NH_2)_2$$

had taken place.[39] The alternative possibility that the product was the ammonium salt of thionyl imide, [NH$_4$]$^+$[OSN]$^-$, was ruled out by a conductimetric titration which seemed to show that the presumed thionyl diamide behaved as a dibasic acid towards the base NH$_2^-$ in liquid ammonia solution. This needs checking, because Schenk's titration sample was not prepared by the above method and may just conceivably have been a different substance, and because his conductivity readings tended to drift.

B. Imidodisulfinamide

Imidodisulfinamide,[4,13] with the structure shown in Fig. 3.2, is derived in principle by elimination of a molecule of ammonia between two molecules of thionyl diamide. It is best prepared by adding undiluted thionyl chloride to liquid ammonia, which produces a very violent reaction with local heating. The surplus ammonia is evaporated off and the imidodisulfinamide extracted from the residue with nitrobenzene.[13] It is a fairly stable amorphous-looking yellow solid. Three hydrogen atoms in the molecule can be replaced by the Zerewitinoff method, and treatment with silver nitrate in aqueous ammoniacal solution precipitates the yellow trisilver salt AgN(OSNHAg)$_2 \cdot$3H$_2$O.[13] A cold aqueous solution of imidodisulfinamide does not decompose quickly, but attempts to recrystallize the compound by evaporation of an aqueous solution at room temperature cause a red color formerly attributed to OSN$^-$ (Section III.H above) to appear, and give rise to a new, well crystallized yellow ammonium salt.[13] This salt is very probably [NH$_4$]$^+$[S$_4$N$_5$O]$^-$ (Chapter 8, Sec. V), though it was originally assigned the formula [NH$_4$]$^+$[S$_3$N$_3$O$_2$]$^-$, with a cyclic anion, on the evidence of a molecular weight determination in aqueous solution.

Imidodisulfinamide with liquid HCl affords an oligomer of thionyl imide (Section III.B above).

C. Organic derivatives of thionyl diamide

Organo-thionylamides $OS(NR_2)_2$ are formed from thionyl chloride and secondary alkylamines in an inert solvent. They are relatively stable colorless liquids or low-melting solids, distillable under reduced pressure without decomposition.[9]

VII. Alkyl and fluoroalkyl sulfimides as reagents for ring synthesis

A. Introduction

The two recently reported compounds to be described in this section are sulfur(IV) imides related under Franklin's system to the organic sulfoxides, R_2SO. One has been used in a ring synthesis and the other holds some promise in this direction.

B. The sulfur imide $Me_2S=NSiMe_3$

This compound has been made by the reaction sequence

$$Me_2SBr_2 + HN(SiMe_3)_2 \longrightarrow [Me_2S\text{---}NHSiMe_3]^+ Br^-$$

$$\downarrow KNH_2$$

$$Me_2S=NSiMe_3$$

It has not been fully described, but its use in the synthesis of the ring compound II (Section VIII below) has been reported.[2a]

C. Bis(trifluoromethyl)sulfimide $(CF_3)_2S=NH$

Though not hitherto employed in ring synthesis, this compound and its lithium derivative resemble in structure the compound just mentioned and might be similarly applicable.

Bis(trifluoromethyl)sulfimide is a colorless liquid more stable than the methyl analog and easily handled in a vacuum line at room temperature.[28] It is obtained from the reaction

$$(CF_3)_2SF_2 + NH_3 \xrightarrow{RNH_2} (CF_3)_2S=NH$$

With butyllithium it readily affords the derivative $(CF_3)_2S=NLi$, a golden-yellow solid which can be kept for a short time at 25°. This is a strong nucleophile which reacts with covalent halides to give a range of other derivatives $(CF_3)_2S=NX$, where X may be, for example, SO_2Cl, SO_2CF_3, POF_2, or $SiMe_3$.[28]

VIII. Mixed P–S rings containing sulfur(IV)

Compounds **I** and **II** contain 4-membered saturated rings in which S(IV) is one component, and can be thought of as either cyclosulfimides or cyclophosph(V)-azanes (cf. Chapter 11, Sec. III.C). Both result from cycloaddition reactions of sulfur(IV) imides. Compound **I** is formed quantitatively from the sulfur diimide $(Bu^tN)_2S$ (Section IV.B above) and the phosphorus(V) compound $(Me_3Si)(Bu^t)NP(S)NSiMe_3$ in benzene at room temperature.[24] Compound **II** arises in good yield from addition of the sulfur(IV) imide $Me_2S=NSiMe_3$ (Section VII.B above) to the phosphorus(V) compound $(Me_3Si)_2NP(NSiMe_3)_2$.[2a]

Compounds **I** and **II** both form colorless crystals soluble in organic solvents. Their structures have been established by n.m.r. and mass spectra, but they are otherwise little known.

References

1. Allegretti, J. M. and Merer, A. J., *Can. J. Phys.* **50**, 404 (1972)
2. Appel, R. and Montenarh, M., *Z. Naturforsch. B* **30b**, 847 (1975)
2a. Appel, R. and Halstenberg, M., *Angew. Chem. Int. Ed. Engl.* **14**, 769 (1975)
3. Armitage, D. A. and Sinden, A. W., *Inorg. Chem.* **11**, 1151 (1972)
4. Becke-Goehring, M., *Adv. Inorg. Chem. Radiochem.* **2**, 169 (1960)
5. Becke-Goehring, M., Schwarz, R. and Spiess, W., *Z. Anorg. Allg. Chem.* **293**, 294 (1958)
6. Bindra, A. P., Elix, J. A. and Morris, G. C., *Aust. J. Chem.* **22**, 2483 (1969)
7. Buckendahl, W. and Glemser, O., *Chem. Ber.* **110**, 1154 (1977)
8. DeKock, R. L. and Haddad, M. S., *Inorg. Chem.* **16**, 216 (1977)
9. Dorlars, A., in *Methoden der Organischen Chemie* (Houben-Weyl), vol. XI/2, Thieme, Stuttgart (1958)

10. Eysel, H. H., *J. Mol. Struct.* **5**, 275 (1970)
11. Fluck, E. and Becke-Goehring, M., *Z. Anorg. Allg. Chem.* **292**, 229 (1957)
12. Fluck, E. and Boeing, H., *Chem. -Ztg. Chem. App.* **94**, 331 (1970)
13. Goehring, M., *Ergebnisse und Probleme der Chemie der Schwefelstickstoffverbindungen*, Akademie-Verlag, Berlin (1957)
14. Goehring, M. and Messner, J., *Z. Anorg. Allg. Chem.* **268**, 47 (1952)
15. Goehring, M. and Voigt, G., *Chem. Ber.* **89**, 1050 (1956)
16. *Gmelins Handbuch der Anorganischen Chemie*, vol. 9, part B, section 3, "Schwefel", Verlag Chemie, Weinheim (1963)
17. Grunwell, J. R., Hoyng, C.F. and Rieck, J. A., *Tetrahedron Lett.* 2421 (1973)
18. Haas, A. and Schott, P., *Chem. Ber.* **101**, 3407 (1968)
19. Hata, T. and Kinumaki, S., *Nature* (London) **203**, 1378 (1964)
20. Heal, H. G., in *Inorganic Sulphur Chemistry*, ed. G. Nickless, Elsevier Amsterdam (1968)
21. Hisatsune, I. C. and Heicklen, J., *Can. J. Chem.* **53**, 2646 (1975)
22. Kirchoff, W. H., *J. Am. Chem. Soc.* **91**, 2437 (1969)
23. Kresze, G. and Wucherpfennig, W., *Angew. Chem. Int. Ed. Engl.* **6**, 149 (1967)
24. Kulbach, N. T. and Scherer, O. J., *Tetrahedron Lett.* 2297 (1975)
25. Lidy, W., Sundermeyer, W. and Verbeek, W., *Z. Anorg. Allg. Chem.* **406**, 228 (1974)
26. May, J. F. and Vallet, G., *Rev. Gen. Elec.* **81**, 255 (1972)
27. Meij, R., Kuyper, J., Stufkens, D. J. and Vrieze, K., *J. Organomet. Chem.* **110**, 219 (1976)
28. Morse, S. D. and Shreeve, J. M., *Inorg. Chem.* **17**, 2169 (1978)
29. Oberhammer, H., *Z. Naturforsch. A* **25a**, 1497 (1970)
30. Pohl, S., Petersen, O. and Roesky, H. W., *Chem. Ber.* **112**, 1545 (1979)
31. Roesky, H. W., *Angew. Chem. Int. Ed. Engl.* **6**, 711 (1967)
32. Roesky, H. W., Holtschneider, G. and Gierek, H. H., *Z. Naturforsch. B* **25b**, 252 (1970)
33. Roesky, H. W. and Holtschneider, G., *J. Fluorine Chem.* **7**, 77 (1976)
34. Roesky, H. W. and Kuhtz, B., *Chem. Ber.* **107**, 1 (1974)
35. Roesky, H. W. and Wiezer, H., *Angew. Chem. Int. Ed. Engl.* **12**, 674 (1973)
36. Roesky, H. W. and Wiezer, H., *Angew. Chem. Int. Ed. Engl.* **13**, 146 (1974)
37. Roesky, H. W., Diehl, M., Bats, J. W. and Fuess, H., *Angew. Chem. Int. Ed. Engl.* **17**, 58 (1978)
38. Ruff, J. K., *Inorg. Chem.* **6**, 2108 (1967)
39. Schenk, P. W., *Monatsh. Chem.* **95**, 710 (1964)
40. Scherer, O. and Hornig, P., *Angew. Chem. Int. Ed. Engl.* **5**, 729 (1966)
41. Scherer, O. J. and Wies, R., *Z. Naturforsch. B* **25b**, 1486 (1970)
42. Scherer, O. J. and Wies, R., *Angew. Chem. Int. Ed. Engl.* **11**, 529 (1972)
43. Scott, W. D. and Lamb, D., *J. Am. Chem. Soc.* **92**, 3943 (1970)
44. Seppelt, K. and Sundermeyer, W., *Angew. Chem. Int. Ed. Engl.* **8**, 771 (1969)
45. Tchir, P. O., *Diss. Abs. Int. B* **33**, 3587 (1973)
46. Verbeek, W. and Sundermeyer, W., *Angew. Chem. Int. Ed. Engl.* **8**, 376 (1969)
47. Weiss, J., *Z. Naturforsch. B* **16b**, 477 (1961)
48. Weiss, J., Ruppert, I. and Appel, R., *Z. Anorg. Allg. Chem.* **406**, 329 (1974)
49. Zinner, G., *Chem. Ber.* **91**, 966 (1958)
50. Zipp, A. P., *J. Inorg. Nucl. Chem.* **36**, 1399 (1974)

4
Imides and amides of sulfur(VI) as source materials for inorganic heterocycles

I. Introduction and structural relationships

The best-known heterocyclics in this area of chemistry are the cyclosulfimides, $(SO_2NH)_x$. They have saturated rings and 3-coordinate nitrogen atoms, a distinction from the sulfanurics described in Chapter 5, which are also based on sulfur(VI). A large part of this chapter is devoted to the cyclosulfimides, to some related linear polymers, and to the principal source material for these compounds, sulfamide. A further section deals with mixed-ring compounds containing the sulfimide group $-SO_2-NR-$ as one component. Some space is also devoted to amido and imido derivatives of sulfur(VI), such as the S,S-dialkyl sulfodiimides, which are of actual or potential value for ring synthesis.

The viewpoint of E. C. Franklin's nitrogen system of compounds, already used in Chapter 3, also has some value here for systematizing the description of compounds and reactions. According to it, certain sulfur(VI) imides and amides which are of interest for heterocyclic chemistry can be regarded as derivatives of sulfur trioxide or sulfuric acid (Fig. 4.1 shows the more important relationships). This point of view is appropriate in a chemical as well as a structural sense; just as OH groups attached to sulfur are acidic, so are the formally related NH and $-NH_2$ groups. In the compounds to be described, NH groups ionize strongly in aqueous solution as a rule, while $-NH_2$ groups ionize more weakly, but nevertheless enough to form salts in strongly basic aqueous solutions or in liquid ammonia. Historically, the silver and ammonium salts have played an important part in the study of the sulfur(VI) imides and amides, which are often unstable as free acids.

The nitrogen system concept has, however, some limitations. Besides those pointed out in Section II of Chapter 3, it takes no account of the existence of nitrogen compounds for which oxygen analogs are unknown or impossible, such as cyclotetrasulfimide, $(HNSO_2)_4$, and the recently prepared[4] ion

$$\begin{bmatrix} HN & & NH \\ & S & \\ HN & \parallel & N \end{bmatrix}^{3-}$$

I. INTRODUCTION

sulfur trioxide (gas) — SO_3

γ-sulfur trioxide

β-sulfur trioxide

sulfimide monomer (postulated reaction intermediate)

cyclotrisulfimide (salts only)

sulfimide linear polymers (salts only)

sulfuric acid

disulfuric acid

chlorosulfuric acid

sulfamide

imidodisulfuric acid (salts only)

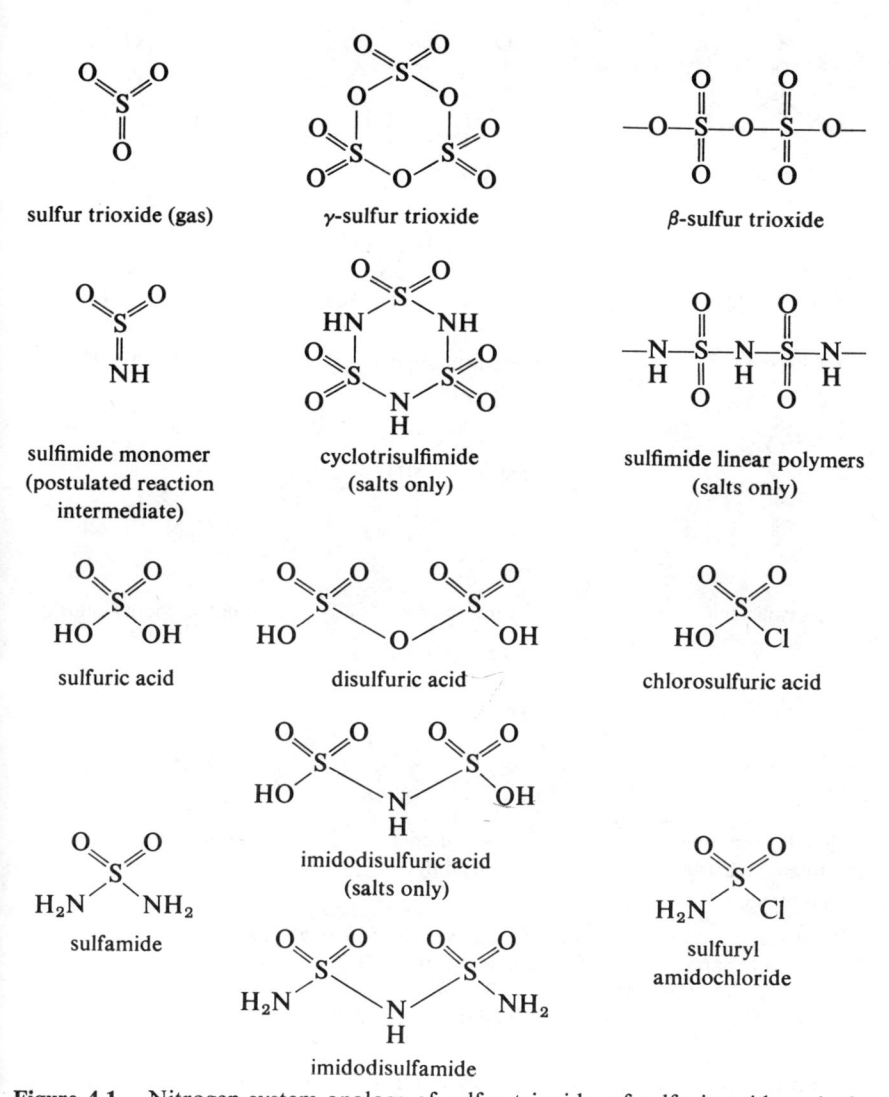

sulfuryl amidochloride

imidodisulfamide

Figure 4.1. Nitrogen-system analogs of sulfur trioxide, of sulfuric acid, and of sulfuric acid derivatives.

The sulfur(vi) imides and amides may also be usefully compared with urea and a group of substances related to it (Fig. 4.2). This parallel was first discussed by Hantzsch at the beginning of the present century. He was asking whether an inorganic analogy existed for the structural isomerism between urea and ammonium cyanate. He correctly concluded that, although sulfamide structurally and chemically resembles urea, the ammonium salt of

CHAPTER 4. IMIDES OF HEXAVALENT SULFUR

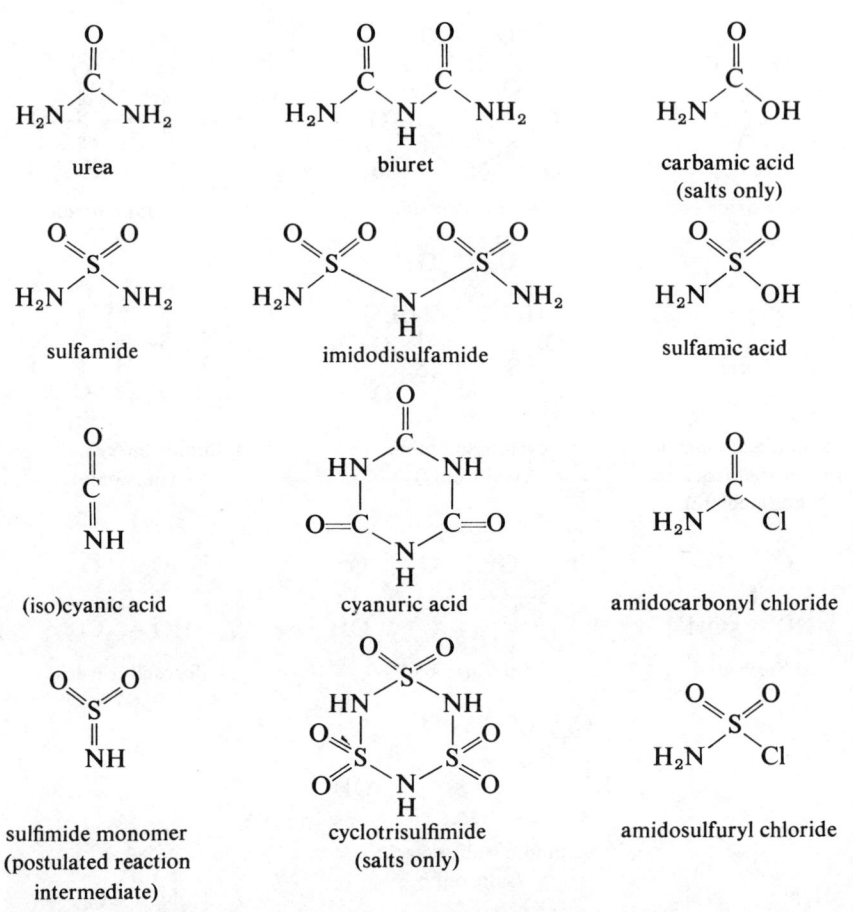

Figure 4.2. Analogous amide derivatives of sulfur(VI) and carbon. The analogy between carbamic and sulfamic acids is imperfect because carbamic acid is not known in the free state, while sulfamic acid (in the crystalline state) has the structure $H_3N \rightarrow SO_3$, not H_2N-SO_2OH as depicted.

sulfimide differs from ammonium cyanate in being trimeric. The compounds described here are covered in a recent volume of *Gmelin*.[19]

II. Sulfamide and its derivatives: monomeric sulfimide

A. Introduction

Monomeric sulfimide, $HN=SO_2$, unlike its sulfur(IV) analog thionyl imide, has not been isolated though there is mass-spectroscopic evidence for its existence.[29] It probably occurs as an unstable intermediate in reactions.[19] An unstable compound of m.p. 57–65°, which is almost certainly its pyridine adduct, py → S(NH)O$_2$ (the nitrogen-system analog of the stable py → SO$_3$), has been obtained from the reaction

$$H_2NSO_2Cl + 2py \xrightarrow{-80°} py{\to}S(NH)O_2 + pyH^+Cl^-$$

as a water-soluble substance decomposing rapidly in aqueous solution, or on heating in the dry state, to H$_2$N–SO$_2$–N=SO$_2$ ← py.[3]

Polymers of sulfimide are easy to make, and have been the main concern of research workers in this branch of chemistry. In this category, a few oligomers of both cyclic and linear polysulfimides have been characterized, and results to date suggest that full homologous series of both kinds may be capable of existence and moderately stable.

The polymeric sulfimides can be regarded as condensation products of sulfamide, SO$_2$(NH$_2$)$_2$, and most of them can be made from sulfamide. We therefore begin with a brief review of the chemistry of sulfamide, and proceed then to the sulfimides, with emphasis on the heterocyclics (HNSO$_2$)$_x$.

B. Sulfamide

This compound, SO$_2$(NH$_2$)$_2$, the diamide of sulfuric acid, is obtainable as a moderately expensive laboratory chemical from specialist suppliers. A much-used method of preparation is to add sulfuryl chloride, mixed with an inert solvent such as petroleum ether, to liquid ammonia.[13,18] Patented variants include the use of sulfur dioxide plus chlorine in place of sulfuryl chloride. The reported yields vary enormously, from practically zero to as high as 90%; with good management 40–50% is typical. The equation

$$SO_2Cl_2 + 4NH_3 \to SO_2(NH_2)_2 + 2NH_4Cl$$

is grossly oversimplified, and the uncertain yields arise from side reactions leading to imidodisulfamide (Fig. 4.1; Table 4.2; Section III.D below) and other longer-chain and cyclic condensation products. In principle, one might hope to minimize these by maintaining a large excess of ammonia, but even with this precaution they are always extensive and seemingly unpreventable. The usual work-up involves an aqueous HCl solution of the crude product, in which the condensation products become hydrolyzed in minutes to hours at room temperature, giving more sulfamide, e.g.[18]

$$\underset{\text{imidodisulfamide}}{HN(SO_2NH_2)_2} + H_2O \to \underset{\substack{\text{sulfamic} \\ \text{acid}}}{H_3N \cdot SO_3} + \underset{\text{sulfamide}}{SO_2(NH_2)_2}$$

This process can be driven to completion by waiting or by briefly heating the HCl solution, but in some reported preparations this was not done and the yields were correspondingly small. Better preparations of sulfamide are known, but from less readily available starting materials. Sulfuryl fluoride, a gas boiling at $-55°$, gives very good yields of sulfamide[16] when reacted with concentrated *aqueous* ammonia (it is inert towards water at ordinary temperatures). Also, sulfamide is formed almost quantitatively by the hydrolysis of sulfuryl diisocyanate:[2,8]

$$SO_2(NCO)_2 + 2H_2O \rightarrow SO_2(NH_2)_2 + 2CO_2$$

Liquid ammonia with sulfuryl chlorofluoride (b.p. 7°) gives good yields of sulfamide, accompanied by the same by-products as from sulfuryl chloride.[16]

Sulfamide forms colorless orthorhombic tablets melting at 92° and unaffected by moist air at room temperature. Its solubility is greater the more polar the solvent, being high in water or liquid ammonia and moderate in organic solvents containing carbonyl or hydroxyl groups. It is stable for long periods in aqueous solution at room temperature, but suffers hydrolysis or condensation in hot aqueous solution (see Section III.D below).

It is only very weakly acidic in aqueous solution, but metal derivatives can easily be prepared, as follows.

C. Metal derivatives of sulfamide

Traube and Reubke in 1923 made stable monoalkali salts of sulfamide from aqueous solution.[18] Dialkali salts cannot be prepared in this way, but a dipotassium salt may result from sulfamide and potassium amide in liquid ammonia.[18] In contrast, sulfamide readily forms a disilver salt, precipitated as a white slightly light-sensitive powder by adding ammonia to aqueous sulfamide plus silver nitrate; but all attempts to make a monosilver salt have failed.[33] The disilver salt can be converted into triply and quadruply substituted

$$\underset{\substack{\text{white,}\\\text{sparingly soluble}}}{\overset{\text{AgN}\diagdown\;\;\diagup\text{OAg}}{\underset{\text{O}\diagup\;\;\diagdown\text{NH}_2}{\text{S}}}} + 3\text{NaOH} \longrightarrow 2\;\underset{\substack{\text{pale yellow,}\\\text{sparingly soluble}}}{\overset{\text{AgN}\diagdown\;\;\diagup\text{ONa}}{\underset{\text{AgN}\diagup\;\;\diagdown\text{OAg}}{\text{S}}}}\; \text{H}_2\text{O} + \text{SO}_2\text{NH}_2(\text{NHNa})$$

$$\text{warm} \Big| \text{5\% AgNO}_3$$

$$\underset{\substack{\text{dark red,}\\\text{explosive when dry}}}{\overset{\text{AgN}\diagdown\;\;\diagup\text{OAg}}{\underset{\text{AgN}\diagup\;\;\diagdown\text{OAg}}{\text{S}}}} + \text{NaNO}_3$$

salts by the reaction sequence shown.[32] The structure of the disilver salt, with one NH_2 group intact, is perhaps unexpected but is supported by i.r. and n.m.r. Moreover, methylation with methyl iodide gives not simply the symmetrical compound $SO_2(NHMe)_2$ (as Traube supposed) but a mixture of all five possible compounds $SO_2(NR^1R^2)(NR^3R^4)$ (R^1, R^2, R^3, R^4 = Me or H).[32] These results suggest that some of the silver salts of the cyclic and linear sulfimide polymers may not have the simple structures hitherto speculatively attributed to them.

On treatment of sulfamide wih mercury(II) acetate or mercury(II) acetamide in aqueous solution, mercurated products of varying composition are precipitated, thought from X-ray evidence to be mixtures of polymers based on the structural units $-NH-SO_2-NH-Hg-$ and $-NH-S(O)-NH-OHg-$.[23]

Bis(trimethylsilyl)sulfamide is readily obtainable by several methods, and can be further silylated to give the tetrakis derivative.[15] The silyl derivatives $SO_2(NMeSiMe_3)_2$[9] and $SO_2(NHSiMe_3)(NMeSiMe_3)$[14] are also known; the former has been employed in the synthesis of mixed sulfimide–phosphazane rings (Section V below).

It seems likely that further heterocyclic syntheses might be effected by reaction of metal derivatives of sulfamide with covalent chlorides, bromides, or iodides. The mercury derivatives, in particular, are easily handled and seem promising for this purpose. The drawback of the ill-defined polymeric structure of the inorganic mercury(II) derivatives could be surmounted by using phenylmercury derivatives; the compound $SO_2(NHHgPh)_2$, which must presumably be monomeric, precipitates from sulfamide and phenylmercury acetate in methanol.[24]

D. *N*-Halogenosulfamides

Several N,N'-dialkyl-N,N'-dichlorosulfamides, $SO_2(NRCl)_2$, have been described,[19] and *N*-chlorosulfamide, $SO_2(NH_2)(NHCl)$, is also known.[18] These compounds are low-melting solids or liquids distillable under reduced pressure. They are potential reagents for ring synthesis (compare Section VI.A below).

E. Organic derivatives of sulfamide and of monomeric sulfimide

Organic mono-, di-, tri-, and tetra- derivatives of sulfamide are easily made, e.g. the N,N'-disubstituted sulfamides from primary amines and sulfuryl chloride. They are generally liquids or solids which can be sublimed or distilled without decomposition and are fairly resistant to hydrolysis.[17,19] An example of the use of one of these compounds in a ring-forming condensation is given below in Section V.C.

CHAPTER 4. IMIDES OF HEXAVALENT SULFUR

Table 4.1. Cyclic oligomers of sulfimide (see text for references).

Compound	Isolated as	Solubilities* Ag salt	Solubilities* NH₄ salt	Ionization constants in water	Other salts reported	Methyl derivative $(O_2SNMe)_x$	Ethyl derivative $(O_2SNEt)_x$	Preparation method in Section III.B
$(O_2SNH)_3$	salts and alkyl derivatives (pure acid unstable)	(trihydrate) 0.1 g per 100 g water at room temp.; more soluble in hot water	very soluble in water at room temp.	2 strong 1 weak	Tl(I), Hg(0), Hg(II), Pb(II), Co(NH₃)₆³⁺, benzidine	colorless prisms, m.p. 120–121°	colorless crystals, m.p. 65–66°	1, 2, 3, 4
$(O_2SNH)_4$	salts and alkyl derivatives (pure acid unstable)	practically insoluble in water	2 g per 100 g water at room temp.	2 strong 2 weak	pyridinium, n-butylammonium, K, Ba	m.p. 212°	dec. 240°	2, 4
$(O_2SNH)_6$	acid and salts	only slightly soluble in water	sparingly soluble in water	4 strong 2 weak	gel-like, sparingly soluble Na, K, Ba salts	not known	not known	4

*Refers to fully substituted salts $(O_2SNM)_x$.

Table 4.2. Linear oligomers derived from sulfamide.

x	Diamides $H_2N(O_2SNH)_xH$ name	preparation	properties	Monosulfonic acids $H_2N(O_2SNH)_xSO_3H$ name	preparation	properties	Disulfonic acids $HO_3SNH(O_2SNH)_xSO_3H$ name	preparation	properties
0	(ammonia)			amidosulfuric acid	see ref. 18	stable crystals, m.p. 205°	imidobis(sulfuric acid)	Sec. III.B.2, and ref. 18	unstable strong acid in water soln.; not isolable; various salts known
1	sulfamide, sulfuryl diamide	see Sec. II.B		sulfamidosulfuric acid	Sec. III.B.5	stable crystals, m.p. 92°	sulfurylamidobis(sulfuric acid)	Sec. III.B.5	known only as salts with py, NH₄, K
2	imidodisulfamide, imidobis(sulfuric acid) diamide	Sec.III.B.1, 3, 4 and ref. 23	stable crystals, m.p. 168–9°	not isolated, but probably formed as moderately stable intermediates in the hydrolysis of the cyclic sulfimides (chromatographic evidence)			imidodisulfamidobis(sulfuric acid)	Sec. III.B.5	known only as pentammonium salt
3	sulfuryl disulfamide	Sec.III.B.1, 3, 4	stable crystals, m.p. 186–7°				mixed acids, $x = 3\text{–}5$ average	Sec. III.B.2	readily hydrolyzed in warm aqueous HCl or NH₃
4		Sec.III.B.1,3	not isolated: chromatographic streaks at low R_f, and impure Ag salts						

Monomeric N-ethyl- and N-benzoyl-sulfimide, $RN=SO_2$, have been obtained as very unstable substances in solution at low temperatures.[19] Alkyl derivatives of trimeric sulfimide (Section III.C below) are, of course, stable and well characterized.

III. Cyclic and linear polymers of sulfimide

A. Introduction

Sulfur trioxide monomer very easily forms the trimer (γ-SO_3) and linear polymers (β-SO_3). Its nitrogen-system analog sulfimide (Fig. 4.1) is known only as polymers. These have been incompletely studied, but a series of cyclic oligomers $(HNSO_2)_x$ is well established, as are three series of linear polymers, with respectively $-NH_2$ end-groups, $-SO_3H$ end-groups, or both (Tables 4.1 and 4.2). The repeating unit in all these compounds is $-(HNSO_2)-$.

The history of these substances goes back to Traube, who reported the cyclic trimer (but not as such) in 1893. This compound was first recognized as trimeric by Hantzsch and Holl in 1901. Since then, the sulfimide polymers have been looked at several times, at long intervals. The most impressive progress has been made as a result of the introduction of paper chromatography to this field by G. Kempe about 1958.[26] Kempe showed that chromatographic R_M values [R_M is defined as $\log(R_f/1 - R_f)$] are an additive function of the constituent groups, for the amides, imides, and sulfonic acids now under discussion. Up to the present, chromatography has been used on these compounds only for identification and the quantitative analysis of mixtures. The way in now open for their chromatographic separation on a preparative scale.

We first describe the preparation of the polymeric sulfimides and then review their properties.

B. The preparation of cyclic and linear sulfimide polymers

Most of the preparative reactions for compounds of this group give mixtures of products. In some cases the mixtures have been only partly analyzed, and in no case are the reasons for variation in the relative yields of different products fully understood. It will be more informative, therefore, to discuss the preparative reactions as such, than to describe routes to particular compounds. The important reactions are as follows.

1. Thermal oligomerization of sulfamide

The thermal decomposition of sulfamide strongly recalls that of urea, which in the molten state gives (among other substances) the cyclic polymer cyanuric acid and the linear polymers biuret and triuret. In both cases the fundamental reaction is the condensation

$$—NH_2 + H_2N— \rightarrow NH_3 + —NH—$$

Ito[25] has made a thorough study by paper chromatography of the products obtained by heating sulfamide at various temperatures and for various periods. The results can be summarized in the diagram:

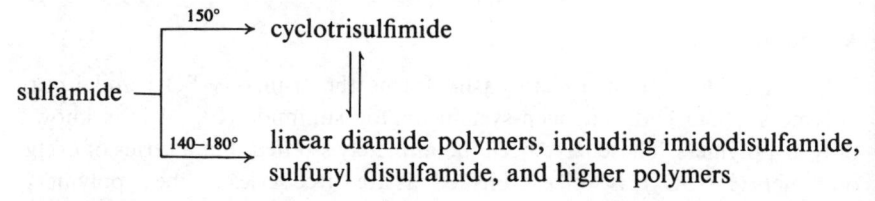

```
                    150°
                  ┌──────→ cyclotrisulfimide
                  │             ⇅
   sulfamide ─────┤
                  │
                  └──────→ linear diamide polymers, including imidodisulfamide,
                   140–180°    sulfuryl disulfamide, and higher polymers
```

The extent to which cyclotrisulfimide is formed via the linear polymers, or more directly from sulfamide, is in dispute,[16,25] but Ito showed that it does form from heated imidodisulfamide and vice versa. He also obtained chromatographic evidence (a streak at low R_f) for polymers higher than sulfuryl disulfamide. His work accurately defines the best conditions for preparing, by this route, cyclotrisulfimide (5 h at 140°, 50% yield) and sulfuryl disulfamide (2–6 h at 160°, 20% yield). Unfortunately the method is not known to give the cyclic tetramer or hexamer.

All the products are obtained as their ammonium salts, the formation of which exactly consumes 1 mol of ammonia for each (acidic) NH group produced, so no ammonia is actually evolved below 220°.

2. Reaction of ammonia with sulfur trioxide

This complicated reaction has been repeatedly investigated, under various conditions, since 1832. By analogy with the trimethylamine–SO$_3$ reaction, one might expect a high yield of the 1:1 adduct amidosulfuric acid, $H_3N \rightarrow SO_3$. In fact, amidosulfuric acid is formed only to the extent of about 10% in the gas-phase reaction and hardly at all in water or nitromethane. A major product with gaseous or aqueous ammonia is actually nitrilotrisulfate, $[N(SO_3)]_3^{3-}$, which may arise from further sulfonation of amidosulfate.[27] Nitrilotrisulfate ion hydrolyzes quickly in acid aqueous solution to imidodisulfate (Section D below), and so was missed in earlier investigations, from which its hydrolysis products imidodisulfate and sulfate were obtained.

Under the following special conditions, the ammonia–SO$_3$ reaction gives rise to condensed sulfimides.[20] Ammonia gas is passed into a concentrated nitromethane solution of sulfur trioxide at 0°. Ammonium trisulfate precipitates, and the solution on evaporation leaves a viscous liquid. The main component of this seems, on the evidence of hydrolysis products and molecular weight, to be linear sulfimide disulfonic acid polymers (Table 4.2) with $x = 3$ to 5. A work-up with aqueous silver nitrate also gave the silver salts of cyclotrisulfimide and cyclotetrasulfimide in about 10% yield.

3. Reaction of ammonia with sulfuryl halides

The formation of sulfimide polymers, as well as sulfamide, by this reaction has been mentioned (Section II.B).

The crude product from sulfuryl chloride and ammonia has been shown by paper chromatography to contain the following compounds (Table 4.2): linear diamides with $x = 1, 2, 3$; monosulfonic acids with $x = 0, 1$; disulfonic acids with $x = 0, 1$; cyclotrisulfimide.[16] Other chromatographic spots at low R_f values probably represent higher diamides, since Ephraim and Michel in 1909 had obtained, from the same crude product, mixtures of silver salts with analyses expected for $x = 4, 5$.

Similar products are obtained from sulfuryl chlorofluoride and ammonia.[16]

4. Condensation of sulfuryl halides with sulfamide and related reactions

Sulfuryl chloride is reluctant to react with sulfamide, but does so in boiling acetonitrile;[12] the solvent plays an essential part, perhaps as an ionizing medium. This is the best route to cyclotetrasulfimide, which can be isolated in 33% yield as its sparingly soluble ammonium salt:

$$2SO_2Cl_2 + 2SO_2(NH_2)_2 \rightarrow (SO_2NH)_4 + 4HCl$$

Under the same conditions, sulfuryl chloride reacts with the monosodium salt of sulfamide[30] giving cyclotetrasulfimide and larger amounts of cyclotrisulfimide. At room temperature, however, the main product is sulfuryl disulfamide, for which this is probably the best method of preparation.[30]

Lehmann and coworkers, in a logical extension of this idea, replaced sulfuryl chloride by imido-bis(sulfuryl chloride), and found the first route to cyclohexasulfimide:[29]

$$2SO_2(NH_2)_2 + 2HN(SO_2Cl)_2 \xrightarrow[\text{reflux}]{\text{acetonitrile}} (SO_2NH)_6 + 4HCl \uparrow$$

The same compound resulted in even better, almost 100%, yield by heating a compound containing the required $-NH_2$ and $-SO_2Cl$ groups in the same molecule, viz. amidosulfuryl chloride:

$$6H_2NSO_2Cl \xrightarrow{120°} (SO_2NH)_6 + 6HCl$$

Amidosulfuryl chloride reacts with pyridine in benzene at room temperature[1] to give the dipyridinium salts of cyclotrisulfimide (in 49% yield) and cyclotetrasulfimide (in 21% yield); and in boiling acetonitrile it reacts with sulfamide to give imidodisulfamide:

$$H_2NSO_2Cl + SO_2(NH_2)_2 \rightarrow HN(SO_2NH_2)_2 + 2HCl$$

Methods of preparing amidosulfuryl chloride and imido-bis(sulfuryl chloride) have been devised relatively recently.[18]

5. Introduction of sulfonic acid end-groups

As indicated in Table 4.2, the terminal $-NH_2$ groups of sulfamide and imidodisulfamide can be sulfonated by the action of molten pyridine–sulfur trioxide adduct, giving pyridinium salts of the sulfonic acids.[31]

C. Properties and derivatives of the cyclic sulfimides

The cyclic sulfimides are water-soluble acids. The first few ionizations of their NH groups are strong, and subsequent ones weaker (Table 4.1).[29] The solubilities of their ammonium and silver salts decrease with increasing ring size, a property made use of in their separation.[18,23] The trisilver salt of the cyclic trimer can be recrystallized from hot water, but silver salts of the tetramer and hexamer are too insoluble for this (Table 4.1).[18,23]

The cyclic hexamer has been prepared free of solvents[29] as a very hygroscopic, water-soluble solid, fairly stable at room temperature, but not recrystallizable without decomposition. The trimer and tetramer, however, have never been isolated and do not seem to be stable in the absence of solvents. An aqueous solution of the trimer, fairly stable at room temperature, has been made by treating the trisilver salt with HCl, but when concentrated, even near room temperature, it decomposed to ammonium hydrogen sulfate.[1] Another approach, treatment of the dipyridinium salt with the hydrogen form of a sulfonic acid ion-exchange resin in liquid SO_2, also failed; the resulting solution yielded imidodisulfamide (Section D below) on evaporation.[1] When the silver salt of the tetramer was treated with HCl in ether, the resulting solution left on evaporation a water-soluble oil containing much SO_4^{2-} formed by decomposition.[12] In spite of these difficulties, it must be emphasized that ammoniacal aqueous solutions of all the cyclic sulfimides are stable for long periods at room temperature, and can be subjected to metathetical reactions and paper chromatography without any change in the rings.

Fully methylated and ethylated derivatives of the trimer and tetramer are easily made by treating the silver salts with the required alkyl iodide. They are stable crystalline solids.[6,12]

Crystallographic studies of the trisilver and trimethyl derivatives of the trimer[22] have confirmed the presence of a 6-membered ring, previously deduced from chemical evidence in conjunction with the experimentally determined molecular weight of the methyl compounds. For the tetramer and hexamer there have been no direct structure determinations, but the sum of other evidence is fairly conclusive: elemental analyses, the molecular weight of the methylated tetramer, and potentiometric and conductimetric titration data which characterize the acid ionizations in number and strength (Table 4.1).

The reactions of the cyclic sulfimides so far studied are salt formation and hydrolysis.

Various salts, other than the silver and ammonium salts mentioned above, have been made by neutralization of the cyclic sulfimides with bases or metathesis with other salts (Table 4.1), and in most cases analyzed with satisfactory results.[12,18,23,29]

Hydrolysis of the sulfimide rings is fast in strongly acid aqueous solutions (minutes to hours at room temperature).[12,28,29] It is not yet possible to make definitive statements about the complex reaction sequences involved, but paper chromatography and sulfate determination have given some clues. In 2M HCl at room temperature the rings are opened and the resulting chains progressively degraded, giving (in times of the order of hours) a mixture of sulfamide and amidosulfate. These are relatively stable substances, their eventual hydrolysis to sulfate taking very much longer. The chains formed by opening the trimer and hexamer rings immediately produce some sulfate as they hydrolyze, but the tetramer chains do not. This probably depends on which point in a chain is most susceptible to hydrolytic attack. The reactions with the trimer may be:

$$(HNSO_2)_3 + H_2O \rightarrow H_2NSO_2HNSO_2HNSO_3H$$

$$H_2NSO_2HNSO_2HNSO_3H + H_2O \rightarrow H_2SO_4 + H_2NSO_2HNSO_2NH_2$$
imidodisulfamide

$$H_2NSO_2HNSO_2NH_2 + H_2O \rightarrow H_2NSO_3H + SO_2(NH_2)_2$$
amidosul-   sulfamide
furic acid

with the second water molecule attacking near the end of the chain. With the hexamer also, sulfuric acid is early produced, and probably in a similar way; chromatography naturally shows more substances, implying more steps in the overall hydrolysis. With the tetramer, which has been less studied, the next step after ring opening probably affords two molecules of sulfamidosulfuric acid, rather than sulfuric acid:

$$H_2NSO_2(HNSO_2)_3OH + H_2O \rightarrow 2H_2NSO_2NHSO_3H$$

The ammonium salts of the tri- and tetra-sulfimide are not appreciably hydrolyzed in hours in neutral solution at 100°, but that of hexasulfimide hydrolyzes considerably in 1.5 h at this temperature.[12,28,29]

D. Properties of the linear polymeric sulfimides

Table 4.2 lists three important kinds of linear sulfimide oligomers grouped according to their end-functions. A few examples of compounds belonging to this general category, but with other types of end-groups, are also known, for example N-chlorosulfamide (Section II.D above), imido-bis(sulfuryl chloride) $HN(SO_2Cl)_2$, and organo-substitution products. On the present rather inadequate evidence, there is no limit to the possible chain length of the diamide series (left-hand side of Table 4.2) or the disulfonic acid series (right-hand side). Less is known about the monosulfonic acids (middle of table). It seems possible that under anhydrous conditions the longer-chain compounds in this series might tend to cyclize by elimination of water; this does not preclude their existence in aqueous solution, for which there is some chromatographic evidence (Section C above).

The three compounds of the diamide series hitherto isolated (Table 4.2) (leaving out the formally first member, ammonia) are stable, colorless, odorless, crystalline solids,[8,18,23,30,34,35] readily soluble in water, insoluble in non-polar organic solvents, and of intermediate solubility in polar organic solvents. The third member, sulfuryl disulfamide, is very hygroscopic.[30,35] The salts of sulfamide ($x = 1$) have been described (Section II.C above). Imidodisulfamide ($x = 2$) is a strong acid in its first dissociation (believed to involve the imido proton), and several monosubstituted salts have been described.[18] Silver nitrate in presence of ammonia, however, precipitates a trisilver salt.[18] The two NH groups of sulfuryl disulfamide are strongly acidic; several salts in which both are substituted have been described, including a very slightly soluble silver salt; aqueous solutions of the dipotassium salt react neutral, showing that the $-NH_2$ groups, like those of sulfamide and imidodisulfamide, are little ionized.[30] However, the compound has been completely methylated, with replacement of all hydrogen atoms, by the prolonged action of diazomethane in ether at room temperature. The hexamethyl derivative[30] is a stable, colorless, odorless solid, m.p. 161°, recrystallizable from boiling water. A dimethyl derivative,[35] and also the monomethyl derivative, of imidodisulfamide have been made similarly.[8,35]

Sulfamide, and the imido-methyl derivatives of imidodisulfamide and sulfuryl disulfamide, all undergo the Kirsanov reaction with phosphorus pentachloride (see Chapter 5, Sec. II.A), that is, replacement of the hydrogen atoms of $-NH_2$ groups by $=PCl_3$,[35] e.g.

$$MeN(SO_2NH_2)_2 + PCl_5 \xrightarrow[24\,h]{CCl_4} MeN(SO_2N\!\!=\!\!PCl_3)_2 + 4HCl \uparrow$$

$-NH-$ groups must first be inactivated by methylation, otherwise unwanted condensation takes place. The resulting trichlorophosphazo derivatives can be converted into triphenylphosphazo or triphenoxyphosphazo compounds by standard methods.[35]

The hydrolytic behavior of the linear diamide polymers is complicated and has not been adequately investigated. In aqueous solution at room temperature the three compounds in Table 4.2 hydrolyze only slowly in absence of added acid, the half-time for 0.25M imidodisulfamide being 17.3 h at 20°.[23] 2M HCl considerably accelerates the process. The hydrolysis products of imidodisulfamide and sulfuryl disulfamide include sulfamide, amidosulfuric acid, and sulfuric acid.[23,30] In hot alkaline solutions, however, condensation rather than chain degradation takes place; thus sulfamide is efficiently converted into imidodisulfamide:[23]

$$2SO_2(NH_2)_2 \xrightarrow[10 \text{ min, } 100°]{0.5M \text{ NaOH}} NH_3 + H_2NSO_2NHSO_2NH_2 \ (69\% \text{ yield})$$

and sulfuryl disulfamide cyclizes to the cyclic trimer, thus:[30]

$$O_2S(NK-SO_2-NH_2)_2 \xrightarrow[\text{boil 5 h}]{30\% \text{ KOH}} NH_3 + (KNSO_2)_3$$

Just as sulfamide and imidodisulfamide can be compared to urea and biuret respectively (Fig. 4.2), so can sulfuryl disulfamide be compared to triuret, $CO(NH-CO-NH_2)_2$. It is interesting that copper(II) ions in alkaline solution give violet colors with both biuret and imidodisulfamide, and steel-blue colors with both triuret and sulfuryl disulfamide.[30]

Little need be said about the properties of the monosulfonic acid series of linear sulfimide polymers (Table 4.2). Amidosulfuric (sulfamic) acid, which can formally be regarded as the first member, actually has the structure $H_3N \rightarrow SO_3$ in the solid state. It is manufactured on a large scale and is adequately described in the literature.[18] The ammonium salt of the second member, sulfamidosulfuric acid, was obtained in 1956 by Appel, Voigt, and Sadek as an ammonolysis product of disulfuryl chloride, $S_2O_2Cl_2$, and the pyridinium salt was prepared by the sulfonation of sulfamide with the molten pyridine–SO_3 adduct.[31] In aqueous solution at room temperature the rate of hydrolysis is negligible at alkaline pH; in 0.5M HCl hydrolysis is almost complete after 2 days. Higher members of the series probably form in the early stages of hydrolysis of the cyclic sulfimides (Section C above); they produce chromatographic spots but have not been isolated in bulk.

As for the disulfonic acid series of linear polymers (Table 4.2), the anions of the individual lower oligomers are moderately stable in cold neutral or alkaline aqueous solutions, and a number of salts have been well characterized.[8,18,31] Mixed acids with $x = 3$ to 5 are said to be obtained in the free state, as an amorphous and very hygroscopic mass, from the reaction of ammonia with excess sulfur trioxide in nitromethane,[20] but acids with $x = 0, 1$, and 2 have not been isolated. All these acids hydrolyze in minutes to hours at room temperature in acid solutions. The first member of the series, imidodisulfuric acid, is the only one which has been much studied; its ammonium salt can readily be obtained from the reaction between excess ammonia and sulfur trioxide (Section B.2 above).[18]

E. Four-membered sulfimide rings

The cyclic sulfimide dimer is unknown, and 4-membered S–N rings are rare. However, the compounds **Ia** and **Ib**, which are derivatives of the cyclic sulfimide dimer, have been described. They are formed by the action of sulfur

$$\text{Ia } (X = F)$$
$$\text{Ib } (X = CF_3)$$

(I)

trioxide on FSO_2NSO [38] and CF_3SO_2NSO [37] (Chapter III, Sec. III.G) respectively. Compound **Ia** is a liquid, m.p. 16.5°, which decomposes above 30°. Compound **Ib** is a white crystalline solid, m.p. 58°. Both are sensitive to moisture. The structures **I** follow from i.r. and mass-spectral data.

Lewis bases L (L = S_4N_4, pyridine, 4-cyanopyridine) cleave the rings of **I** giving adducts $XSO_2NSO_2 \cdot L$.[37]

IV. Mixed six-membered rings containing the sulfimide group

A. Introduction

The compound $N_3S_3O_4H$ (**II**) and the two isomeric compounds $N_3P_2SCl_4O_2Me$ (**III** and **IV**) have structural components in common with the N-methyl

(II)   (III) (R = Me, Et, Pr, Bu)   (IV)

sulfimides (Section III.C above). Rings **III** and **IV** also contain phosphazene segments (Chapter 12). These rings are of interest in containing a possible path of π-electron delocalization, like the nitrides S_4N_4 and S_4N_2 (Chapter 6) and phosphazenes respectively, but interrupted at the 3-coordinate nitrogen atom.

B. The compound $N_3S_3O_4H$

The preparation of this compound has been mentioned in Chapter III, Sec. IV.C. It crystallizes in bright yellow rods, m p. 139°(dec.). It can be vaporized without decomposition. The structure **II** has been suggested on i.r., mass-spectral, and n.m.r. evidence, but needs confirmation.[14] If this structure is correct, this would be the acid from which the well characterized anion $S_3N_3O_4^-$ (Chapter 8, Sec. III.A) is derived.

C. Preparation and properties of $N_3P_2SCl_4O_2Me$ isomers

Kirsanov reported in 1952 that sulfamide (Chapter 4, Sec. II) reacts smoothly with phosphorus pentachloride as follows:

$$SO_2(NH_2)_2 + 2PCl_5 \rightarrow SO_2(N=PCl_3)_2 + 4HCl$$

This reaction is analogous to the reaction of PCl_5 with amidosulfuric acid (Chapter 5, Sec. II.A). The product $SO_2(N=PCl_3)_2$ contains all the components of Me-**III** above, in the correct sequence, except the *N*-methyl group. Becke-Goehring, Bayer, and Mann supplied this by reaction with heptamethyldisilazane:[7]

$$MeN(SiMe_3)_2 + O_2S(NPCl_3)_2 \xrightarrow[50-60°]{benzene} 2Me_3SiCl + O_2S(NPCl_2)_2NMe$$

(isomer **III**)
white needles, m.p. 177°

The principle of this reaction is the same as that of syntheses from bis(trimethylsilyl) sulfur diimide (Chapter 3, Sec. IV.C), namely, the strong thermodynamic drive for formation of trimethylchlorosilane.

Isomer **IV** above was made in 1974 by Roesky and Grosse-Böwing, on similar principles, by the following reaction sequence:[36]

$ClSO_2N=PCl_2-N=PCl_3 + MeN(SiMe_3)_2 \longrightarrow$
(see Section B)

$Me_3SiCl + ClSO_2N=PCl_2-N=PCl_2-NCH_3-SiMe_3$

$\downarrow$ cyclizes $-Me_3SiCl$ | reflux 20 h in tetrachloroethane dilute solution

isomer **IV**
white crystals, m.p. 80°

The structures of isomers **III** and **IV** have been adequately characterized by i.r., ^{31}P and ^1H n.m.r., and mass spectra. Me-**III** can be fluorinated or aminated.[19]

D. The cyclic anion $O_2S[NP(NH_2)_2]_2N^-$

The compound $O_2S(N=PCl_3)_2$ (Section C above) reacts with liquid ammonia[7] in a similar way to its reaction with heptamethyldisilazane, forming the ring of isomer **III** above. However, the four terminal chlorines are aminated at the same time. The product is the ammonium salt $[NH_4]^+O_2S[NP(NH_2)_2]_2N^-$, a water-soluble compound from which potassium and silver salts have been prepared by metathesis. On heating to 135° *in vacuo*, the ammonium salt loses ammonia and leaves a polymer thought to contain linked rings of the type of III.

V. Mixed four-membered rings containing the sulfimide group

A. Introduction

All the compounds to be described in this section contain saturated SN_2P rings, with sulfur in the +6 oxidation state as $-SO_2-$, and phosphorus in the +5 state. They can be regarded equally well as cyclosulfimides or as cyclophosphazanes (compare Chapter 11). Phosphazane chemistry affords many examples of 4-membered N_2P_2 rings and also examples of P–N spiro structures quite like those to be mentioned here. The entire known chemistry of the present compounds was developed by Becke-Goehring and her collaborators in the late 1960s.

B. The one-ring compound $(MeN)_2SO_2PF_3$

This compound, with structure V, was reported in 1968 from the reaction [10,19]

$$SO_2(NMeSiMe_3)_2 + PF_3Cl_2 \xrightarrow[-20°]{\text{benzene}} (MeN)_2SO_2PF_3 + 2Me_3SiCl$$

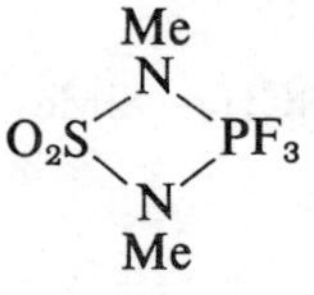

<pre>
        Me   Me F Me   Me Cl Me
         N   N | N   N | N
       /   \   / | \   / | \
  O₂S   PF₃   O₂S   P   SO₂   O₂S   P   SO₂
       \   /   \   /   \   /
         N   N   N   N   N
        Me   Me   Me   Me   Me
        (V)   (VI)   (VII)
</pre>

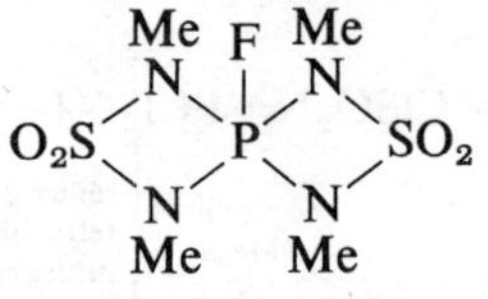

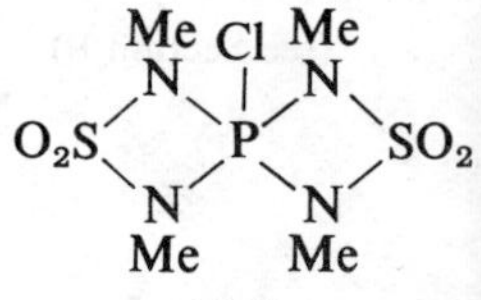

Compound V results in 50% yield in this way from the trimethylsilyl derivative, but could not be obtained from the parent sulfamide $SO_2(NHMe)_2$ and PF_3Cl_2 in presence of base.[10] It forms colorless moisture-sensitive crystals of m.p. 54°, readily soluble in benzene and other indifferent solvents. The structure shown is based on an ebullioscopic molecular weight, and proton, ^{19}F, and [31]P n.m.r. data.

C. Spiro compounds

Treatment of V with a further 1 mol of the silyl compound $SO_2(NMeSiMe_3)_2$ gives the two-ring spiro compound VI.[10] This requires more drastic conditions than the preparation of **V**, namely 24 h at 60–70°. Compound **VII** is made as follows, without the use of a silylated intermediate:[11]

$$2SO_2(NHMe)_2 + PCl_5 \xrightarrow[\text{reflux}]{\text{pyridine, CCl}_4} \textbf{VII} + 4HCl$$

and can readily be fluorinated to **VI** by means of silver(I) fluoride.[11] Compounds **VI** and **VII** form colorless moisture-sensitive crystals, soluble in benzene or chlorinated alkanes but almost insoluble in light petroleum or cyclohexane.

No X-ray crystallographic data are available for **VI** or **VII**, but the spiro structures seem fairly well established by n.m.r. and i.r., which show the presence of a 5-coordinate phosphorus atom and sulfonyl groups.[11]

A phenyl analog of **VII** can be made similarly to **VII**.

The preparation of the three-ring spiro compound **VIII** resembles that of **VII** just mentioned, except that the cyclophosphazane **IX** (Chapter 11, Sec.

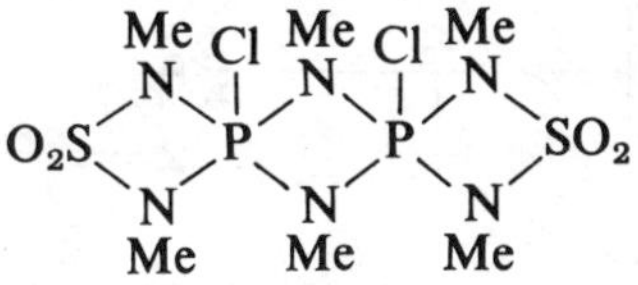

(VIII) 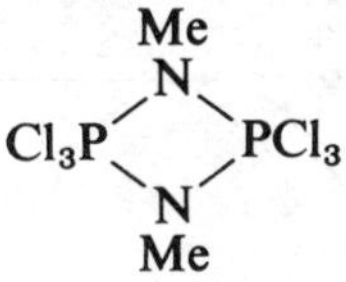   (IX)

III.C) replaces PCl_5 and supplies the middle ring of **VIII**.[11] Physically, **VIII** resembles **VII**. The structure shown is supported by the known structure of **IX** (X-ray crystallography), by ^{31}P n.m.r. which shows the two phosphorus atoms of **VIII** to be chemically equivalent and 5-coordinate, and by the crystallographically proven structure of the closely related spirophosphazane (formula **XXIV** in Chapter 11).

VI. Further imido derivatives of sulfur(VI), used or potentially usable in heterocyclic synthesis

A. The S,S-dialkylsulfodiimides

These reagents have been developed by Appel and coworkers; a bibliography is available in ref. 5.

The S,S-dialkylsulfodiimides have structure **X**. They are prepared by

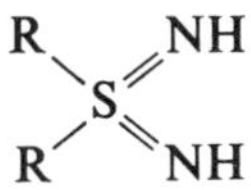

(X) (R = Me, Et) 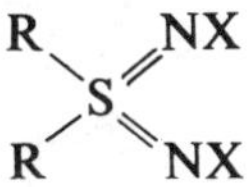   (XI) (X = Cl, Br, I, SnMe₃)

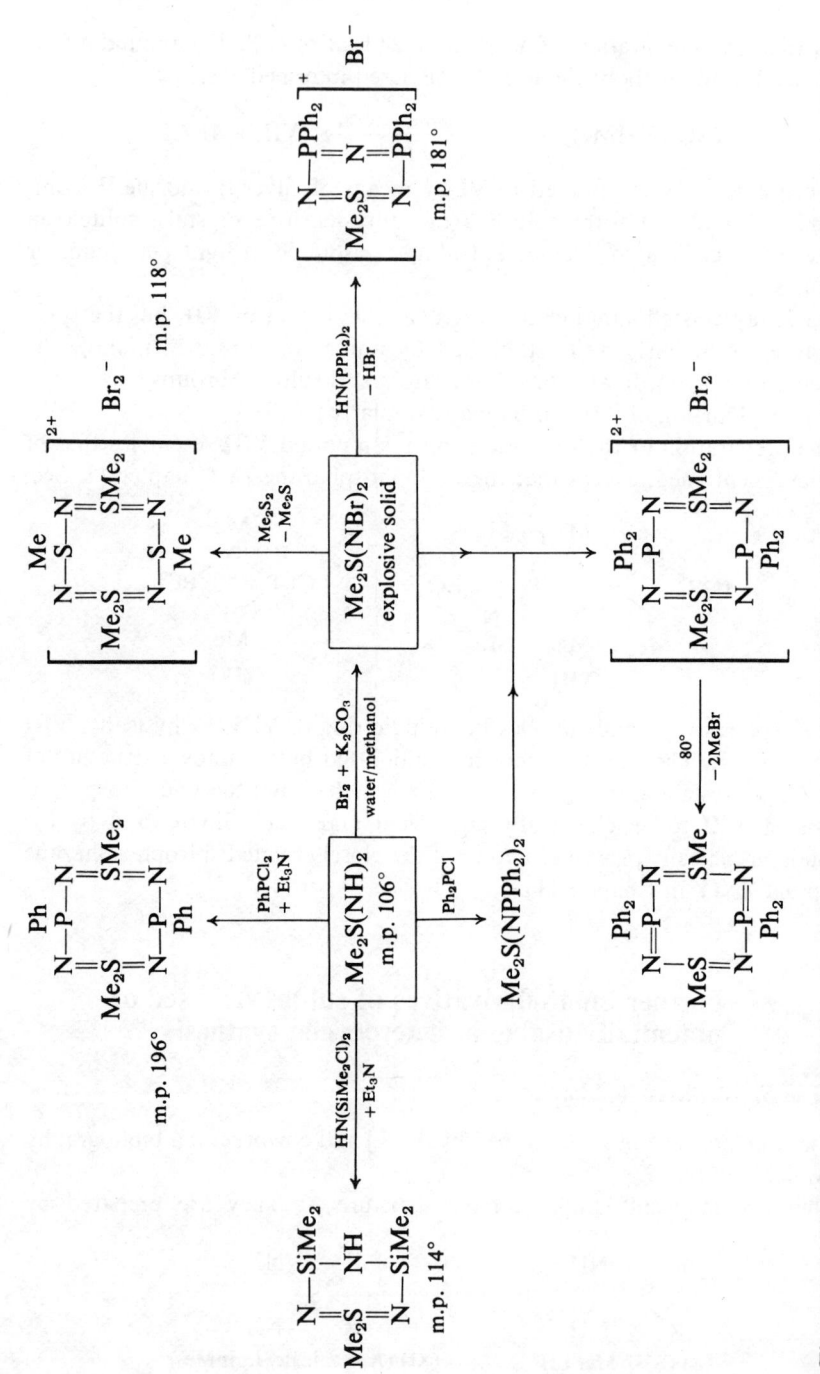

Figure 4.3. Syntheses of inorganic heterocycles from S,S-dimethyl sulfodiimide.[5,21] Further inorganic and organic examples may be found in these references. See also Chapter 7, Sec. II.G.

chloramination of dialkyl sulfides, as in the following example:

$$Me_2S + 2NH_2Cl + NH_3 \xrightarrow[20°]{Pr^iOH} [Me_2S(NH)NH_2]^+Cl^- + NH_4Cl$$

$$[Me_2S(NH)NH_2]^+Cl^- + Na_2CO_3 \xrightarrow{water} Me_2S(NH)_2 + NaCl + NaHCO_3$$

The methyl and ethyl compounds form colorless hygroscopic crystals, m.p. 106° and 47° respectively. They are readily soluble in water and also in organic solvents including benzene. They are weakly basic and form crystalline hydrochlorides.

The dihalogeno derivatives **XI** can be made by treatment of **X** with the halogen in aqueous sodium carbonate solution. The dichloro derivatives are dangerously explosive; the dibromo derivatives are explosive but more easily handled. Both the parent imides **X** and the dibromo derivatives **XI** have been employed in heterocyclic syntheses; Fig. 4.3 shows some inorganic examples. It will be noted that the ring-forming reactions of the imide itself are simple heterofunctional condensations, whereas those of the dibromo compound entail conversion of positive into negative halogen accompanied by oxidation of $P(III)$ to $P(V)$ or $S(I)$ to $S(IV)$.

The derivative **XI** with X = $SnMe_3$ can be made by transamination of **X** with Me_3SnNMe_2.[21] It forms colorless crystals, m.p. 28°. It may have some potential for ring synthesis (compare Chapter 3, Sec. IV.D) but it has been found[21] that replacement of one of the $SnMe_3$ groups by an electron-donating group deactivates the other.

B. The mercurial $Hg[NS(O)F_2]_2$

This compound is a potential but so far unexploited reagent for heterocyclic synthesis. It is a sulfur(VI) analog of the sulfur(IV) compounds $Hg(NSO)_2$ and $Hg(NSF_2)_2$ (Chapter 3, Sec. III.E). It has been made by the reaction sequence

$$(Me_3Si)_3N + SOF_4 \xrightarrow[80°]{autoclave} Me_3SiN{=}S(O)F_2 + 2Me_3SiF$$

$$2Me_3SiN{=}S(O)F_2 + HgF_2 \xrightarrow[85°]{autoclave} Hg[N{=}S(O)F_2]_2 + 2Me_3SiF$$

It forms colorless moisture-sensitive crystals. It can be used to replace chlorine atoms in such compounds as $HN(SO_2Cl)_2$ by $[NS(O)F_2]$ groups,[14] and might perhaps be made to yield $[NS(O)F_2]$ radicals by, for example, reaction with iodine.

References

1. Appel, R. and Berger, G., *Z. Anorg. Allg. Chem.* **327**, 114 (1964)
2. Appel, R. and Gerber, H., *Chem. Ber.* **91**, 1200 (1958)
3. Appel, R. and Helwerth, R., *Angew. Chem. Int. Ed. Engl.* **6**, 952 (1967)

CHAPTER 4. IMIDES OF HEXAVALENT SULFUR

4. Appel, R. and Ross, B., *Angew. Chem. Int. Ed. Engl.* **7**, 546 (1968)
5. Appel, R. and Eichenhofer, K.-W., *Chem. Ber.* **104**, 3859 (1971)
6. Becke-Goehring, M., *Adv. Inorg. Chem. Radiochem.* **2**, 169 (1960)
7. Becke-Goehring, M., Bayer, K and Mann, T., *Z. Anorg. Allg. Chem.* **346**, 143 (1966)
8. Becke-Goehring, M. and Fluck, E., in *Developments in Inorganic Nitrogen Chemistry*, ed. C. B. Colburn, Elsevier, Amsterdam (1966)
9. Becke-Goehring, M. and Wunsch, G., *Liebigs Ann. Chem.* **618**, 43 (1958)
10. Becke-Goehring, M. and Weber, H., *Z. Anorg. Allg. Chem.* **365**, 185 (1969)
11. Becke-Goehring, M. and Wald, H.-J., *Z. Anorg. Allg. Chem.* **371**, 88 (1971)
12. Bencker, K., Leiderer, G. and Meuwsen, A., *Z. Anorg. Allg. Chem.* **324**, 202 (1963)
13. Brauer, G., *Handbook of Preparative Inorganic Chemistry*, 2nd ed., vol. 1, p. 482, Academic Press, New York and London (1963)
14. Buckendahl, W. and Glemser, O., *Chem. Ber.* **110**, 1154 (1977)
15. Buss, W., Krannich, H. J. and Sundermeyer, W., *Z. Naturforsch.* B **30b**, 842 (1975)
16. Cueilleron, J. and Monteil, Y., *Bull. Soc. Chim. France* 888, 892 (1966)
17. Dorlars, A., in *Methoden der Organischen Chemie* (Houben-Weyl), vol. XI/2, Thieme, Stuttgart (1958)
18. *Gmelins Handbuch der Anorganischen Chemie*, Teil 3B, "Schwefel", Verlag Chemie, Weinheim (1963)
19. *Ibid.*, Ergänzungswerk zur 8 Auflage, Band 32, Teil 1, "Schwefelstickstoffverbindungen", Springer-Verlag, Berlin, Heidelberg, New York (1977)
20. Goehring, M., *Ergebnisse und Probleme der Chemie der Schwefelstickstoffverbindungen*, Akademie-Verlag, Berlin (1957)
21. Hänssgen, D. and Appel R., *Chem. Ber.* **105**, 3271 (1972)
22. Hazell, A. C., *Acta Cryst.* **B30**, 2721, 2724 (1974)
23. Heal, H. G., in *Inorganic Sulphur, Chemistry*, ed. G. Nickless, Elsevier, Amsterdam (1968)
24. Heal, H. G. and Edwards, S., unpublished work
25. Ito, Yukio, *Nippon Kagaku Kaishi* 320 (1972)
26. Kempe, G., *Z. Anal. Chem.* **180**, 9 (1961)
27. Lehmann, H.-A., Beyer, D. and Schneider, W., *Z. Anorg. Allg. Chem.* **337**, 22 (1965)
28. Lehmann, H.-A. and Kempe, G., *Z. Anorg. Allg. Chem.* **306**, 273 (1960)
29. Lehmann, H.-A., Schneider, W. and Hiller, R., *Z. Anorg. Allg. Chem.* **365**, 157 (1969)
30. Meuwsen, A. and Papenfuss, T., *Z. Anorg. Allg. Chem.* **318**, 190 (1962)
31. Meuwsen, A. and Reichelt, H., *Angew. Chem.* **71**, 162 (1959)
32. Nachbaur, E. and Popitsch, A., *Angew. Chem. Int. Ed. Engl.* **12**, 339 (1973)
33. Nachbaur, E., Popitsch, A. and Burkert, P., *Monatsh. Chem.* **105**, 822 (1974)
34. Nannelli, P., Failli, L. and Moeller, T., *Inorg. Chem.* **4**, 558 (1965)
35. Nara, K., Nakagaki, M., Manabe, O. and Hiyama, H., *Kogyo Kagaku Zasshi* **69**, 20 (1966)
36. Roesky, H. W. and Grosse-Böwing, W., *Z. Anorg. Allg. Chem.* **406**, 260 (1974)
37. Roesky, H. W., Aramaki, M. and Schönfelder, L., *Z. Naturforsch.* B **33b**, 1072 (1978)
38. Schmidt, K.-D., Mews, R. and Glemser, O., *Angew. Chem. Int. Ed. Engl.* **15**, 614 (1976)

5
Sulfanuric halides and related compounds

I. Introduction: structural relationships

This chapter deals with a group of unsaturated ring systems (Chapter 1, Sec. II) containing nitrogen and sulfur(vi). The presence of unsaturation, or in other words, of 2-coordinate doubly bonded nitrogen atoms (–N=) in the ring, separates these compounds from those covered in Chapter 4.

The range of compounds described in the present chapter is defined by the characteristic grouping

$$\begin{array}{c} O \diagdown \quad X \\ S \\ \diagup \quad \diagdown N \\ | \end{array}$$

(see Fig. 5.1). The X in the parent compounds is chlorine, but many derivatives exist with X = fluorine or an alkyl, aryl, or amino group. The defining group may be combined in the ring with others of the same kind (Fig. 5.1, **II** and **III**), or with sulfur in a different oxidation state (**V** to **VIII**), or with phosphazene [NPX$_2$] groups (**IX, X, XI**). All rings in this category so far studied in detail are 6-membered, but there has been one preliminary report of an 8-membered ring, and the analogy of the phosphazenes (Chapter 12) suggests the possibility of many 8-membered sulfanuric ring compounds and perhaps larger rings too. In spite of a burst of activity in the last decade, the study of the sulfanurics is still in its early stages and obviously capable of large development. The sulfanurics are likely to get a good deal of attention because of their structural and chemical interest, and because they are fairly stable inorganic covalent compounds readily amenable to study by such methods as vapor-phase chromatography.

The energy difference between unsaturated sulfanuric rings and their saturated isomers (derivatives of cyclic trisulfimide; Chapter 3) seems to be

(I)
Cyanuric chloride

(II)
Sulfanuric chloride trimer

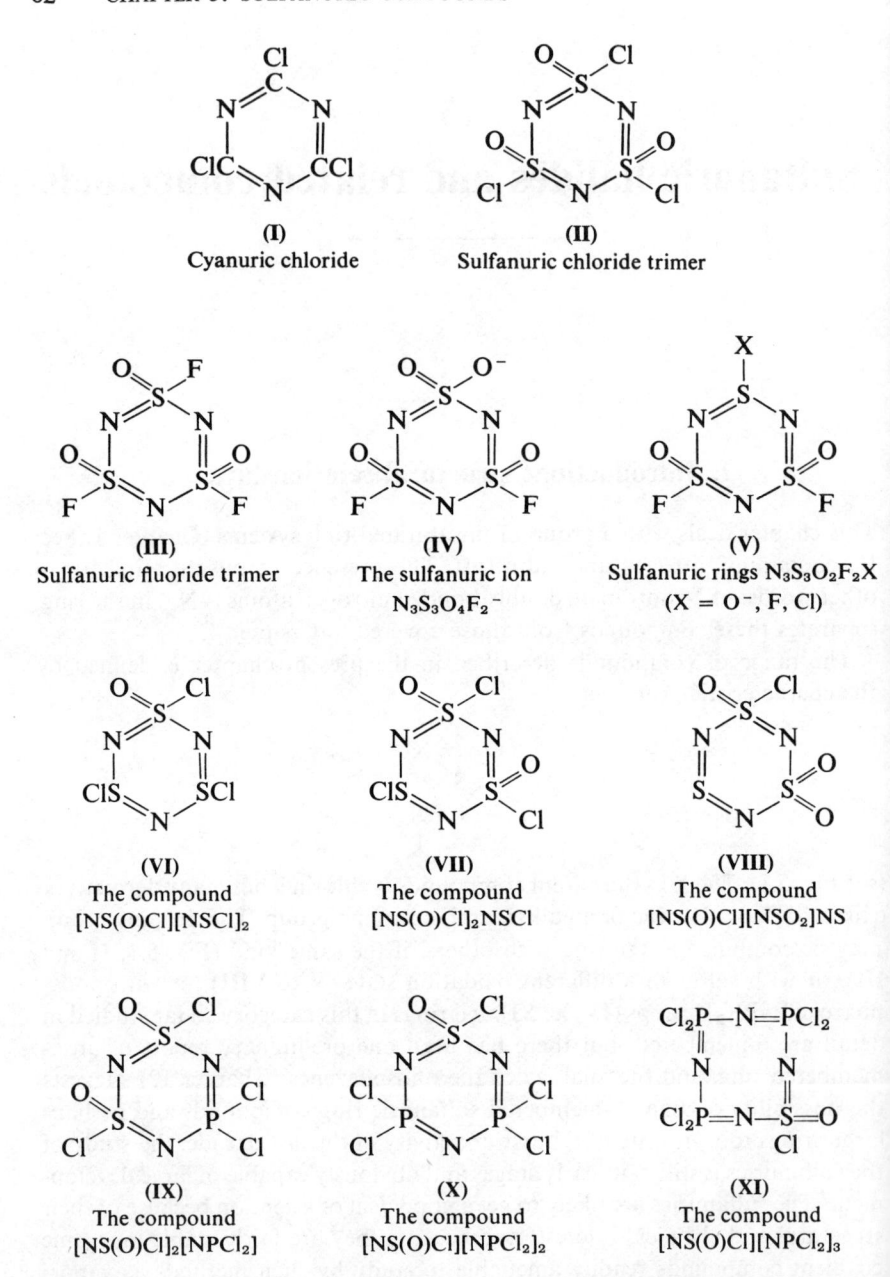

(III)
Sulfanuric fluoride trimer

(IV)
The sulfanuric ion $N_3S_3O_4F_2^-$

(V)
Sulfanuric rings $N_3S_3O_2F_2X$ (X = O^-, F, Cl)

(VI)
The compound [NS(O)Cl][NSCl]$_2$

(VII)
The compound [NS(O)Cl]$_2$NSCl

(VIII)
The compound [NS(O)Cl][NSO$_2$]NS

(IX)
The compound [NS(O)Cl]$_2$[NPCl$_2$]

(X)
The compound [NS(O)Cl][NPCl$_2$]$_2$

(XI)
The compound [NS(O)Cl][NPCl$_2$]$_3$

Figure 5.1. Structural formulas of sulfanuric halides (II to IV); S–N heterocycles containing sulfanuric and also S(IV) segments (V to VIII); and mixed sulfanuric–phosphazene halides (IX to XI). Cyanuric chloride (I) is shown for comparison.

relatively small in some cases, since reactions starting with one type of ring can give products with rings of the other type (see Sections II.G and H below).

It is interesting that the [NPX$_2$] group (Fig. 5.1, **IX** and **X**) is in a loose sense isoelectronic with [NS(O)X]; however, as formulas **V** to **VIII** show, it is by no means a condition of stability of these rings that they be composed of isoelectronic groups.

The common name for the compounds [NS(O)X]$_3$ (X = Cl, F), sulfanuric halides, is short, unambiguous, and understood everywhere, but it has little systematic basis. It was first used by the discoverer of [NS(O)Cl]$_3$, A. V. Kirsanov, about 1950, by analogy with cyanuric chloride, (NCCl)$_3$. This latter compound is now known to have the structure **I** in Fig. 5.1, so the structural analogy is sound;[34] it is the same analogy as exists between sulfamide and urea and their derivatives (Fig. 4.2). The names, however, are misleading. Cyanuric chloride is so named as if it were the acid chloride of cyanuric acid, but it is not; cyanuric acid in the crystalline state actually has the structure shown in Fig. 4.2.[11] Sulfanuric acid is unknown and seems unlikely to be made (Section II.G below).

The sulfanurics are covered in a recent volume of *Gmelin*.[14]

II. The sulfanuric halides

A. Preparation of the sulfanuric chlorides

Sulfanuric chloride, or trithiazyl oxychloride, [NS(O)Cl]$_3$ (Fig. 5.1, **II**), is best prepared by a recent procedure of Klüver and Glemser:[20]

$$SOCl_2 + NaN_3 \xrightarrow{\text{acetonitrile}} \tfrac{1}{3}[NS(O)Cl]_3 + N_2 \uparrow$$
$$35\%$$

the reagents being mixed at $-35°$ and then allowed to warm up slowly.

Kirsanov's original method of preparation[17,27,28] is still useful, and interesting for its mechanism and by-products. About 1910, Ephraim and Gurewitsch had obtained from amidosulfuric acid and phosphorus pentachloride what they wrongly took to be an addition compound, $H_2NSO_2Cl \cdot PCl_3$. Kirsanov showed that this product was really the new compound trichlorophosphazosulfuryl chloride:

$$2PCl_5 + H_3NSO_3 \xrightarrow[80°]{CCl_4} 3HCl + POCl_3 + Cl_3P{=}NSO_2Cl$$
$$75\%$$

a readily hydrolyzable solid, m.p. 35–36°. Its pyrolysis yields sulfanuric chloride:

$$3Cl_3P{=}NSO_2Cl \xrightarrow{130-150°} [ClS(O)N]_3$$
$$17\%$$

The sulfanuric chloride is produced by either method as a stable α-form mixed with a less stable β-form, and there is evidence of other even less stable forms. Recrystallization of the crude product by cooling a hot solution in n-heptane gives mainly the less soluble α-isomer, rhombic prisms with m.p. 144–145°. For the more soluble β-isomer, the mother-liquor from the crystallization is evaporated *in vacuo* and the residue sublimed *in vacuo*. The β-form sublimes at room temperature, whereas the α-form requires a temperature of about 80° for sublimation.[28]

The second stage of Kirsanov's preparation has, for obscure reasons, given somewhat different results in the hands of different investigators. In 1963 van de Grampel and Vos obtained the first sample of $[NS(O)Cl]_2[NPCl_2]$ from it (Section IV.B below), and the same compound was again found as a minor product by Clipsham *et al.*[10]

One may speculate that the first product of pyrolysis of $Cl_3P=NSO_2Cl$ is the monomer **A**, which then trimerizes to sulfanuric chloride. This monomer,

$$N\equiv S\begin{matrix}O\\Cl\end{matrix} \qquad \left[\begin{matrix}N\diagup S\diagdown O\\ \|\\O\end{matrix}\right]^{-} \qquad N\diagup S\diagdown O\atop Cl$$

(A)   **(B)**   **(C)**

which has not been isolated, is isomeric with the known compound Cl–N=S=O (Chapter 3, Sec. III.E). Goehring and coworkers in 1953 reported another preparation of sulfanuric chloride, based upon the generation of the monomer by a different route.[28] They passed ammonia into a petroleum ether solution of sulfuryl chloride to give (as they argued) the intermediate **B**; thionyl chloride was included in the mixture with the object of converting this intermediate into **C**. Sulfanuric chloride was in fact obtained, though in poor yield.

A third way of making sulfanuric chloride is from trithiazyl trichloride, $N_3S_3Cl_3$ (Chapter 7), which already contains the required ring system but has sulfur in the (IV) oxidation state. Oxidation is effected with sulfur trioxide at 150°/20 atm. Because of the relative inconvenience of the starting materials, this method has not been much investigated.[28]

B. Physical properties and structure of sulfanuric chlorides

α-Sulfanuric chloride, the thermodynamically stable isomer under ordinary conditions, forms colorless rhombs melting at 144–145°. The β-isomer forms needles with m.p. 44–46°. It is stable in the solid state or in non-polar solvents, but changes in about 1 hour to the α-isomer in acetonitrile.[28]

The α-isomer is insoluble in water and soluble in most organic solvents;

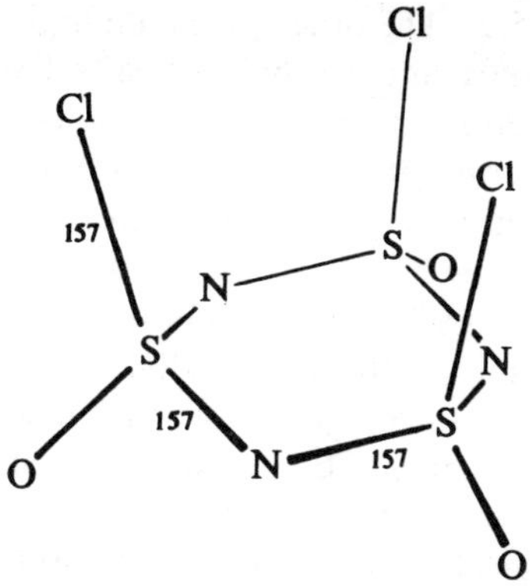

Figure 5.2. Molecular structure of α-sulfanuric chloride (after ref. 15); bond lengths are in pm.

reported solubilities at 25° range from (g compound per 100 g solvent) 1.56 in n-heptane to 22.50 in benzene.[28,37] The β-isomer is more soluble, but no figures have been given for it.

Measurements in benzene at 25° show that the α-form has a much larger molecular dipole moment (3.88 D) than the β-form (1.91 D). This may explain why, at 0.005 Torr, the β-form sublimes at room temperature while the α-form only does so at 80°.[37]

The infrared spectra of the two forms are quite similar, with some difference in relative band intensities.[37]

The molecular structure of the α-form, as determined by X-ray diffraction, is shown in Fig 5.2. It resembles that of S_3O_9. The 6-membered S–N ring has a chair form, with all the chlorine atoms in axial positions on one side of the ring. The configuration around each sulfur atom is nearly tetrahedral.[15]

The structure of the β-form has not been determined. The obvious suggestion of a boat-form ring is not, according to Vandi et al.,[37] easily reconciled with the dipole moment. This and n.q.r. data[10] suggest that the β-form has also a chair ring, but with two axial and one equatorial chlorine atoms. The β → α transformation might well be favored, in polar solvents, by the coordination of a molecule of solvent to one of the sulfur atoms.

C. The cyclic trimeric sulfanuric fluorides

The trimeric sulfanuric fluorides (Fig. 5.1, **III**) were first prepared in 1964 by Seel and Simon,[32] by treatment of the α-chloride with potassium fluoride in carbon tetrachloride at 145°. Two isomers $[NS(O)F]_3$ were obtained. One gave a singlet ^{19}F n.m.r. signal; Seel and Simon therefore suggested a *cis* structure with the three fluorine atoms in equivalent positions on one side of the ring (compare Fig. 5.2). The other gave a multiplet in agreement with the theory for AB_2 compounds, and was provisionally described as a *trans* isomer with one fluorine differently situated from the other two. The *cis* compound boils at 138.4° and the *trans* at 130.3°. There are relatively small differences between their i.r. spectra. The isomers were separated by vapor-phase chromatography.

Moeller and Ouchi soon afterwards found that the fluorination with potassium fluoride could be more conveniently done in refluxing acetonitrile.[26] In this way, followed by fractional distillation, they made the pure *cis* isomer in sufficient quantity for the study of the reactions, but still in the relatively small yield of 23%. Lin *et al.* have recently reported that SbF_3 fluorinates sulfanuric chloride at 110° with an 80% yield of the trifluoride.[21] Chlorofluorides are also produced (Section F below).

Sulfanuric fluoride trimer is also reported to arise from the thermal decomposition of $Cs^+[N=SF_2=O]^-$.[13]

There is no report of an X-ray structure determination on either of the sulfanuric fluoride trimers. However, the n.m.r. data can be explained on the postulate that the "*cis* isomer" has a similar structure to α-$[ClS(O)N]_3$ (Fig. 5.2), while the "*trans* isomer" has the oxygen and fluorine atoms on one sulfur interchanged. This is consistent with the observation that the singular fluorine atom in the "*trans* isomer" is less shielded from solvent effects in n.m.r. than the other two fluorines.[13] Moreover, the two forms are unlike enough in structure to give a simple eutectic in the freezing-point diagrams, without solid solutions.[33]

D. The cyclic sulfanuric fluoride tetramer

$[NS(O)F]_4$ has been reported, without details, to result from thermolysis of $Hg(NSOF_2)_2$ (Chapter 4, Sec. VI.B).[21]

E. High polymeric sulfanuric fluorides

Compounds of this type, believed by their discoverers to be linear polymers $[-N=S(O)F-]_n$, have been reported from two reactions. In the preparation of $[NS(O)F]_3$ from the chloride and KF in acetonitrile, distillation of the product solution leaves a yellow sticky resinous residue that sets to a yellowish-white powder on cooling.[26] A similar product results when the reaction product of ammonia with thionyl tetrafluoride, SOF_4, is heated to drive off ammonium fluoride.[31] Little more is known about these substances; even analytical data are incomplete. The second product is reported to decompose to oligomers on heating, and to dissolve slowly in boiling water.

F. Mixed sulfanuric chlorofluorides

The fluorination of [NS(O)Cl]$_3$ with SbF$_3$ at 110° gives, besides the trifluoride, the partially fluorinated compounds [NS(O)Cl]$_2$[NS(O)F] and [NS(O)Cl][NS(O)F]$_2$ in about 10% yield.[21] They have been isolated by fractional distillation. The monofluoride boils at 105–110°/20 Torr and the difluoride at 168–175°/760 Torr. Each occurs as three isomers, which have been separated by gas chromatography and characterized by ^{19}F n.m.r. Until the ring shape is determined it will not be possible to choose the best terminology for the isomers, but they have been provisionally labelled in *cis–trans* terms. For example, two of the difluoro isomers give ^{19}F singlets, indicating F atoms in equivalent positions on the same side of the ring, and the third gives a doublet (AB spectrum) which might result from one F atom on the same side as the chlorine and the other on the opposite side.[21]

G. Reactions and derivatives of the sulfanuric halides

α-Sulfanuric chloride decomposes abruptly and exothermically ($\Delta H = -823$ kJ mol^{-1}) at 250–285° in a sealed evacuated tube, emitting an orange flash.[25] It can be exploded by shock, but is not very shock-sensitive and is not considered dangerous to handle, given ordinary care. The products of decomposition include sulfur oxides and nitrogen. The *cis* fluoride [NS(O)F]$_3$ is more stable, suffering little decomposition when led as a vapor through a nickel tube at 350°.[32]

The cyclic sulfanuric halides resemble trimeric sulfur trioxide in structure and to a considerable extent in bonding. In both, each sulfur atom, in the oxidation state +6, is surrounded tetrahedrally by four very electronegative ligands. Consequently the sulfanuric halides, like sulfur trioxide, typically behave as Lewis acids in their reactions, and nearly all the reactions investigated have been with Lewis bases.

Uncharged Lewis bases (B) without reactive hydrogen atoms undergo simple addition to sulfanuric chloride, giving liquid or solid adducts [BNS(O)Cl]$_3$ with one molecule of base for each sulfur atom. Here B can be pyridine, isoquinoline, trimethylamine, triphenylphosphine, triphenylarsine, or triphenylstibine.[6]

Uncharged Lewis bases with reactive hydrogen atoms react with sulfanuric chloride or fluoride with elimination of hydrogen halide and substitution of the halogen. Ring-cleavage accompanies substitution; the inadequate data suggest that the chloride is more prone to cleavage than the fluoride, and that there is less cleavage at low temperatures. The reactions with water, methanol, ammonia, and amines will be discussed in turn.

α-Sulfanuric chloride is insoluble in, and only slowly hydrolyzed by, cold water.[28] Acid hydrolysis gives imidodisulfamide, sulfuric acid, and hydrochloric acid,[8] showing that the ring is cleaved. In aqueous silver nitrate, however, α-sulfanuric chloride gives silver chloride and trisilver salt of cyclotrisulfimide (Chapter 4, Sec. III.C).[16] The hypothetical sulfanic acid, $[NS(O)OH]_3$, and the equally hypothetical cyclotrisulfimide, $[HNSO_2]_3$, are isomeric acids giving the same cyclic anion, $[NSO_2]_3^{3-}$, which is present in the trisilver salt just mentioned. This anion, then, or protonated forms of it, must be early products of the hydrolysis of α-sulfanuric chloride, giving rise secondarily to imidodisulfamide and sulfuric acid. The trimeric sulfanuric fluoride is more resistant to hydrolysis than the chloride, being unaffected by cold water, but decomposes to the extent of 50% in 7 hours' boiling with water.[32] Boiling aqueous NaOH decomposes it more quickly according to the equation

$$[NS(O)F]_3 + 7OH^- \rightarrow 3F^- + 2H_2NSO_3^- + SO_4^{2-}$$

Seel and Simon state that boiling water does not destroy the rings, but no evidence for this statement has been published.[32] Methanol in the presence of trimethylamine gives a ring anion with one fluorine replaced by oxygen (Section H below).

With ammonia, α-sulfanuric chloride has been stated (without details) to give polymers of the melam or melem type containing intact rings linked together through nitrogen atoms.[7] Since, however, its reaction with dimethylamine, which could not bridge in this way, also gives insoluble polymers,[6] it seems more likely that the rings are destroyed in both cases.

With the organic base morpholine, α-sulfanuric chloride undergoes ring cleavage when refluxed in n-heptane, giving N,N'-morpholido sulfamide and unidentified products.[37] But in benzene slightly below room temperature, in presence of triethylamine as a hydrogen chloride acceptor, two isomeric trimorpholido derivatives of the intact ring are obtained (Table 5.1).[12] The trifluorides suffer relatively little ring cleavage when treated with amines, and so have been more investigated than the α-chloride as a starting point for the preparation of amino derivatives. The main findings with them are as follows.[43] The simplest situation arises with secondary amines. Dimethylamine substitutes one fluorine atom by NMe_2 even at $-40°$ in weakly polar solvents; the second fluorine requires $+60°$ and a polar solvent, while the third is only substituted at 80° in the absence of solvent; i.e. the reactivity falls off with increasing substitution. Steric as well as electronic effects may be responsible, since no more than two fluorines are replaceable by NEt_2 even under drastic conditions. With primary amines, the first fluorine atom is readily replaced below room temperature (Table 5.1), and all three can be substituted at 90° in the absence of a solvent, but the formation of diamino derivatives has not been detected. Monosubstitution by ammonia takes place readily at $-20°$. The $-NH_2$ and $-NHR$ derivatives have acidic hydrogen atoms. Thus they form 1:1 adducts (presumably salts) with excess ammonia or amine, and precipitate slightly soluble salts from aqueous solution on addition of tetraphenylphosphonium cation (Table 5.1). The derivative $(NSO)_3F_2NH_2$ serves as starting material for several other derivatives by substitution of the amino hydrogens.[22,43]

Negatively charged Lewis bases B^- attack the sulfanuric halides, driving out halide ions and forming B-substituted sulfanurics. The fluorination of α-sulfanuric chloride (Section C above) by potassium fluoride is an example. This same principle has been employed for the preparation of alkyl and aryl derivatives (Table 5.2) by reaction of the sulfanuric halides with organometallics (acting, at least in a formal sense, as sources of carbanions). In no case has trisubstitution of the halide been thus achieved. A large excess of dimethylmercury effected only monosubstitution,[5] and even at 83° diphenylmercury would not replace the chlorine atom in $(NSO)_3ClPh_2$.[24] This is in accordance with the expectation that the ring would be deactivated towards nucleophilic attack by replacement of the halogen by less electronegative groups. All these mono- and di-substituted products should exist as isomers, and there is infrared evidence that the diphenyl halides listed in Table 5.2 were obtained as mixtures of isomers.[5]

Trisubstitution with phenyl groups has been effected only by a Friedel–Crafts type reaction (Table 5.2) in which the sulfanuric ring has been activated towards the nucleophile benzene by withdrawal of a fluoride ion by $AlCl_3$.[26]

H. The sulfanuric anion $N_3S_3O_4F_2^-$

Methanolysis of trimeric sulfanuric fluoride, in presence of methylamine at −20° in ether, seems initially to replace one fluorine by –OMe, but the products rearrange to the salt $[N_3S_3O_4F_2]^-NMe_4^+$, with the anion **IV** shown in Fig. 5.1. The anion is stable for months in aqueous solution. Ion-exchange gives the free acid, as a crystalline monohydrate, and from this can be made a silver salt and other salts.[41] Reaction of the silver salt with methyl iodide gives a ring methylated on one nitrogen atom (i.e. containing a saturated segment) whereas with Me_3SiCl or Me_3SnCl the Me_3Si or Me_3Sn group attaches itself to an exocyclic oxygen, leaving the ring fully unsaturated.[42]

The reaction of trimeric sulfanuric fluoride with methanethiol is not completely analogous to that with methanol; one sulfur atom is reduced to S(IV), giving the anion **V** (X = O⁻) in Fig. 5.1 (see also Section III.B below).

J. Isomerism of sulfanuric derivatives

Discussion of this subject has been hampered by lack of information on the shapes and conformational rigidity of the rings. Positions on opposite sides of

Table 5.1. Amino derivatives of the sulfanuric halides.

Derivative	Preparative reaction	Description	Ref.
$(NSO)_3F_2NH_2$	$[NS(O)F]_3$ + 3 mol ammonia in ether at $-20°$	m.p. 132°, colorless crystals	43
$(NSO)_3F_2NH^-Ph_4P^+$	$(NSO)_3F_2NH_2 + Ph_4P^+Cl^-$ in water	m.p. 141°, colorless crystals	43
$(NSO)_3F_2NHR$			
R = methyl	$[NS(O)F]_3$ + 3 mol amine in ether at $-10°$	m.p. 69°, colorless crystals	43
ethyl		m.p. 73°, colorless crystals	43
phenyl	$[NS(O)F]_3$ + 1 mol aniline + 2 mol triethylamine in ether at $-5°$	m.p. 82.5°, colorless crystals	39
$(NSO)_3F(NHR)_2$			
R = n-octyl	$[NS(O)F]_3$ + amine in n-heptane at $-23°$	involatile colorless liquid	5
$(NSO)_3(NHMe)_3$	$[NS(O)F]_3$ + 100 mol amine, 8 d at 90°	m.p. 226°, colorless crystals	43
$(NSO)_3(NHEt)_3$		m.p. 108°, colorless crystals	43
$(NSO)_3F_2NMe_2)(NHMe)$	$(NSO)_3F_2NMe_2$ + 13 mol NH_2Me in MeCN, 81°	m.p. 71°, colorless solid	43
$(NSO)_3F(NMe_2)(NMe)^-Ph_4P^+$	$(NSO)_3F(NMe_2)(NHMe) + Ph_4P^+Cl^-$ in water	m.p. 138°, colorless solid	43
$(NSO)_3F_2NR_2$			
R = methyl	$[NS(O)F]_3$ + 2 mol amine in ether at $-40°$	m.p. 61.5°, colorless crystals	39
ethyl		b.p. 62°/0.01 Torr, liquid	43
$(NSO)_3F_2(NX)$			
HNX = piperidine	$[NS(O)F]_3$ + 1.62 mol amine in MeCN at 40°	m.p. 44°	39
$(NSO)_3F(NR_2)_2$			
R = methyl	$[NS(O)F]_3$ + 12 mol amine, 3 d at 60°	m.p. 52°, colorless crystals	43
ethyl	$[NS(O)F]_3$ + 8 mol amine in MeCN at 40°	m.p. 68°, colorless crystals	5, 39
$(NSO)_3F(NX)_2$			
HNX = piperidine		2 isomers; m.p. 114°, 134°	26
pyrrolidine		2 isomers; m.p. 127°, 148°	26
cis-$[NS(O)F]_3$ + 8 mol amine in MeCN at 20°		2 isomers; m.p. 146°, 195°	26

Derivative	Preparative reaction	Description	Ref.
$(NSO)_3(NMe_2)_3$	$(NSO)_3F(NMe_2)_2$ + 84 mol amine, 4 d at 80°	m.p. 238°	43
$(NSO)_3(NX)_3$ HNX = morpholine	α-$[NS(O)Cl]_3$ + excess amine in benzene or MeCN at room temperature	2 isomers; m.p. 171°, 196°	12
$(NSO)_3Ph_2NEt_2$		m.p. 127°	23
$(NSO)_3Ph_2NX$ HNX = piperidine morpholine	$(NSO)_3Ph_2Cl$ + amine in benzene or MeCN at 20°	m.p. 151.5° m.p. 140°	23 23

Table 5.2. Alkyl and aryl derivatives of the sulfanurics.

Derivative	Preparative reaction	Description	Ref.
$(NSO)_3Cl_2Me$	α-$[NS(O)Cl]_3$ + excess Me_2Hg in benzene at 0–20°	white crystals, m.p. 148°	5
$(NSO)_3F_2Ph$	cis-$[NS(O)F]_3$ + 1 mol PhLi in ether at −70°	m.p. 95°	26
$(NSO)_3FPh_2$	cis-$[NS(O)F]_3$ + 2 mol PhLi in ether at −70° or $[(NSO)_3ClPh_2]$ + KF refluxed in moist acetonitrile	m.p. 119°	26
$(NSO)_3Ph_3$	reflux cis-$[NS(O)F]_3$ 2 d with benzene and $AlCl_3$	white solid, m.p. 107° 2 isomers; m.p. 148°, 177°	24 26
$(NSO)_3Cl(C_6H_3C_{-2})_2$	$(NSO)_3ClPh_2$ + Cl_2 in CCl_4 at 20° in presence of iodine	white crystals, m.p. 72°	5
$(NSO)_3ClPh_2$	α-$[NS(O)Cl]_3$ + 2 mol Ph_2Hg in benzene at 30°, 67 h	m.p. 120°	24

a planar ring would be geometrically equivalent and could be fully described in *cis–trans* terms. On a chair-form ring, however, there are non-equivalent positions which have sometimes been described in *cis–trans* terms but which for full characterization would need to be called axial or equatorial. Actually only two X-ray structure determinations have been carried out on sulfanuric compounds, the α-trichloride[15] and the monophenyl difluoride;[1] both show chair-form rings. It seems reasonable to assume that the chair form is a general feature of sulfanuric rings. If so, the axial and equatorial sites are not equivalent, and there are four conceivable isomers of a sulfanuric trihalide and eight of a mono- or di-substituted halide. Obviously only a small fraction of the conceivable isomers have been found in experimental studies hitherto. Some, however, may be incapable of existence for steric reasons, and others (as the example of the β-chloride suggests) may transform so readily into thermodynamically more stable forms that their isolation is impracticable.

^{19}F n.m.r. spectroscopy has been an important source of structural information on the sulfanurics.[9,26,39] It has clearly shown the nature of the structural difference between the *cis* and *trans* trifluorides (Section C above). Spectra of various substituted fluorides have been reported;[9,43] a tentative interpretation was suggested,[9] on the hypothesis that the chemical shifts of axial and equatorial fluorines differ characteristically. Using this argument, the monophenyl derivative was assigned a structure with fluorines and phenyl all in equatorial positions.[9] However, an X-ray diffraction study on (presumably) the same isomer has shown the fluorines axial and the phenyl equatorial.[1] Whether the structures are really different in the crystalline state and in solution, or whether the n.m.r. data need reinterpreting is not clear at the moment.

III. Mixed sulfanuric rings with other sulfur groups

A. Introduction

In 1968 Schläfer and Becke-Goehring drew attention to the bond-theoretical interest of sulfur–nitrogen rings containing sulfur atoms in two oxidation states, +6 and +4. There is evidence pointing to "aromatic stabilization" of some sulfur–nitrogen rings (Chapters 6 and 9); how is this stabilization affected by mixing the oxidation states? The compounds described in this Section III form part of the gradually accumulating experience that will in time enable such matters to be better understood.

B. The anion $N_3S_3O_3F_2^-$ and compounds $N_3S_3O_2XF_2$ (X = F, Cl)

This ion (Fig. 5.1, V, X = O$^-$) contains one sulfur(IV) atom along with two sulphur(VI) atoms. It is formed as its trimethylammonium salt by the action of methanethiol[41] or phenylhydrazine[43] on trimeric sulfanuric fluoride, in presence of trimethylamine. The thiol or phenylhydrazine behaves here as a reducing agent.

Several salts of $N_3S_3O_3F_2^-$ have been described.[41]

Neutral compounds $N_3S_3O_2XF_2$ (Fig. 5.1, V, X = F, Cl) isoelectronic with this anion have also been reported without details.[41]

C. Preparation and properties of [NS(O)Cl][NSCl]$_2$ and [NS(O)Cl]$_2$[NSCl]

Trisulfur dinitrogen dioxide [bis(thionylimino)sulfur] (Chapter 3, Sec. III.F), when mixed at $-80°$ with excess liquid chlorine and allowed to warm to room temperature, undergoes the following reaction:[30]

$$3S(NSO)_2 + 5Cl_2 \rightarrow 2[NS(O)Cl][NSCl]_2 + 2SOCl_2 + SO_2$$

The product, [NS(O)Cl][NSCl]$_2$, can be crystallized from benzene as nearly colorless crystals, m.p. 110°, and is soluble in most organic solvents. The evidence for the ring structure **VI** shown in Fig. 5.1 is chemical. When oxidized with SO_3, the compound behaves like $(NSCl)_3$ (Chapter 7, Sec. II.C), yielding trimeric sulfanuric chloride.[30] The results of alkaline hydrolysis can also be understood in terms of the structure shown.[30] Fluorination, and n.m.r. evidence of the structure of the fluoride product, also confirm it. The chloride reacts readily with silver difluoride in carbon tetrachloride at room temperature,[30] to give [NS(O)Cl][NSF]$_2$, a colorless low-melting solid. In methylene chloride this gives a single ^{19}F n.m.r. signal with a chemical shift showing clearly the attachment of fluorine to sulfur(IV) and not sulfur(VI).[30]

Chlorination of $S(NSO)_2$ also affords the related compound **VII** in Fig. 5.1.[44]

D. Preparation and properties of [NS(O)Cl][NSO$_2$][NS]

This compound (Fig. 5.1, **VIII**) contains in its ring one typical sulfanuric group, an =SO_2 group with sulfur also in the (VI) oxidation state, and an NS group in which the sulfur atom is (IV). It has been obtained by the cyclization-elimination:[29]

$$3ClSO_2NSO \xrightarrow[20°]{u.v.} 3SO_2 + Cl_2 + [NS(O)Cl][NSO_2][NS]$$

It was isolated by vacuum sublimation at 60–70°/0.01 Torr as a solid melting at 105–108°, soluble in benzene and acetonitrile, and readily hydrolyzed. The structure shown in Fig. 5.1 is based on i.r. and mass-spectral evidence.

IV. Sulfanuric rings containing a saturated segment

Attention has already been drawn to the isomerism between the saturated cyclosulfimide ring and the unsaturated sulfanuric ring (Chapter 4, Sec. I; and Section II.G above). The silver salt of anion **IV** in Fig. 5.1 reacts readily with methyl iodide at room temperature to give a monomethyl derivative which has been shown by ^{19}F and proton n.m.r. to have the structure **XI** with a saturated

ring segment, and not **XII** with a normal sulfanuric ring. The compound is a white sublimable solid, m.p. 78°. An analogous ethyl derivative can be made similarly.[40]

V. Mixed sulfanuric–phosphazene rings

A. Introduction

The sulfanuric chloride group [NS(O)Cl] is, in a loose sense, isoelectronic with the phosphazene chloride group [NPCl$_2$]. Phosphazene rings are the more stable and now have a very extensive chemistry (Chapter 12). Here again, mixed rings containing both types of grouping may help to fill out the picture and throw light on the factors responsible for stability, or lack of it, in inorganic heterocycles. Three mixed sulfanuric–phosphazene chlorides have been made.

B. Preparation and properties of [NS(O)Cl][NPCl$_2$]$_2$ and [NS(O)Cl]$_2$[NPCl$_2$]

The mixed sulfanuric–phosphazene compound [NS(O)Cl]$_2$[NPCl$_2$] was first reported in 1962 (see ref. 10) from attempts to make sulfanuric chlorides by the Kirsanov method. Its preparation in this way proved not to be reproducible. Later, Baalmann and coworkers[2] worked out well defined and logical conditions for the preparation of the two mixed sulfanuric–phosphazene chlorides, as follows. It was known that pyrolysis of the compounds [Cl$_3$P=N=PCl$_3$]$^+$[PCl$_6$]$^-$ and [Cl$_3$P=N=PCl$_2$–N=PCl$_3$]$^+$[PCl$_6$]$^-$ (Chapter 12, Sec. VI.B) gives rise to the cyclic trimer and tetramer (NPCl$_2$)$_{3,4}$, presumably because these compounds act as a source of $NPCl_2$ monomer. Also, just as PCl_5 (really $PCl_4{}^+PCl_6{}^-$) reacts with amidosulfuric acid to generate $Cl_3P=N-SO_2Cl$ (Section II.A above), so either of the above $PCl_6{}^-$ salts will react with amidosulfuric acid on heating to give phosphazene chains with SO_2Cl end-groups, viz. $Cl_3P=N=PCl_2-NSO_2Cl$ and $Cl_3P=N=PCl_2-N=PCl_2=NSO_2Cl$. When these latter substances are pyrolyzed (in the same manner as $Cl_3P=N-SO_2Cl$; Section II.A above), the mixed phosphazene–sulfanuric ring compounds (Fig. 5.1, **IX** and **X**) are obtained, with larger quantities of $[NPCl_2]_3$ and $[NPCl_2]_4$. The yield of mixed rings can be greatly increased[2] by pyrolyzing the same compounds in admixture with $Cl_3P=N-SO_2Cl$, which acts as a source of $NS(O)Cl$ monomer, and is further improved by adding an $AlCl_3$ catalyst.[19]

The mixed-ring compounds are colorless crystalline solids, soluble in organic solvents and conveniently crystallizable from hot n-heptane.[2] $[NS(O)Cl][NPCl_2]_2$ melts at 60–62° and $[NS(O)Cl]_2[NPCl_2]$ at 96.5°. Both compounds appear to be stable in moist air but are susceptible to hydrolysis when in solution in organic solvents.[2,10]

An X-ray structure determination has shown the second of these compounds to have a chair-form ring very like that of α-$[NS(O)Cl]_3$, with three chlorines in axial positions (see ref. 15). Its ^{35}Cl n.q.r. spectrum is consistent with this structure, which has been described as *cis*.[3]

C. Preparation and properties of [NS(O)F]₂[NPCl₂]

Fluorination of $[NS(O)Cl]_2[NPCl_2]$ with an excess of the powerful fluorinating agent AgF_2 in boiling CCl_4, or with SbF_3 at 85°, causes replacement of the two chlorines on sulfur, but not those on phosphorus.[19] The product, $[NS(O)F]_2[NPCl_2]$, occurs in the ratio 4:1 as *cis* and *trans* isomers recognized by their different ^{19}F n.m.r. spectra. The *cis* isomer forms colorless needles melting at 56.5°, and slowly hydrolyzed by atmospheric moisture.

D. Reactions and derivatives of the mixed sulfanuric–phosphazene halides

Some nucleophilic substitution reactions of these halides have been examined. As just mentioned, fluorination of the chloride occurs exclusively on the sulfur atoms. In contrast, amination[4] often starts preferentially on phosphorus. For example, monosubstituted *P*-amino derivatives result from reaction of the chlorides in ether at room temperature[19] with stoichiometric amounts of ammonia, several primary alkylamines, and dimethylamine. However, a tendency to substitute on sulfur develops on passing from n-butylamine through s-butylamine to t-butylamine, probably because of increasing steric hindrance at phosphorus.[35] The fluoride reacts similarly with ammonia and with various silylamines, yielding P-amino derivatives, which have been characterized by [19]F n.m.r. spectroscopy as mixtures of *cis* and *trans* isomers. The amino derivatives are low-melting crystalline solids or liquids distillable under reduced pressure.[19] *cis*-[NPCl$_2$][NS(O)F]$_2$ readily undergoes replacement of one or both chlorines by –NCS when treated with KSCN in acetonitrile below room temperature.[18]

Friedel–Crafts phenylation of these mixed halides take place mainly on sulphur.[36]

E. The eight-membered ring [NPCl$_2$]$_3$[NS(O)Cl]

A compound believed on mass-spectrometric and i.r. evidence to have structure XI in Fig. 5.1 was reported in 1974 by Voswijk and van de Grampel.[38] It was obtained in the amount of 0.2 g from a deliberate modification of the standard method for mixed sulfanuric–phosphazene rings (Section B above), designed to improve the yield of larger rings. Sulfamide was allowed to react over several days at room temperature with [Cl$_3$P=N–PCl$_2$=N–PCl$_3$]$^+$PCl$_6^-$, the products being subsequently separated by sublimation and recrystallization. [NPCl$_2$]$_3$[NS(O)Cl] crystallized from pentane as colorless needles, m.p. 58–60°. It has not been further reported on.

References

1. Arrington, D. E., Moeller, T. and Paul, I. C., *J. Chem. Soc. A* 2627 (1970)
2. Baalmann, H. H., Velvis, H. P. and van de Grampel, J. C., *Recl. Trav. Chim. Pays-Bas* **91**, 935 (1972)
3. Baalmann, H. H. and van de Grampel, J. C., *Recl. Trav. Chim. Pays-Bas* **92**, 716 (1973)
4. Baalmann, H. H. and van de Grampel, J. C., *Z. Naturforsch. B* **33b**, 964 (1978)
5. Banister, A. J. and Bell, B., *J. Chem. Soc. A* 1659 (1970)
6. Banister, A. J., *MTP Int. Rev. Sci.: Inorg. Chem., Ser. Two.* **3**, p. 41, Butterworth, London (1975)
7. Becke-Goehring, M., in *Developments in Inorganic Polymer Chemistry*, ed. M. Lappert and G. J. Leigh, Elsevier, Amsterdam (1962)
8. Becke-Goehring, M. and Fluck, E., in *Developments in Inorganic Nitrogen Chemistry*, vol. 1, ed. C. B. Colburn, Elsevier, Amsterdam (1966)
9. Chang, T. H., Moeller, T. and Allen, C. W., *J. Inorg. Nucl. Chem.* **32**, 1043 (1970)
10. Clipsham, R., Hart, R. M. and Whitehead, M. A., *Inorg. Chem.* **8**, 2431 (1969)
11. Coppens, P. and Vos, A., *Acta Cryst.* **B27**, 146 (1971)
12. Failli, A., Kresge, M. A., Allen, C. W. and Moeller, T., *Inorg. Nucl. Chem. Lett.* **2**, 165 (1966)
13. Glemser, O. and Mews, R., *Adv. Inorg. Chem. Radiochem.* **14**, 333 (1972)

REFERENCES

14. *Gmelins Handbuch der Anorganischen Chemie*, Ergänzungswerk zur 8 Auflage, Band 32, Teil 1, "Schwefelstickstoffverbindungen", Springer Verlag, Berlin-Heidelberg-New York (1977)
15. Hazell, A. C., Wiegers, G. A. and Vos, A., *Acta Cryst.* **20**, 186 (1966)
16. Hazell, A. C., *Acta Chem. Scand.* **26**, 2542 (1972)
17. Kirsanov, A. V., *Zhur. Obshchei Khim.* **22**, 81, 88 (1952)
18. Klei, E. and van de Grampel, J. C., *Z. Naturforsch. B* **31b**, 1035 (1976)
19. Klingebiel, U., Lin, T.-P., Buss, B. and Glemser, O., *Chem. Ber.* **106**, 2969 (1973)
20. Klüver, H. and Glemser, O., *Z. Naturforsch. B*, **32b**, 1209 (1977)
21. Lin, T.-P., Klingebiel, U. and Glemser, O., *Angew. Chem. Int. Ed. Engl.* **11**, 1095 (1972)
22. Lin, T.-P. and Glemser, O., *Chem. Ber.* **109**, 3537 (1976)
23. Maricich, T. J and Khalil, M. H., *J. Chem. Soc. Chem. Commun.* 195 (1977)
24. McKenney, R. L. and Fetter, N. R., *J. Inorg. Nucl. Chem.* **30**, 2927 (1968)
25. McKenney, R. L. and Fetter, N. R., *J. Inorg. Nucl. Chem.* **34**, 3569 (1972)
26. Moeller, T. and Ouchi, A., *J. Inorg. Nucl. Chem.* **28**, 2147 (1966)
27. Moeller, T. and Dieck, R. C., *Preparative Inorganic Reactions* **6**, 63 (1971)
28. Moeller, T., Chang, T.-H., Ouchi, A., Vandi, A. and Failli, A. *Inorg. Synth.* **13**, 9 (1972)
29. Roesky, H. W., *Angew. Chem. Int. Ed. Engl.* **10**, 266 (1971)
30. Schläfer, D. and Becke-Goehring, M., *Z. Anorg. Allg. Chem.* **362**, 1 (1968)
31. Seel, F. and Simon, G., *Angew. Chem.* **72**, 709 (1960)
32. Seel, F. and Simon, G., *Z. Naturforsch, B* **19b**, 354 (1964)
33. Seel, F., Velleman, K. and Heinrich, E., *Z. Anorg. Allg. Chem.* **382**, 61 (1971)
34. Thomas, D. W., Bates, J. B., Bandy, A. and Lippincott, E. R., *J. Chem. Phys.* **53**, 3698 (1970)
35. van den Berg, J. B., Klei, E., de Ruiter, B., van de Grampel, J. C. and Kruk, C., *Recl. Trav. Chim. Pays-Bas*, **95**, 206 (1976)
36. van den Berg, J. B. and van de Grampel, J. C., *Z. Naturforsch. B*, **34b** 27 (1979)
37. Vandi, A., Moeller, T. and Brown, T. L., *Inorg. Chem.* **2**, 899 (1963)
38. Voswijk, C. and van de Grampel, J. C., *Recl. Trav. Chim. Pays-Bas* **93**, 120 (1974)
39. Wagner, H., Mews, R., Lin, T.-P. and Glemser, O., *Chem. Ber.* **107**, 584 (1974)
40. Wagner, D.-L., Wagner, H. and Glemser, O., *Z. Naturforsch. B*, **30b**, 279 (1975)
41. Wagner, D.-L., Wagner, H. and Glemser, O., *Chem. Ber.* **108**, 2469 (1975)
42. Wagner, D.-L., Wagner, H. and Glemser, O., *Chem. Ber.* **109**, 1424 (1976)
43. Wagner, D.-L., Wagner, H. and Glemser, O., *Chem. Ber.* **110**, 683 (1977)
44. Weiss, J., Mews, R. and Glemser, O., *J. Inorg. Nucl. Chem.* Herbert H. Hyman Memorial Volume, p. 213 (1976)

6
Formally unsaturated sulfur nitrides and sulfur nitride ions

I. Introduction and plan of treatment

The compounds dealt with in this chapter are called unsaturated because double bonds must be included in their structural formulas if these are to be written with normal valencies of sulfur and nitrogen. This is in contrast to the nitrides described in Chapter 2, which can be formulated with single bonds only. It is useful to make the distinction, because the unsaturation strongly affects molecular structure and reactivity. However, unsaturated S–N rings cannot usually be well represented by single canonical formulas, as is so often possible with unsaturated carbon compounds. The S–N bond multiplicities are non-integral and may differ fractionally between different bonds in a molecule or ion.

In the chemistry of inorganic heterocyclics, the unsaturated sulfur nitrides and sulfur nitride ions are exceptionally important both historically and in current research. Tetrasulfur tetranitride, first prepared about 1835,[39] is one of the longest-known and most-studied of all inorganic heterocycles, yet its chemistry still challenges the imagination. It and its numerous derivatives continually stimulate new experiments and new ideas about structure, bonding, and reactivity. It is opening up the mental horizons of the inorganic chemist as benzene did for the organic chemists of the last century.

It is not immediately obvious how best to organize the subject-matter of this chapter and Chapters 7 and 8 in order to achieve a clear exposition. To classify the compounds according to the oxidation numbers or covalencies of sulfur, as in earlier chapters, and as an inorganic chemist would naturally tend to do, would be awkward because some of them contain several non-equivalent sulfur atoms. Nor is it possible to use an organic chemist's kind of treatment, because the sulfur–nitrogen compounds do not possess stable skeletons that persist unchanged through reaction after reaction. So every writer on the subject has organized it in his own way. Rational classifications based on molecular structures are being attempted and show promise,[11,76] but cannot yet be fully evaluated. In this book the compounds are introduced in an order based on simple structural features and likely to be helpful to a practising chemist; no deep theoretical justification is claimed for this order of treatment. In order to reduce the mass of material to manageable chunks, a somewhat arbitrary decision has been made; the present chapter deals only with S–N rings having no attached exocyclic atoms or groups. The unsaturated S–N ring systems with such groups are dealt with in Chapters 5, 7, and 8. This decision, regrettably, breaks a few useful cross-connections, but on balance seems about as good as any alternative.

This account begins with tetrasulfur tetranitride, which in the laboratory has been the usual starting point for most of the other compounds in this chapter and Chapter 7. There follows an account of three other nitrides with the same empirical formula $(SN)_n$, all obtainable from S_4N_4 by depolymerization and polymerization. The remaining sections of this chapter deal with the unsaturated and probably cyclic nitride S_4N_2, and with a variety of ring and cage sulfur nitride structures, cationic, anionic, and neutral, which can be made from S_4N_4. The non-cyclic ions NS^+ and NS_4^- are relevant to this area of chemistry and so are briefly discussed at the end.

Tetrasulfur tetranitride is indeed an enormously interesting compound both in its own right and as an intermediate for the production of other interesting compounds. Unfortunately it is explosive. One would like to find other ways into this area of chemistry which do not require its use. Such ways are being found, and will be emphasized as appropriate in this account. For example, the easily prepared and relatively safe compound $[S_3N_2Cl]^+Cl^-$ (Chapter 7, Sec. III.B) is being increasingly used as a versatile intermediate.

The problems of nomenclature in this area of chemistry have been discussed in Chapter 1. At the moment one cannot foresee what system the IUPAC is likely to recommend for the future, whereas regular use of *Chemical Abstracts* (Ring Index) systematic names would make this chapter almost unreadable. Accordingly, compounds will be referred to either by their common trivial names or by formulas.

II. Thiazyl polymers, $(SN)_n$, and thiazyl monomer

A. Tetrasulfur tetranitride

1. Preparation

Tetrasulfur tetranitride is formed in small amounts in many reactions.[38] One cannot work for long in this area of chemistry without finding a few of its distinctive orange crystals in a beaker, or seeing its strong bands in an infrared spectrum (Fig. 6.1).

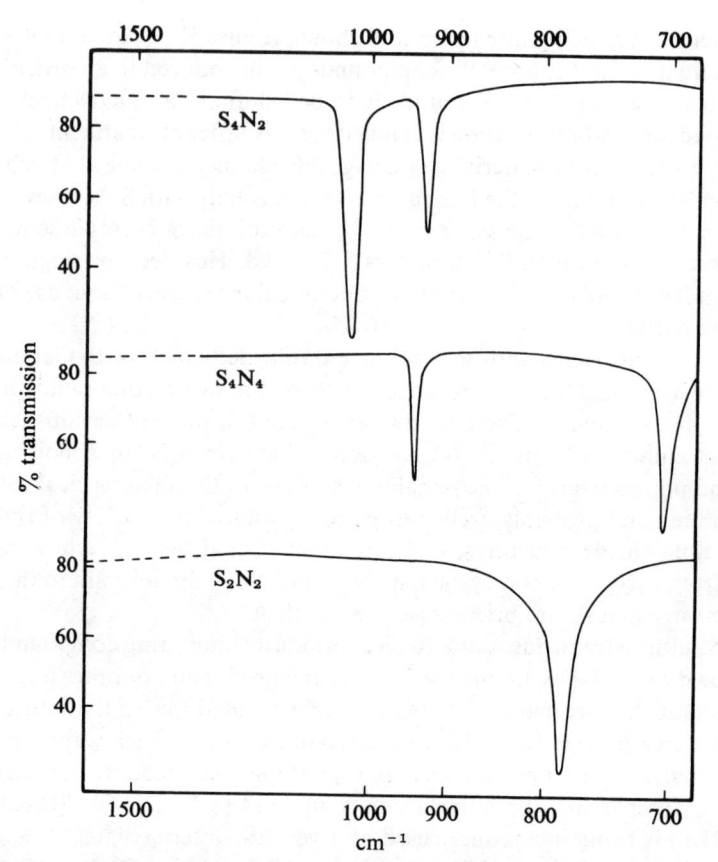

Figure 6.1. Infrared spectra of CS_2 solutions of sulfur nitrides.[46] 2–5% solutions, 0.2 mm cell; broken lines are in regions of strong solvent absorption.

The nitride is usually prepared by the reaction of a chlorosulfane with ammonia. Even concentrated aqueous ammonia gives some S_4N_4,[4] as Gregory observed as early as 1835.[38] For good yields, however, the reaction must be effected in an inert organic solvent. The procedure has been improved empirically over the years. In the most widely used technique,[90] S_2Cl_2 in CCl_4 is first treated with plenty of chlorine, and then ammonia gas passed in. The object of chlorination is to raise the oxidation state of the sulfur, since ideally, for S_4N_4, one should start with sulfur(+3). It cannot be assumed that all of the S_2Cl_2 actually changes into SCl_2, since this reaction is known to be slow in the absence of a catalyst. However, it is well established that the yield of S_4N_4 after chlorination, about 19 g from 100 g S_2Cl_2, is roughly twice that obtained from S_2Cl_2 without chlorination.[90] The chlorination is a nuisance and introduces a risk, the possible formation of explosive NCl_3, which has not been properly evaluated. There is much to be said for dispensing with chlorination and accepting the smaller yield. Another possibility would be to use commercial SCl_2 instead of S_2Cl_2; there is no recent information about this, but earlier reports[38] indicate yields comparable to those from the chlorination procedure.

The mixture becomes warm as ammonia is passed in. Its temperature should be maintained between 20° and 50°. Below 20° the yield of S_4N_4 falls and a considerable amount of S_7NH (Chapter 2) is formed instead.

When reaction is complete the mixture consists of a bulky precipitate of NH_4Cl, sulfur, and S_4N_4, with supernatant solvent containing relatively little S_4N_4. The precipitate is filtered off, washed free of NH_4Cl with water, and air-dried.

S_4N_4 is a powerful explosive, liable to detonate when struck, ground, or suddenly heated. The purest samples are said to be the most sensitive.[90] The crude mixture with sulfur, prepared as just described, will not explode when struck and may safely be stored. Small quantities of pure crystalline S_4N_4 can be made as required by soxhlet-extraction with dioxan from this mixture.[90] The compound should never be stored in large quantities in a pure state. There is one report[46] of an explosion of 0.5 g, which clearly shows the risk.

The sequence of reactions in the above preparation is probably:[46]

(i) $\quad 2S_2Cl_2 + 4NH_3 \rightarrow NSCl + 3NH_4Cl + 3S$

(ii) $\quad 2NSCl + S_2Cl_2 \rightarrow [S_3N_2Cl]^+Cl^- + SCl_2$

(iii) $\quad 3[S_3N_2Cl]^+Cl^- + S_2Cl_2 \rightarrow 2[S_4N_3]^+Cl^- + 3SCl_2$

(iv) $\quad [S_4N_3]^+Cl^- + 4NH_3 + 2SCl_2 \rightarrow S_4N_4 + 3NH_4Cl + S_2Cl_2$

The evidence is as follows. The thiazyl chloride monomer, NSCl, produced in (i) is a gas and can be driven out of the reaction mixture at an early stage by a stream of nitrogen and identified by i.r.[46] It is known from other experiments[52] to undergo reaction (ii) with S_2Cl_2. Reaction (iii) has been separately studied,[52] and the thiotrithiazyl chloride (see Section III.C below) formed in (iii) can be isolated in large quantities from the reaction mixture if the flow of ammonia is stopped at the right stage.[15] Finally, reaction (iv) has been studied separately.[15]

The only really useful alternative synthesis of S_4N_4 is to pass S_2Cl_2 vapor through pellets of ammonium chloride at 170°.[51] The yields are small (about 3.3 g S_4N_4 from 100 g S_2Cl_2) but the apparatus is simple and an ammonia cylinder is not required. S_4N_4 can also be made from S_2Br_2 and ammonia.[69]

2. General description

Tetrasulfur tetranitride forms orange crystals which melt at 178–187° with decomposition. Its vapor pressure reaches 1 Torr at about 130°.[42] As electron-diffraction studies have shown,[46] the vapor is largely S_4N_4 molecules below

40°, but at 130° there is extensive dissociation to S_2N_2 (Section B below) and other products.[85]

The nitride is endothermic ($\Delta H_f^\circ = 460$ kJ mol^{-1})[13] and, as explained above, explosive.

It can be kept in moist air at 20° without apparent decomposition. It is insoluble in, and little affected by, water.

It dissolves without decomposition in many organic solvents, but it is not extremely soluble in any of them; the best solvents are dioxan and 1,1,2,2-tetrachloroethane, and benzene and chloroform are also fairly good. Solubilities from 20° to 40° in thirteen solvents have been reported[43] (cf. ref. 46).

Other important physical properties are described in ref. 46.

In the laboratory the nitride can be reliably identified by its characteristic i.r. spectrum in CS_2 solution (Fig. 6.1). An X-ray powder diffraction pattern has been published,[41] but it is dangerous to powder the compound.

3. Molecular and electronic structure

The structure shown in Fig. 6.2(a) has been established for the vapor by electron diffraction and for the crystal by X-ray diffraction.[46] The symmetry is D_{2d}. The nitrogen atoms form a square and the sulfur atoms a bisphenoid. The molecule is often described as a ring, but it is enlightening to think of it as a cage (see Chapter 9). An argument for the cage viewpoint is the short S–S distances (broken lines in the figure) which according to molecular orbital calculations[81,89] result from weak bonds of order ~0.3. The S–N bond lengths

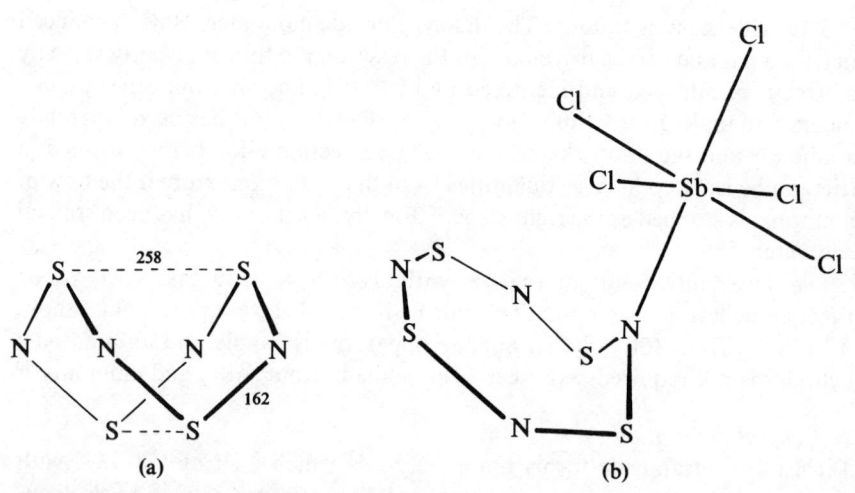

Figure 6.2. (a) Molecular structure of S_4N_4.[46] (b) Molecular structure of $S_4N_4 \cdot SbCl_5$;[46] bond distances are in pm.

are all short and equal (for discussion see Sec. V.C of Chapter 9); the deduction of a bond order from this length is controversial.[67] The electronic structure of S_4N_4 has recently been studied experimentally by means of photoelectron spectroscopy,[81] with the conclusion that about one-half of an electronic charge is transferred to each nitrogen atom from sulfur. CNDO calculations have given for the first time an adequate quantitative explanation of all the u.v. and photoelectron spectroscopic data on S_4N_4.[81]

4. Reactions

The reactions of S_4N_4 are unusually diverse and are still being investigated. We discuss them here in the following order: dissociation; addition; oxidation of S_4N_4; its reduction; its behavior with Lewis acids; its behavior with Lewis bases.

Above 130°, at ordinary pressures, S_4N_4 decomposes exothermically to its elements.[46] If heated quickly, it explodes. However, in the vapor state at low pressures, between 130° and 300°, it dissociates to S_2N_2 (Section B below).[55,85]

Under mild conditions (e.g. refluxing in ether) S_4N_4 fails to react with cyclohexene and 3-methyl-1,3-butadiene,[24] but readily adds two molecules of *trans*-cyclooctene.[63] Certain strained, rigid bicyclic olefins are similarly added to give white crystalline adducts which appear to dissociate slowly to their components in solution;[24] these olefins displace each other from their S_4N_4 adducts in the following order of reactivity: C_7H_8 (norbornadiene) > C_7H_{10} (norbornene) > C_8H_{10} (5-methylnorbornene) > $C_{10}H_{12}$ (dicyclopentadiene). There is controversy about the structure of the adducts; thus chemical and spectroscopic evidence point to $S1,S3$ addition of cyclooctene while n.m.r. evidence has been cited for $N1,N3$ addition of norbornene.

There are only old reports[38,39] of the reaction of S_4N_4 with oxygen or sources of oxygen, and there is a need for work which meets modern standards. The nitride burns vigorously when slowly heated in air, and detonates if heated quickly.[38] In concentrated sulfuric acid or disulfuric acid at room temperature, a radical is formed by oxidation, which from its e.s.r. spectrum and other evidence was once thought[5,53] to be $S_2N_2{}^+$ but is now known to be $S_3N_2{}^+$ (Section III.F below, and ref. 35). Oxidation of S_4N_4 by AsF_5 gives a solid salt of this ion, while SbF_5 and $SbCl_5$ oxidize S_4N_4 to $S_4N_4{}^{2+}$ (Section III.F below).

S_4N_4 is oxidized by NO_2 in CCl_4 to nitrosyl disulfate, $(NO)_2S_2O_7$, and by chloramine-T in acidified dioxan to sulfuric acid and ammonia.[46]

The reactions of S_4N_4 with halogens have been considerably investigated. At $-78°$, with diluted fluorine, it gives 81% of $(NSF)_3$ (**I**) and up to 12% of $(NSF)_4$ (**III**) (Chapter 7, Sec. II). At ordinary temperatures, smaller acyclic nitrogen–sulfur–fluorine molecules are formed.[31,57] The action of chlorine on a suspension of S_4N_4 in CCl_4 gives $(NSCl)_3$ (II) (Chapter 7, Sec. II.C). With liquid bromine at 70°, S_4N_4 gives[94] orange crystalline $[S_4N_3]^+Br_3^-$ (compare refs. 64, 87, and 97); an old report[39] of the formation of $(SNBr)_x$ is clearly unsound. Reaction of S_4N_4 with iodine is slow at room temperature, but in boiling inert solvents a dark red amorphous material analyzing fairly accurately as $(SN)_3I$ is precipitated.[64]

The radical bis(trifluoromethyl) nitroxide, $(CF_3)_2NO$, oxidizes S_4N_4 at room temperature to the white crystalline compound IV.[46]

$$(CF_3)_2NO-S-N-S-ON(CF_3)_2$$
$$| \quad \quad \quad |$$
$$N \quad \quad \quad N$$
$$| \quad \quad \quad |$$
$$(CF_3)_2NO-S-N-S-ON(CF_3)_2$$
(IV)

The reduction of S_4N_4 has been examined under various conditions. In acetonitrile solution at $-25°$, it is reduced at a platinum or mercury cathode to a yellow substance which has been pretty definitely identified by e.s.r. as the radical-anion $S_4N_4^-$. This product decomposes with a half-life of about 8 seconds at $-30°$,[93] giving $S_3N_3^-$, which also arises (with $S_4N_5^-$) from reduction of S_4N_4 by potassium in dimethoxyethane[22] (see also Sections III.H and IV.B below). The nitride can easily be reduced to sulfur imides. With tin(II) chloride in boiling methanol–benzene, there is an immediate reaction, adding hydrogen to each nitrogen atom of the S_4N_4 ring and so producing the cyclic imide $S_4(NH)_4$ (Chapter 2) in high yield.[38,39,46] In contrast, hydrazine adsorbed on silica gel slowly reduces the nitride, in benzene at 46°, to a mixture of cyclic sulfur imides $S_{8-n}(NH)_n$ (Chapter 2).[33] Hydrogen iodide in CCl_4 completely reduces S_4N_4 to H_2S and NH_3.[46]

S_4N_4 above 100° is a strong dehydrogenating agent for hydrocarbons. With aralkyl hydrocarbons, significant yields of thiadiazole derivatives are obtained,[17] e.g. 9,10-dihydrophenanthrene gives phenanthrene and phenanthro-[9,10-c]-1,2,5-thiadiazole.

S_4N_4 is a good Lewis base, with donor nitrogen. However, little is known about its ability to add the simplest of Lewis acids, the proton. Some of the more obvious proton donors with which it might be tested are unsuitable because it is so easily oxidized or reduced. It is known that S_4N_4 forms a conducting solution in anhydrous HF;[31] this may well be a case of simple proton addition. A further relevant observation is that dry HCl precipitates from a solution of S_4N_4 in CCl_4 a dark red substance which with more HCl changes into $[S_4N_3]^+Cl^-$;[56] the red substance may be $[S_4N_4H]^+Cl^-$.

S_4N_4 reacts in a simple manner with a number of Lewis acid halides of main-group elements, and with sulfur trioxide, forming adducts which often have the 1:1 formula and sometimes other stoichiometries. These are listed in Table 6.1. The main points about them are as follows. They are usually made by bringing the components together in an inert solvent. The second molecule of BCl_3, however, is added only with difficulty, in liquid SO_2 at $-40°$. Only three true adducts of S_4N_4 have been characterized by X-ray crystallography, namely $S_4N_4 \cdot SbCl_5$ [Fig. 6.2(b)] and the structurally similar $S_4N_4 \cdot BF_3$[46] and $S_4N_4 \cdot SO_3$.[34] In all, the S_4N_4 ring is intact but somewhat flattened, and the structure is clearly that of a covalent donor-acceptor complex. The same may hold for other adducts in Table 6.1 in their solid state; there is some i.r. evidence for this.[3,95] However, $S_4N_4 \cdot SbCl_5$, $S_4N_4 \cdot BCl_3$, and eight other adducts tested were all strong electrolytes in nitrobenzene or acetonitrile,[71] indicating that the covalent structure very easily goes over to ions, such as $[S_4N_4 \cdot BCl_2]^+Cl^-$, in a polar solvent. Paul and coworkers, who made the conductance measurements, also considered it likely that some of the adducts were ionic in the solid state, and identified i.r. bands attributed to the anion in the following solid compounds among others: $S_4N_4 \cdot 2BCl_3$, which they formulated $[S_4N_4 \cdot BCl_2]^+BCl_4^-$; and $S_4N_4 \cdot BCl_3 \cdot SO_3$, formulated as $[S_4N_4 \cdot BCl_2]^+SO_3Cl^-$. In the absence of structural information, it should not be too readily assumed that reported adducts of S_4N_4 are really such. It is now known, for example, that AsF_5 slowly oxidizes S_4N_4 to $S_3N_2^+$ in SO_2 solution,[35] and it seems possible that the adduct $S_4N_4 \cdot 4SbF_5$[3,46] may be identical with the recently characterized compound $[S_4N_4]^{2+}[SbF_6]^-[Sb_3F_{14}]^-$ or a mixture containing it.[36]

The reactions of S_4N_4 with S_2Cl_2, S_2Br_2, and Se_2Cl_2 are not simple additions (see Section III.C below).

Many reactions of S_4N_4 with the compounds of transition and post-transition metals give rise to complexes, some but not all of which contain intact molecules of S_4N_4. These reactions have been well reviewed[3,11,70,92] and only a few points are mentioned here. Some of the reactions, in inert solvents, give adducts believed to contain flattened out but otherwise intact S_4N_4.[3] With iron, cobalt, and nickel carbonyls in benzene, S_4N_4 gives strongly colored crystalline compounds $M(SN)_4$, of unknown constitution but monomeric in some solvents. $Mo(CO)_6$ gives the explosive $Mo(SN)_5CO$.[96] The ligands here may be not S_4N_4 but unidentate SN or bidentate S_2N_2. The same holds for the complexes $CuCl_2 \cdot S_2N_2$ and $CuBr_2 \cdot S_2N_2$, obtained from S_4N_4 and the

Table 6.1. Adducts of tetrasulfur tetranitride with main-group halides and with sulfur trioxide.

Group III	Group IV	Group V	Group VI
B	C	N	O
$S_4N_4 \cdot BF_3$, dark red crystals (95)			
$S_4N_4 \cdot BCl_3$, red-orange crystals (71, 95)			
$S_4N_4 \cdot 2BCl_3$, brown crystals (71)			
$S_4N_4 \cdot BBr_3$, impure orange-brown solid (95)	Si	P	S
		$S_4N_4 \cdot SnCl_4 \cdot POCl_3$	
$S_4N_4 \cdot BCl_3 \cdot SbCl_5$ (95)			
$S_4N_4 \cdot BCl_3 \cdot SO_3$, orange liquid (71)		As	†$S_4N_4 \cdot SO_3$, brick-red solid (34)
$S_4N_4 \cdot PhBCl_2$, orange solid (3)		*$S_4N_4 \cdot AsF_5$ (33, 59)	
Al	Ge	Sb	Se
$S_4N_4 \cdot AlCl_3$, dark red crystals (25)		$S_4N_4 \cdot 2SbF_5$, white crystals (35, 46)	‡$S_4N_4 \cdot Se_2Cl_2$, olive powder (46)
$S_4N_4 \cdot AlBr_3$, orange crystals (25)		*$S_4N_4 \cdot 4SbF_5$, green solid (3, 35, 46)	$S_4N_4 \cdot SeCl_4$, yellow solid (3, 71)
$S_4N_4 \cdot 2AlBr_3$, orange solid (3)	Sn	$S_4N_4 \cdot SbCl_5$, red crystals (3, 35, 71, 85)	$S_4N_4 \cdot SeCl_4 \cdot SO_3$, reddish-orange liquid (71)
$S_4N_4 \cdot AlCl_3 \cdot SbCl_5$, brown powder (25)	$S_4N_4 \cdot \tfrac{1}{2}SnCl_4$, red crystals (3, 10, 46)	*$S_4N_4 \cdot 2SbCl_5$, orange solid (35, 71)	
	$S_4N_4 \cdot \tfrac{1}{2}SnBr_4$, dark red solid (3)	*$S_4N_4 \cdot SbCl_5 \cdot SO_3$, reddish-brown liquid (71)	Te
Ga	$S_4N_4 \cdot SnCl_4 \cdot POCl_3$, brown crystals (3)		$S_4N_4 \cdot TeCl_4$, red solid (3, 71)
$S_4N_4 \cdot 2GaCl_3$, red solid (3)		$S_4N_4 \cdot 2SbBr_3$, amorphous (46)	$S_4N_4 \cdot TeCl_4 \cdot BCl_3$, brown solid (71)
In		$S_4N_4 \cdot 2SbI_3$ (46)	$S_4N_4 \cdot TeCl_4 \cdot SO_3$, red-orange liquid (71)
$S_4N_4 \cdot 2InCl_3$, red solid (3)			$S_4N_4 \cdot TeCl_4 \cdot SbCl_5$, red-brown solid (71)
Tl	Pb	Bi	$S_4N_4 \cdot TeBr_4$, orange powder (45)

* In liquid SO_2, S_4N_4 is oxidized by AsF_5, SbF_5, and $SbCl_5$[35], so the status of these adducts needs reexamination.
† Excess SO_3 oxidizes S_4N_4 to $S_3N_2O_5$ (Chapter 8, Sec. VI).
‡ This adduct is probably spurious (cf. Section III.C of this chapter).

halide in dimethylformamide. Molybdenum, rhenium, and osmium complexes with an NS ligand have been made in another way, by the action of S_8 or S_2Cl_2 on a nitride ligand.[18] In some other reactions of complex formation described in the literature, the solvents participate chemically in obscure ways. The nickel, cobalt, and palladium complexes **V**, **VI**, and **VII**, obtained by heating methanol or dimethylformamide solutions of the metal halide with S_4N_4,[44] are colored crystalline compounds with definite melting points; their structures have been well established crystallographically and otherwise. The hydrogen atoms in the NH groups must come from the solvent; their presence

was recognized by infrared spectroscopy and their substitution reactions have been studied.[88,92] With the chlorides of various main-group metals and transition metals in thionyl chloride, S_4N_4 yields complicated products consisting largely of the chlorometallates of the ions $S_4N_3^+$, $S_3N_2Cl^+$, and $S_5N_5^+$.[7]

Tetrasulfur tetranitride is readily attacked by nucleophiles, always with opening of the ring. These reactions are a fruitful source of novel linear and cyclic S–N structures. Perhaps the least interesting is hydrolysis, in which the ring is totally destroyed.[46] It has been studied under heterogeneous and homogeneous conditions. The time required in a homogeneous, aqueous-organic solvent medium containing 0.5M alkali is about 2 h at room temperature. Reaction in acid solutions is generally slower, but the rate increases with acidity. The nitrogen is always completely converted into ammonia, confirming the physical-structural deduction that there are no N–N bonds in the molecule, and justifying the usual assignment of oxidation number -3 to nitrogen and $+3$ to sulfur. On this view of the sulfur oxidation state, reasonable mechanisms have been suggested to explain the complex mixture of sulfur oxidation numbers found in the end-products of hydrolysis. In the presence of excess sulfite ion, the overall reaction of "hydrolysis" becomes relatively simple:

$$S_4N_4 + 6H_2O + 2H_2SO_3 \rightarrow 2H_2S_3O_6 + 4NH_3$$

but it is not clear whether the sulfite ion attacks the S_4N_4 ring itself or its primary hydrolysis products. Treatment of S_4N_4 with aqueous ammonia gives similar products and also sulfamate, probably formed via trithionate.[46]

S_4N_4 and S_2N_2 (Section B below) both dissolve in liquid ammonia to give (after a few minutes' standing in the case of S_4N_4) the same red solution, which on evaporation leaves a cinnabar-red solid originally formulated as $S_2N_2 \cdot NH_3$ or $S_4N_4 \cdot 2NH_3$.[46] This substance is a mixture, containing much $[NH]_4^+[S_4N_5]^-$ (Section IV.B below),[84] and with a strong absorption band at 360 nm which arises from the recently described $S_3N_3^-$ ion[22] (see also Section III.H below). The ammonia solution has conductivity appropriate to a 1:1 electrolyte;[46] it serves as the source of some interesting metal complexes. Thus addition of lead nitrate or iodide precipitates green $Pb(NS)_2 \cdot NH_3$, which has

$$H_3N \rightarrow Pb \begin{array}{c} N = S \\ | \\ S - N \end{array}$$

(VIII)

been shown by X-ray diffraction to contain the chelate structure **VIII**. Thallium, silver, and copper salts originally formulated $M(NS)_2$ can be made similarly, but since the copper salt can only be made from Cu(I) these salts may really be MS_2N_2H, derived from the +1 metal.[92] Air oxidation of a pyridine solution of "$S_2N_2 \cdot NH_3$" gives $S_4N_5O^-$ (Chapter 8, Sec. II.C).

S_4N_4 is attacked by primary and secondary amines.[11] With secondary amines, $S_3N_3^-$ and $S_4N_5^-$ (Sections III.H and IV.B below) are formed.[22] Benzylamine at room temperature in absence of a solvent, gives benzylideneimine polysulfides, $(ArCH=N-)_2S_x$, especially the tetrasulfide.[83]

Controlled ring-opening of S_4N_4 by organic nucleophiles has given a variety of compounds containing unsaturated chains of alternating sulfur and nitrogen atoms. Information on chains of this type is summarized in Table 6.2. The longest are the quasi-infinite chains believed to exist in the $(SN)_x$ polymer (Section IV.C below), which conducts electricity well and is regarded as a metal. When S_4N_4 reacts with diaryldiazomethanes (presumably through a nucleophilic attack by the methylene carbon atom), almost the whole of the S_4N_4 ring is preserved in the products, which have the general formula $(R_2C=N-S-N=)_2S$. They are strongly colored, and when the organic groups are polycyclic aromatics (fluorenylidene etc.), they are pigments with photoconductivity and presumably a high degree of electron delocalization extending through the aromatic rings and the S–N chain.[40,91] The central S–N bonds of the bis-diphenylmethylene compound[49] are considerably shorter (154.9 pm) than the outer ones (165.6 and 168.5 pm) and the chain is not fully planar. Another series of colored crystalline compounds with shorter S–N chains, $(Ar-S-N=)_2S$, results from nucleophilic attack on S_4N_4 by the carbanions of Grignard reagents (Table 6.2). The central $-N=S=N-$ group in the bis-(p-chlorophenyl) compound resembles in shape that in the compound just previously discussed; again there are short central and longer outer S–N

II. THIAZYL AND ITS POLYMERS   109

Table 6.2. Compounds containing unsaturated chains of alternate sulfur and nitrogen atoms

Compound	Description	Source	Reference
$(SN)_x$	brass-like fibrous crystals, conductor	polymerization of S_2N_2	Chapter 6 Sec. IV.C
$[Ar-S-N=S-N=S-N-S-Ar]^+Cl^-$			58
Ar = p-tolyl	metallic golden crystals	arylsulfenyl chloride + $(Me_3SiN)_2S$ in benzene	
Ar = p-chlorophenyl	steel-blue crystals		
$R_2C=N-S=N-S=N-S-N=CR_2$			
R = phenyl	orange needles, m.p. 168°, giving dichroic solutions in organic solvents	diphenyldiazomethane + S_4N_4 in ether	49, 50
R_2 = fluorenylidene	greenish-black needles with metallic reflex, m.p. 225° dec., giving violet solutions in organic solvents	reflux fluorenyldiazomethane + S_4N_4 in benzene	49
R_2 = various polycyclic aromatics	patented pigments for electrophotographic plates and electrophoretic imaging		40
$Me_3Si-N=S=N-S-N=S=N-SiMe_3$	orange-yellow needles, m.p. 69°	$(Me_3SiN)_2S + SCl_2$	Chapter 3, Sec. IV.C
$Ar-S-N=S=N-S-Ar$			
Ar = phenyl	brownish platelets with gold sheen, m.p. 105°	add ArMgBr in ether to S_4N_4 or $[S_3N_2Cl]^+Cl^-$ in benzene	6, 91
Ar = p-chlorophenyl	red-brown needles, m.p. 160°		68, 91
Ar = p-bromophenyl	dark brown crystals		91
Ar = p-methoxyphenyl	orange-brown platelets, m.p. 125°		91
$R_2N-S-N=S=N-SiMe_3$			
R = methyl	red oily liquid, b.p. 29°/0.01 Torr	$Me_3SiNR_2 + S_4N_4$ in refluxing CH_2Cl_2	77
R = ethyl	red oily liquid, b.p. 37°/0.01 Torr		
$Et-S-N=S=N-H$	very unstable red oil	$EtMgBr + S_4N_4$ see Chapter 3, Sec. IV	68
$R-N=S=N-R$	substituted sulfur(iv) diimides, yellow to red oils		

110   CHAPTER 6. UNSATURATED SULFUR NITRIDES

bonds.[68] Nucleophilic attack on S_4N_4 by trimethylsilylamines gives compounds (Table 6.2)[77] with the same central $-N=S=N-$ group. The $-N=S=N-$ group without any extra sulfur or nitrogen atoms is found in the sulfodiimides (Chapter 3, Sec. IV), and the compound $(Me_3Sn-N=)_2S$ containing it has been made by nucleophilic attack on S_4N_4 by tris(trimethylstannyl)amine.[79]

The attack of nucleophiles on S_4N_4 in some cases affords new S–N ring systems, viz. S_3N_3 or S_3N_2, as follows.

The thiophile triphenylphosphine attacks S_4N_4 in dimethylformamide solution,[46] removing one sulfur atom and forming compound **IX**, red monoclinic crystals. The molecular structure has been confirmed by X-ray diffraction (Fig. 6.3);[48] its most notable feature is the unusual 6-membered sulfur–nitrogen ring, planar except for the sulfur atom with the ligand attached, which lies considerably out of plane. The corresponding cyclohexyl compound has also been made.[46] With phenyldichlorophosphine in excess, S_4N_4 is totally broken down, forming cyclic chlorophosphazenes (Chapter 12) and phenyldichlorophosphine sulfide.[46] [See also note on p. 129.]

(IX)

Attack on S_4N_4 by the thiophile cyanide ion gives $S_4N_5^-$ (Section IV.B below), NCS⁻, and a brick-red compound containing CN but so far not fully characterized.[22]

At 25° in methyl chloride, S_4N_4 with trifluoroacetic anhydride or trichloroacetic anhydride affords the compounds $CX_3CON=S_3N_2$,[86] containing the S_3N_2 ring. Derivatives of this ring are more fully discussed in Chapter 7, Sec. IV.B.

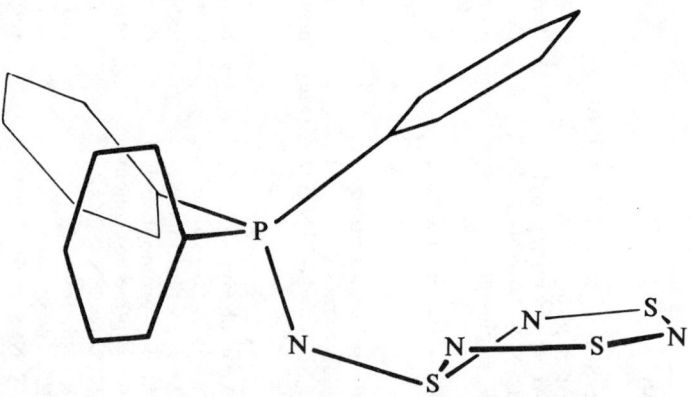

Figure 6.3. Molecular structure of the compound $Ph_3PN_4S_3$.[48]

B. Disulfur dinitride, S_2N_2

Disulfur dinitride, or thiazyl dimer, is best prepared by passing the vapor of S_4N_4 at pressures below 10^{-3} Torr through silver wool at 220°.[62,85] The volatile dinitride is condensed out in a cold trap. The preparation works at temperatures up to 300°, but the best yields (60%) and safer working are achieved at the lower temperature. The silver is converted into silver sulfide. Its function is not clear, but it is well established that silver wool gives better results than quartz wool[85] (cf. ref. 46).

S_2N_2[46,62] forms large colorless crystals with a repulsive iodine-like smell, which soon begin to polymerize to $(SN)_x$ even at 0°,[62] developing a blue-black color. S_2N_2 can be sublimed at 10^{-2} Torr at room temperature.[46,62] It is endothermic ($\Delta H_f^\circ \approx 230$ kJ mol^{-1})[23] and detonates with friction or shock or on heating above 30°. It is insoluble in water but soluble in many organic solvents. It can be recrystallized from ether. It is occasionally formed in solution from the decomposition of other sulfur–nitrogen compounds, and can then be recognized by its i.r. spectrum (Fig. 6.1).

An X-ray single-crystal study has shown the molecule of S_2N_2 to be an almost exact square, with alternating S and N atoms and practically equal bond lengths averaging 165.4 pm.[27] I.r. data[23] are consistent with this. The bond lengths indicate delocalized partial multiple bonding[27] (Chapter 1, Sec. II; Chapter 9, Sec. V.B). In the crystal, polymerization to $(SN)_x$ (below) is facilitated by short intermolecular N–S contacts along the a axis (Fig. 6.4).

Rapid and quantitative polymerization to S_4N_4 takes place when traces of alkalis or potassium cyanide are added to solutions of S_2N_2 in organic solvents.[46] However, when dry purified S_2N_2 is stored for some days *in vacuo*, nearly 100% conversion into polythiazyl $(SN)_x$ (Section C below) is observed. In moist air, S_2N_2 changes quickly into a mixture of S_4N_4, hydrolysis products, and a little $(SN)_x$.[46] When heated to 250° in a sealed tube it decomposes quantitatively into its elements.[46]

The reactions of S_2N_2 with Lewis acids have recently been examined. Adducts of S_2N_2 can be formed, but in some circumstances adducts of S_4N_4 or $(SN)_x$ are obtained. When S_2N_2 is gradually added to $SbCl_5$ in dichloromethane at room temperature,[46] the following sequence of reactions takes place:

$$S_2N_2 + 2SbCl_5 \rightarrow S_2N_2(SbCl_5)_2$$
(yellow crystalline precipitate)

$$S_2N_2 + S_2N_2(SbCl_5)_2 \text{ (dissolves)} \rightarrow S_2N_2 \cdot SbCl_5$$
(orange, crystallizes on evaporation)

$$S_2N_2 \cdot SbCl_5 + S_2N_2 \rightarrow S_4N_4 \cdot SbCl_5$$
(orange-red precipitate)

Any further S_2N_2 added is quickly changed into S_4N_4, apparently because of catalysis by some unidentified component of the mixture. The adduct $S_2N_2(SbCl_5)_2$ has been shown by X-ray crystallography to have the structure X, with two donor nitrogen atoms and a planar S_2N_2 ring almost the same in

$$Cl_5Sb \leftarrow N \underset{S}{\overset{S}{\cdots}} N \rightarrow SbCl_5$$

(X)

dimensions as the free S_2N_2 molecule.[46] Mono- and di-adducts are also formed with BCl_3 at $-78°$.[46] The diadduct easily loses BCl_3. The monoadduct readily polymerizes to a brown solid, $[(S_2N_2)(BCl_3)]_x$, which almost certainly does not contain intact S_2N_2 rings. It resembles, and may be identical with, the adduct of $(SN)_x$ with BCl_3 (Section C below). With excess BF_3,[46] S_2N_2 gives only the known $S_4N_4 \cdot BF_3$ (Section A above). This may be a result of the weakness of BF_3 as a Lewis acid; even in the presence of excess BF_3, there would be a relatively high concentration of free S_2N_2 in equilibrium with the presumed primary adduct $S_2N_2 \cdot BF_3$, permitting the following secondary reaction to take place quickly:

$$S_2N_2 + S_2N_2 \cdot BF_3 \rightarrow S_4N_4 \cdot BF_3$$

The nucleophilically catalyzed dimerization of S_2N_2 has been mentioned above. There is little further information on its behavior with nucleophiles. With liquid ammonia[46] and aqueous alkali it gives the same products as S_4N_4.

C. Thiazyl polymer, $(SN)_x$

This compound was first observed in 1910 as a blue film with a bronzy reflex, accompanying the pyrolysis of S_4N_4 vapor.[38,55] Until 1975 it was seldom prepared, but since that year there has been intense interest in its electronic properties, resulting in a spate of publications which have come largely from American university laboratories (supported by the National Science Foundation) and from the firms of General Electric and Xerox Webster. This recent work has firmly established the remarkable fact that $(SN)_x$ is a highly anisotropic metal, the best example of this unusual class of substances, and the only one exhibiting superconductivity. Its properties have been well reviewed.[55]

Analytically pure $(SN)_x$ has only recently been made. The best procedure is to grow large crystals of S_2N_2 (Section B above) by sublimation *in vacuo*, and to let these stand for some days at or near room temperature.[55] From each crystal of S_2N_2 there results a mass of parallel fibers of $(SN)_x$, which look bright gold when viewed from the side and blue-black at the ends. The crystals are soft and malleable. The compound can be sublimed by heating *in vacuo* and condensed as fully oriented epitaxial films on e.g. Mylar, Teflon, and polyethylene.[55]

During the polymerization of S_2N_2 to $(SN)_x$, an e.s.r. signal caused by radicals develops and passes through a maximum, finally disappearing when polymerization is complete.

X-ray diffraction data show that $(SN)_x$ consists of nearly planar chains of alternating sulfur and nitrogen atoms (Fig. 6.4). These chains are readily formed from the square molecules of S_2N_2 by ring-opening accompanied by relatively small atomic displacements, the crystallographic *a* axis of S_2N_2 becoming the *b* axis of $(SN)_x$ (Fig. 6.4).[29] The $(SN)_x$ structure is defective because of the lack of long-range coherence in the polymerization process.

The S–N bond lengths in the chains of $(SN)_x$ correspond to a bond order of about 1.5, but are not exactly equal, alternating between 159 and 163 pm.[29] X-ray photoelectron spectroscopy shows that each nitrogen atom receives a charge transfer of about 0.5 e from sulfur.[82]

$(SN)_x$ is the best example known of a very anisotropic metal, and for this reason is now under intensive study by solid-state physicists. It is a metallic conductor ($\sigma = 10^3$ ohm^{-1} cm^{-1}) in the direction of the S–N chains[55] and a

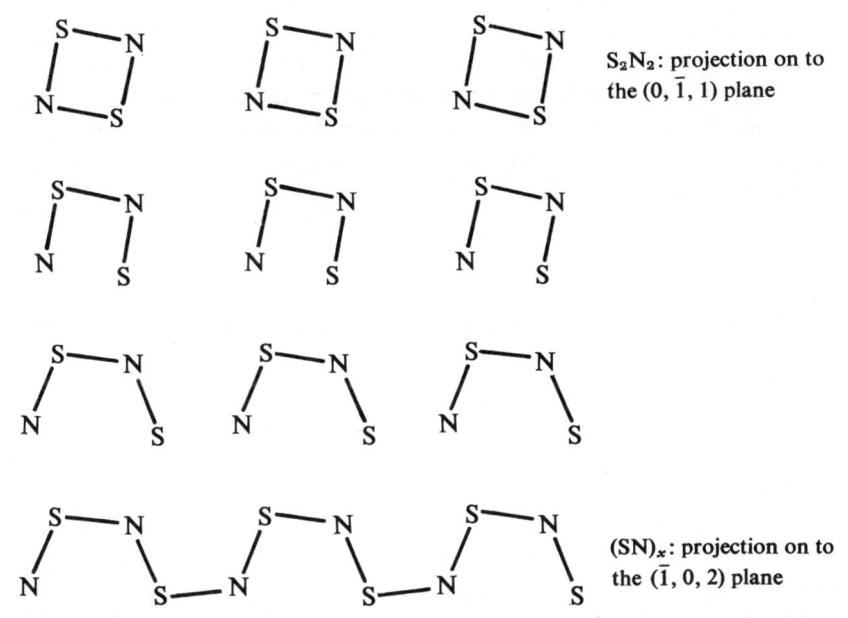

Figure 6.4. Stages in the polymerization of S_2N_2 to $(SN)_x$; the *a* axis of S_2N_2 becomes the *b* axis of $(SN)_x$ (after ref. 29).

much poorer conductor in perpendicular directions. It is unique among conducting polymers in becoming superconducting at low temperatures (below 0.26 K).[55] There are theoretical grounds for believing that electronic coupling *between* chains may play an essential part in this behavior. Light polarized parallel to the chains is reflected metallically, whereas with transverse polarization the crystals appear dark blue-black. The specific heat at low temperatures is consistent with this picture.

By mass spectrometry, and by the matrix isolation technique,[55] the vapor of $(SN)_x$ undergoing sublimation *in vacuo* has been shown to contain much S_4N_4 (probably a linear form), some S_2N_2, a little SN, and more than one unidentified species. $(SN)_x$ is not explosive. It ignites in air near 130°,[38] and at 140° in a sealed evacuated tube decomposes to sulfur, nitrogen, and perhaps other materials. When pure it is hardly affected by oxygen or moisture in 7 days, but tarnishes if any S_2N_2 is present as an impurity. Bromine vapor converts it into $(SNBr_{0.4})_x$, an even better conductor.[55]

D. Thiazyl monomer, SN

Because of its tendency to polymerize, this compound has only a transient existence in the gas phase, and cannot be isolated as a solid or liquid. It is formed by the action of an electric discharge on a mixture of sulfur vapor and nitrogen, from various mixtures of gaseous nitrogen and sulfur compounds subjected to photolysis or electric discharges,[30,46] or by thermal decomposition of S_7NH[30] or S_4N_4.[85] It has been characterized by the emission spectra of its excited states. With the help of photoelectron spectroscopy, a potential-energy diagram has been constructed for the (ground) NS ($X^2\Pi$) state and the first ionized state NS^+ ($X^1\Sigma^+$),[30] and may be accepted with some confidence.

III. Other unsaturated sulfur–nitrogen ring molecules and ions

A. Introduction

The ease with which the S_4N_4 cage can be opened has been demonstrated in Section II.A.4, where a number of examples showed the basic structure of alternating sulfur and nitrogen atoms, or part of it, surviving into the products. Among the substances so obtainable are five unsaturated ring systems containing only sulfur and nitrogen, namely, the neutral molecule S_4N_2, the ions $S_4N_3^+$, $S_5N_5^+$, and $S_6N_4^{2+}$, and the radical-ion $S_3N_2^+$. We shall also describe the ring ions $S_4N_4^{2+}$ and $S_3N_3^-$, obtained respectively by oxidation and reduction of S_4N_4, and report the limited evidence for $S_3N_2^{2+}$.

B. Tetrasulfur dinitride, S_4N_2

Everyone who has worked with the compounds described in this chapter and Chapter 2 is familiar with a deep-red color seen in CS_2 solutions of many crude reaction products. The color is often due to traces of tetrasulfur dinitride, S_4N_2. This compound was first isolated in an impure state in 1896, but was not properly characterized until 1951.[46] Before then, impure samples were often formulated S_5N_2.[38]

The traditional way to prepare S_4N_2 is by heating S_4N_4 with sulfur, in CS_2, to 100–120° in an autoclave.[14,39] The yield is only about 10%, and the mechanism is not well understood.[39] The solvent must participate, because much thiocyanogen polymer is formed, and some of the sulfur in the S_4N_2 comes from the solvent.[39,46] This preparation is not very convenient, because it calls for large quantities of the explosive S_4N_4 and requires an autoclave. Better methods are now available[47,66] to anyone prepared to go to the trouble of setting up a column for gel-permeation chromatography; such a column can be used repeatedly without re-packing. In one such method, the starting material is heptasulfur imide, S_7NH (Chapter 2), a safe and stable compound which can easily be made and stored in bulk. The imide is converted into its mercury(II) derivative, $Hg(S_7N)_2$, by treatment with mercury(II) acetate in methanol solution. The mercury derivative decomposes in 3 days on storage at room temperature. S_4N_2 forms in the decomposing mass in 64% yield, and can be isolated pure from the CS_2 extract of products by one pass through the g.p.c. column.[47] A second method is to react S_2Cl_2, in CS_2, with 10 molar aqueous ammonia at 5°. Again the products are separated by g.p.c. Although the percentage yield is smaller (about 8%), this method has the great advantage of using commercial chemicals off the shelf.[66] From a practical standpoint, other reported[1,38] methods of formation of S_4N_2 are now of little interest.

Although different isomeric structures for the S_4N_2 molecule can be imagined, only one substance with this formula has ever been found in the chromatographic analysis of products from three methods of preparation.[46,47]

Tetrasulfur dinitride forms opaque red-grey needles which melt at 23° to a dark red liquid. It has a distinct unpleasant smell and can be sublimed at about 1 Torr at room temperature. Sublimation, and chromatography on silica gel or polystyrene gel, are good methods of purification.[46] It is insoluble in, and slowly hydrolyzed by, water. It dissolves readily without decomposition in many organic solvents such as CS_2, benzene, and hexane.[46]

The solid or liquid nitride decomposes in a few hours at room temperature. Decomposition becomes explosive at 100°.[38,46] The compound keeps much better in solution in CS_2.

The molecular formula S_4N_2 has been established by analysis and cryoscopic determination of the molecular weight in benzene.[46] Determination of the molecular structure has been more difficult. An astonishing number of Lewis structures can be written for the formula S_4N_2. Excluding unlikely features such as 3-membered rings and adjacent charges of the same sign, the number is of the order of sixty. No X-ray diffraction data are yet available. However, the following physical evidence makes it possible to rule out the great majority of hypothetical structures at once, and then to select **XIa** as a likely structure among the few remaining contenders.[46]

(XIa) (XIb) (XIc) (XId)

First, the ^{14}N n.m.r. spectrum contains a single resonance 105 ppm upfield from aqueous nitrate ion, indicating that the two nitrogen atoms are equivalent and doubly bonded. Most of the hypothetical structures have either non-equivalent nitrogen atoms or nitrogen atoms at bridgeheads (i.e. singly bonded), so probably should not be considered. Secondly, the dipole moment, 1.75 ± 0.28 D, is definitely non-zero, and there are coincidences between Raman and infrared spectral frequencies, so centrosymmetric structures (including planar and chair form **XIc**) can be ruled out. Thirdly, linear structures with equivalent nitrogen atoms cannot be reconciled with the mass-spectral fragmentation pattern.[46] This leaves only **XIa**, **XIb**, and boat-form **XIc** to be discussed. There is no frequency in the vibrational spectrum high enough for an N=N bond. Hence **XIb** can be ruled out. Finally, there are four fundamental S-N stretching vibrations,[46] the number expected for **XIa**; boat-form **XIc** would give only three. The physical evidence is thus consistent with **XIa**. This structure is also supported by chemical evidence. Hydrolysis[46] converts all the nitrogen in the compound into ammonia; no hydrazine or molecular nitrogen is formed as would perhaps be expected from **XIb**. Detailed studies of the hydrolysis products in the presence and absence of HSO_3^- have shown[46] that the sulfur in S_4N_2 behaves as if it were in three oxidation states.

$$S_4N_2 \rightarrow S(+4) + S(+2) + 2S(0) + 2N(-3)$$

This can be understood in terms of **XIa**, by assigning reasonable oxidation states to the different sulfur atoms as shown; but it is less easy to reconcile with **XIc**, in which all the sulfur atoms must have the same oxidation number, or with **XId**.

Structure **XId** does not seem to fit the evidence (as at present interpreted) so well as **XIa**, but recent findings on the stability of the S_3N_2 ring (Chapter 7, Sec. IV; Chapter 8, Sec. IV) strongly suggest **XId** as a possibility; it would be the thio analog of the oxide S_3N_2O.

It is noteworthy that S_4N_2 has an entirely different structure from N_2O_4. N_2O_4 is in equilibrium with the paramagnetic NO_2. S_4N_2, in contrast, gives no evidence of dissociation to radicals. It is diamagnetic, with susceptibility near the sum of the Pascal constants.[46]

Apart from the hydrolysis just described, the chemistry of S_4N_2 has not been adequately investigated.

S_4N_2 appears to be less strongly basic than S_4N_4 or S_2N_2, undergoing no reaction with BCl_3 in CS_2 at room temperature,[46] and does not hydrogen-bond to phenol. It is oxidized by $SbCl_5$, giving $S_4N_4 \cdot SbCl_5$ (Section II.A.4 above).[46] It forms a 1:1 adduct with dicyclopentadiene,[2] red crystals of m.p. 129° (dec.), the i.r. spectrum of which has been interpreted as evidence of addition across the nitrogen atoms.

Its sensitivity to oxidative ring-opening is also apparent in its reactions with chlorine and bromine, which still have to be fully worked out.[46] Simple halogenation products containing the intact S_4N_2 ring do not appear to be formed; reaction with chlorine produces instead S_4N_4, $[S_4N_3]^+Cl^-$, and $[S_6N_4]^{2+}Cl^-_2$.

Like other sulfur nitrides, S_4N_2 is easily reduced by HI in anhydrous formic acid:[46]

$$S_4N_2 + 6H^+ + 6I^- \rightarrow 3I_2 + 4S + 2NH_3$$

Reduction with hydrogen and palladium, potassium borohydride, sodium dithionite, lithium aluminum hydride, or hydrazine[46] gives a mixture of cyclic sulfur imides with 8-membered rings (Chapter 2).

C. The thiotrithiazyl ion, $S_4N_3^+$, and its salts

Impure $[S_4N_3]^+Cl^-$ was first made about 1880 by Demarçay,[38] by the reaction

$$3S_4N_4 + 2S_2Cl_2 \xrightarrow[\text{reflux}]{CHCl_3} 4[S_4N_3]^+Cl^-$$

By 1957 the pure chloride was well known.[39] The ionic constitution of this compound was recognized, and other salts had been prepared by metathesis from the chloride. But the planar 7-membered ring structure of the ion was first definitely established in a crystallographic study of the nitrate by J. Weiss in 1962,[11] and has been confirmed by ^{15}N n.m.r. and by crystallographic studies of other salts.[11]

The chloride results in 90–92% yield from the reaction above, when the reactants are refluxed for 1 hour in carbon tetrachloride.[39] Good yields are obtained if $SOCl_2$ is substituted for SCl_2.[38] In practice, however, it is better to avoid making S_4N_4 first; this is possible in two ways. Either, the S_4N_4 synthesis from S_2Cl_2 and ammonia can be stopped at an intermediate stage (Section II.A.1 above), giving ~35% yield, roughly 27 g product from 100 g S_2Cl_2,[15] or, for quantities of the order of 8 g from very simple equipment, Jolly and

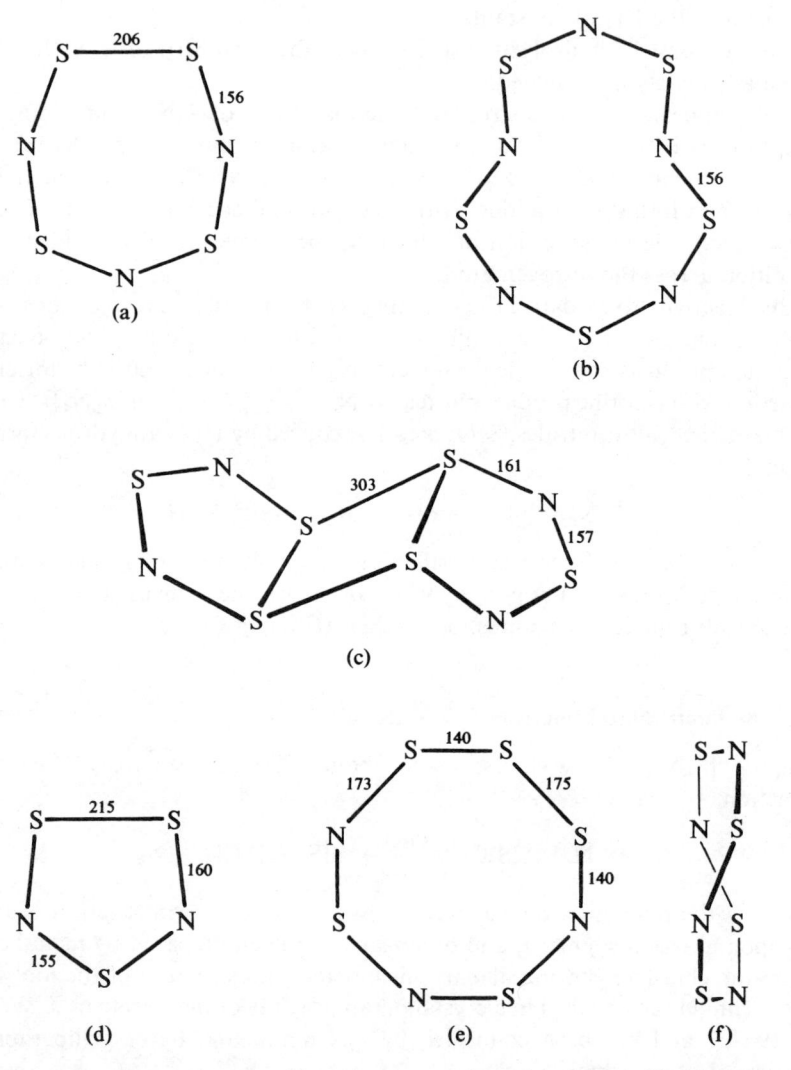

Figure 6.5. Structures of sulfur–nitrogen cations (a) $S_4N_3^+$;[54] (b) $S_5N_5^+$;[10] (c) $S_6N_4^{2+}$;[9] (d) $S_3N_2^+$;[35] (e) and (f) face view and side view of the ion $S_4N_4^{2+}$, in the form that occurs in $[S_4N_4][SbCl_6]_2$;[36] bond distances are in pm.

Maguire's method[52] can be recommended; $[S_3N_2Cl]^+Cl^-$ (Chapter 7, Sec. III.B) is first made by refluxing S_2Cl_2 with NH_4Cl, and then quantitatively converted *in situ* by the reaction:

$$3[S_3N_2Cl]^+Cl^- + S_2Cl_2 \xrightarrow[\text{reflux}]{CCl_4} 2[S_4N_3]^+Cl^- + 3SCl_2$$

The chloride can be recrystallized from thionyl chloride.

Other salts are obtainable by metathesis from $[S_4N_3]^+Cl^-$. The bromide and iodide are precipitated by reaction in ice-cold water with KBr or KI,[73] but water is of limited usefulness as a medium since $[S_4N_3]^+Cl^-$ decomposes fairly quickly in it.[38] The salts of several strong mineral acids, such as the nitrate and fluorosulfate, have been made from the chloride and the appropriate concentrated acid. The bromide, thiocyanate, etc. can be made in anhydrous formic acid. Reaction with the appropriate metal chloride in thionyl chloride gives chlorometallates with anions such as $AlCl_4^-$, $FeCl_4^-$, and $SbCl_6^-$.[11,45]

The chloride, $[S_4N_3]^+Cl^-$, has been studied much more than any other salt. It forms small shining golden-yellow tablets. It is stable in dry air and only slowly affected by moist air. On heating in air, it explodes with a blue luminescence, producing a red smoke. It has not been reported to explode during normal handling.[11,15,38,45]

As expected from its ionic constitution, it is insoluble in non-polar organic solvents such as CCl_4 and CS_2.[38] It is fairly soluble, without decomposition, in hot thionyl chloride,[11] and in anhydrous formic acid.[15] Most other solvents decompose it.

Other salts are also, for the most part, yellow well crystallized substances.[11,38,45] The bromide decomposes exothermically to its elements at 179°.[98]

X-ray diffraction studies of three salts of $S_4N_3^+$ have shown the presence of the planar 7-membered ring cation depicted in Fig. 6.5(a).[11,54] The six S–N bond lengths are nearly equal and only slightly longer than the presumed double-bond distance.[67] The S–S bond is about equal in length to the formally single bonds in S_8. The ring is seen as an aromatic one (Chapter 9, Sec. V.B) with delocalized π bonding interrupted at the S–S bond.

As expected, $[S_4N_3]^+Cl^-$ is a strong 1:1 electrolyte in formic acid.[15]

The metathetical conversion of the chloride into other salts has been mentioned above. Apart from this, there is rather little recent or reliable information on the reactions of the chloride, and practically nothing on those of other salts.

On heating alone, or on boiling a suspension of it in benzene or chloroform, the chloride gives some S_4N_4.[39] Controlled pyrolysis of the chloride or bromide gives S_4N_4 and some thiazyl halide monomer[74] (Chapter 7, Sec. II.F). The chloride does not react with chlorine, even in presence of an iodine catalyst.[38] Like S_4N_4, it can be reduced by hydrogen iodide in formic acid:

$$S_4N_3^+ + 13HI \rightarrow 5I_2 + 3NH_4^+ + 3I^- + H^+ + 4S$$

A few of its reactions with Lewis acids have been examined. In thionyl chloride it reacts with SO_2 to give $S_3N_2O_2$ (Chapter 3, Sec. III.F).[38] An adduct is formed at 0° with $2SO_3$.[38] With $CoCl_2$ in thionyl chloride, the chlorocobaltate $[S_4N_3^+]_2[Co_2Cl_6]^{2-}$ is formed.[72]

The chloride is easily decomposed by nucleophiles, but the reactions have not been fully elucidated. Water, while dissolving it slightly, converts much of the sample into a black solid. A similar solid, made by shaking the chloride with ice-cold aqueous sodium acetate, has been thought to be the hydroxide $[S_4N_3]^+OH^-$, because it gives the chloride again in 85% yield when refluxed with thionyl chloride,[38,45] but this does not amount to proof of its identity. Hot aqueous HCl or alkali destroys the ring and gives ammonia with a mixture of sulfur oxidation states.[39] Two nucleophilic reagents are known to enlarge the ring of $S_4N_3^+$ to S_4N_4. One of them, ammonia, does this in the course of the S_4N_4 synthesis[15] (Section II.A.1 above). The reaction goes smoothly, with high yields of S_4N_4, in a solvent such as CCl_4.[15] In the absence of a solvent, however, $[S_4N_3]^+Cl^-$, after freely absorbing gaseous ammonia, explodes.[39] The chloride dissolves in liquid ammonia with a violet-red color; evaporation of the solution usually leads to an explosion.[39] Ammonia gas diluted with nitrogen is absorbed only up to the composition $[S_4N_3]^+Cl^- \cdot NH_3$, giving a black product of unknown constitution.[39] The other reaction of ring-enlargement is with aluminum azide:[45]

$$[S_4N_3]^+Cl^- + Al_{1/3}N_3 \xrightarrow[40°]{\text{tetrahydrofuran}} N_2 + S_4N_4 + Al_{1/3}Cl$$

It is nearly quantitative.

D. The cyclopentathiazenium cation, $S_5N_5^+$, and salts

This interesting ion was first reported in 1969 by Banister and Dainty.[11] The well crystallized pale yellow tetrachloroaluminate, $[S_5N_5]^+[AlCl_4]^-$, was obtained during a study of the reactions of S_4N_4 with Lewis acid chlorides in $SOCl_2$. These reactions (Section II.A.4 above) give complicated mixtures which include salts of $S_4N_3^+$ and $S_3N_2Cl^+$ (Chapter 7) as well as $S_5N_5^+$ salts.[7] It was later found that high yields of relatively pure $S_5N_5^+$ salts could be obtained by including $(NSCl)_3$ (formula II above; see also Chapter 7, Sec. II.C) in the reaction mixture. $(NSCl)_3$ may react with the Lewis acid chloride, generating SN^+, which then inserts into the S_4N_4 ring to give $S_5N_5^+$. Independent experiments have shown that $[SN]^+[MF_6]^-$ (M = As, Sb) does give high yields of $S_5N_5^+$ with S_4N_4.[61]

$S_5N_5^+$ is now recognized to be a 14π aromatic species, and its discovery strongly stimulated efforts to rationalize the electronic structures of unsaturated S–N rings (Chapter 9).

The yellow $AlCl_4^-$ salt (m.p. 181° dec.), the dark orange $FeCl_4^-$ salt (m.p. 181° dec.), and the yellow $SbCl_6^-$ salt (m.p. 188° dec.) are all made as follows.[8] At room temperature, $(NSCl)_3$ is dissolved in $SOCl_2$ and the appropriate metal chloride added. An adduct precipitates. On now adding S_4N_4, the adduct reacts with it and the corresponding $S_5N_5^+$ salt crystallizes out.[8] The AsF_6^- and SbF_6^- salts, and the yellow salt with the anion $[(SnCl_5)OPCl_3]^-$,[10] can be made by a related technique.

The reineckate and the tetraphenylborate are prepared from the tetrachloroferrate by metathesis in formic acid solution.

An X-ray diffraction study of the tetrachloroaluminate has shown the presence of a nearly planar ring of alternating sulfur and nitrogen atoms, with two re-entrant angles [Fig. 6.5(b)].[10] All the S–N bond lengths are short enough unequivocally to indicate multiple bonding (Chapter 9). Earlier reports of a heart-shaped $S_5N_5^+$ cation are now attributed to a misinterpretation.[12]

Little has been published on the properties of $S_5N_5^+$ salts. There are data on the electronic absorption spectrum.[11]

E. The $S_6N_4^{2+}$ cation and salts

The chloride $[S_6N_4]^{2+}Cl_2^-$ is a black powder with a green reflex, which leaves a carmine-red trace when rubbed on paper.[38] It was originally described by Demarçay in 1880 as the first product of the S_4N_4–S_2Cl_2 reaction. On refluxing in CCl_4 these reagents give yellow $[S_4N_3]^+Cl^-$ (Section C above), but in the cold, without a solvent or in nitromethane, black-green $[S_6N_4]^{2+}Cl_2^-$ is formed.[38,39] An ionic structure, with $S_6N_4^{2+}$, was reported without details at a conference[16] in 1969, but the complete structure of a salt was first announced in 1974 by Banister, Clarke, Rayment, and Shearer.[9] Since there is some doubt about what the structure means in terms of bonds (see below) it would be premature to name the ion.

The purest samples of $[S_6N_4]^{2+}Cl_2^-$ have been made by thermal decomposition of $[S_3N_2Cl]^+Cl^-$ (Chapter 7) *in vacuo* at 90°.[9,52] An alternative method, reduction of $[S_3N_2Cl]^+Cl^-$ with formic acid at room temperature, gives a good yield but a less pure product.[9]

Among the few known salts of $S_6N_4^{2+}$ is the chlorodisulfate, prepared by adding the chloride to chlorosulfuric acid and leaving the solution to crystallize.[9]

The X-ray diffraction study of the chlorodisulfate[9] has revealed the curious ion shown in Fig. 6.5(c), with three fused rings in a chair conformation. Each ring is planar. The S–S bonds holding the two S–N rings together and forming the middle ring are very long, though well below typical van der Waals distances of 370 pm. Since they are considerably longer than the S–S bonds in S_4N_4, which are believed to be of order ~ 0.3, there is perhaps some slight doubt whether the middle ring exists at all as a bonded structure. The S–N bonds have short lengths typical of the unsaturated rings discussed in this chapter, and unquestionably have a π component. N.q.r. shows all the chlorine in the chloride to be ionic.[11]

The chloride and chlorodisulfate both form deep-green, nearly black, crystals. The chloride is insoluble in organic solvents but slightly soluble in liquid SO_2. It is decomposed by water but keeps well in dry air.[9,52]

When heated *in vacuo*, the chloride decomposes without melting at 120–140°,[52] giving SCl_2, NSCl, and $[S_4N_3]^+Cl^-$. It has not been observed to explode, but since it has been little investigated it should be handled with care.

With water it dissolves partially to give a yellow-brown unstable solution.[38] There is no reliable information on the course of hydrolysis. It dissolves in concentrated sulfuric acid, evolving HCl; the solutions have been used for recording the visible-u.v. spectrum of the ion,[9] but decompose quickly on standing.

Attempts to prepare other salts from the chloride by reaction with Lewis acid chlorides in SO_2, $SOCl_2$, and organic solvents were unsuccessful or gave impure products.[9]

F. The cations $S_3N_2^+$ and $S_4N_4^{2+}$

These ions result from the oxidation of S_4N_4 under different conditions.

The ion $S_6N_4^{2+}$ [Fig. 6.5(c)] can be seen as two planar S_3N_2 rings held together by long S–S bonds. Recently two compounds have been reported which contain well separated $S_3N_2^+$ rings, and therewith the separate existence and stability of this radical-cation is demonstrated. Deep red-brown crystals of $[S_3N_2]^+[AsF_6]^-$ can be made by the oxidation of S_4N_4 with AsF_5 in liquid SO_2;[35] by X-ray diffraction they have been shown to contain planar S_3N_2 rings almost identical in dimensions with those in $S_6N_4^{2+}$ [Fig. 6.5(d)]. Treatment of S_4N_4 with $(CF_3SO_2)_2O$ in CH_2Cl_2 gives blackish-brown needles of $[S_3N_2]^+[CF_3SO_3]^-$.[80] Both compounds give five-line 1:2:3:2:1 e.s.r. spectra.

Recently the remarkable discovery has been made that oxidation of S_4N_4 in liquid SO_2 by means of two other Group V halides, SbF_5 and $SbCl_5$, gives salts of another new cation $S_4N_4^{2+}$.[36] This exists in three forms, one of them a somewhat warped near-plane, present [Fig. 6.5(e),(f)] in the crystals of $[S_4N_4]^{2+}[SbCl_6^-]_2$, as so prepared. The other forms are planar. The ion may

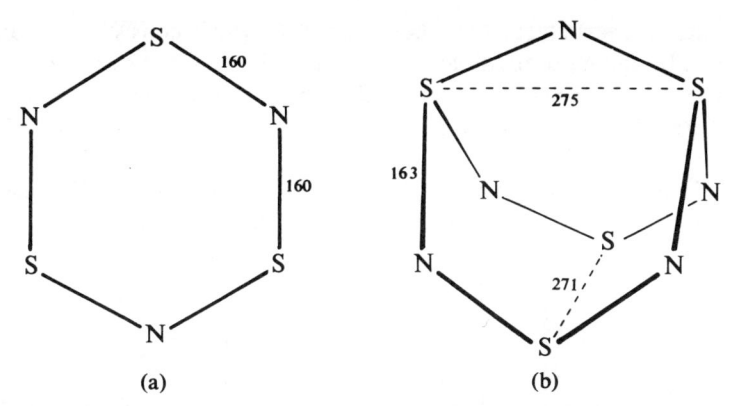

Figure 6.6. Structures of sulfur–nitrogen anions: (a) $S_3N_3^-$;[21] (b) $S_4N_5^-$;[32] bond distances are in pm.

represent a new Hückel-type aromatic species (Chapter 9, Sec. V.B) the existence of which had been postulated earlier.

G. The cation $S_3N_2^{2+}$

This ring ion has been supposed, on i.r. evidence, to occur in the compound $S_3N_2Cl_2 \cdot 2SbCl_5$, which consists of brownish crystals and is made by direct combination of $[S_3N_2Cl]^+Cl^-$ (Chapter 7, Sec. III.B) with $SbCl_5$.[78] It would be isoelectronic with $S_3N_2Cl^+$ and with S_3N_2O (Chapter 8, Sec. IV), and would be a Hückel aromatic species. Although its existence seems very likely, there is still a need for X-ray crystallographic evidence.

H. The anion $S_3N_3^-$

The reduction of S_4N_4 by cesium or tetraalkylammonium azides in ethanol, or by potassium metal in dimethoxyethane, gives explosive salts of the yellow anion $S_3N_3^-$, a nearly planar 6-membered ring[21] [Fig. 6.6(a)]. This anion has an ultraviolet absorption maximum at 360 nm, and is probably the source of a peak at this wavelength in the spectra of S_4N_4 solutions in liquid ammonia. $S_3N_3^-$ may be regarded as a 10π aromatic species (Chapter 9, Sec. V.B).

IV. Sulfur–nitrogen cages based on tetrasulfur tetranitride

A. Introduction

Reference has been made to the short S–S distances in the molecule of S_4N_4 (Section II.A.3 above; and see Chapter 9, Sec. V.C). Recently three new

CHAPTER 6. UNSATURATED SULFUR NITRIDES

sulfur–nitrogen structures have been reported which consist essentially of S_4N_4 with bridging groups added across one of these S–S distances. They are the ions $S_4N_5^-$ and $S_4N_5^+$, in which the bridge is a nitrogen atom, and the sulfur nitride S_5N_6, in which it is an –N=S=N– group (formulas **XI**, **XII**, and **XIII**). In addition, the ion $S_4N_5O^-$ (Chapter 8, Sec. II.C) is known, with essentially the same structure as $S_4N_5^-$ plus an oxygen ligand. The electron counts of these compounds are discussed in Chapter 9, Sec. V.C.

(XI) $S_4N_5^-$ (XII) $S_4N_5^+$ (XIII) S_5N_6

B. The anion $S_4N_5^-$

The yellow t-butylammonium and sodium salts of this ion were first described in 1975 by Scherer and Wolmershäuser. The simplest preparation, due to Bojes and Chivers,[19] is

$$NaN_3 + S_4N_4 \xrightarrow[20°]{\text{ethanol}} Na^+[S_4N_5]^- + N_2$$

The sodium salt explodes violently at about 180°.

The structure of **XI** [Fig. 6.6(b)] has been established by X-ray diffraction.[32] The "bridged" sulfur atoms at the top have almost the same spacing as in the S–N–S groups elsewhere in the molecule.

Although only recently characterized, the $S_4N_5^-$ ion has for long played an important but unappreciated part in experimental studies of sulfur–nitrogen compounds. It is formed in large amounts when $(NSCl)_3$ or S_4N_4 dissolve in liquid ammonia, and when S_4N_4 is attacked by a variety of other nucleophiles.[22,84] In these circumstances it may well arise from reaction of $S_3N_3^-$ first formed with S_4N_4.[22]

C. The cation $S_4N_5^+$

This ion, formula **XII**, is present in golden-brown crystals of S_4N_5Cl formed by the reaction at room temperature, in carbon tetrachloride, of equimolar amounts of $(NSCl)_3$ (Chapter 7, Sec. II.C) and $(Me_3SiN)_2S$,[27] or by chlorine oxidation of $S_4N_5^-$.[28] The most interesting feature of its X-ray structure is a considerable lengthening of the unbridged S–S distance (dashed line in formula **XII**) compared with S_4N_4 or $S_4N_5^-$.[27] The significance of this is discussed in Chapter 9, Sec. V.C.

D. The sulfur nitride S_5N_6

This new nitride was reported independently in 1978 by Chivers and Proctor[28] and by Roesky.[75] It is obtained by oxidation of $S_4N_5^-$ (Section B above). The salt $[Bu_4N]^+[S_4N_5]^-$ is treated with bromine in methylene chloride at 0°. Evaporation of the filtered reaction mixture gives S_5N_6 as orange crystals in 73% yield. [See also note on p. 129.]

S_5N_6 is stable for long periods under an inert atmosphere but immediately blackens in air. It sublimes without decomposition at $45°/10^{-2}$ Torr, and decomposes at 130°. It sometimes explodes during grinding. It is soluble in methylene chloride and in carbon disulfide. Its u.v.-visible and i.r. spectra have been reported.

The molecular structure **XIII** has been determined in two X-ray crystallographic studies.[28,75] The S_4N_4 "basket" is bridged by an $-N=S=N-$ "handle". The bridged S–S distance is lengthened relative to S_4N_4 but the unbridged S–S distance (dashed line in the formula) is shorter than in S_4N_4.

V. Unsaturated non-cyclic ions

A. The thiazenium ion, SN^+

Compounds containing this ion were first reported by Glemser and Koch[37] in 1971 as resulting from the reaction

$$NSF + XF_5 \xrightarrow[20°]{\text{gas phase}} [SN]^+[XF_6]^- \quad (X = As, Sb)$$

Their structure was deduced from the Raman spectra, which contained known bands of the XF_6^- ion and a cationic N–S stretching vibration corresponding to a bond order of about 2.6 (3 expected for SN^+). The adiabatic first ionization potential of SN is so high (8.87 eV) as to suggest that ionic compounds of SN^+ could only form with a few anions. These SN^+ salts can be used in the synthesis of $S_5N_5^+$ (Section III.D above), and other applications in synthesis are being examined.[60,61]

B. The perthionitrate anion, NS_4^-

The recent discovery of this ion, and the elucidation of its behavior, by Chivers and collaborators represent a major advance in our understanding of sulfur–nitrogen chemistry.

NS_4^- is the intensely blue compound observed in the standard synthesis of sulfur imides in dimethylformamide (Chapter 2, Sec. III.B). One route to

NS_4^- is equilibration of the anion of heptasulfur imide according to the equation

$$S_7N^- \rightleftharpoons NS_4^- + \tfrac{3}{8}S_8 \qquad K_i = 0.13 \pm 0.06 \text{ in DMF at room temp.}$$
(pale yellow)   (dark blue)

S_7N^- results from the dissolution of S_7NH in basic solvents (dimethylformamide, dimethyl sulfoxide, hexamethylphosphoramide) or from treatment of S_7NH with strong bases in tetrahydrofuran. Near room temperature it rapidly equilibrates as above. The tetra-n-butylammonium salt of NS_4^- is obtained as a blue-black powder, m.p. 49°, by allowing the corresponding salt of S_7N^- (precipitated at $-80°$) to stand at room temperature.[26] NS_4^- is also produced by rearrangement of the anions of the hexasulfur diimides (Chapter 2, Sec. III.B.4).[26] A second good route to NS_4^- is the reaction of sodium azide with S_8 in hexamethylphosphoramide or dimethylformamide.[20]

The vibrational spectrum of NS_4^- appears to support the structure **XIV**.

$$S\!=\!\overset{+}{N}\!\!\diagup^{\!\!\!S-S^-}_{\!\!\!S^-}$$

(XIV)

It resembles the spectrum of CS_4^{2-}.[26] [See also note on p. 129.]

Aqueous HCl with solutions of NS_4^- in dimethylformamide or hexamethylphosphoramide yields sulfur and cyclic sulfur imides, principally S_7NH (Chapter 2, Sec. III.B.1). Bojes and Chivers postulate that these imides are formed via their anions, which are thought to exist in equilibrium with NS_4^- in the solution prior to hydrolysis.[20]

References

1. Adkins, R. R. and Turner, A. G., *J. Chromatogr.* **110**, 202 (1975)
2. Adkins, R. R. and Turner, A. G., *Inorg. Chim. Acta* **25**, 233 (1977)
3. Alange, G. G. and Banister, A. J., *J. Inorg. Nucl. Chem.* **40**, 203 (1978)
4. Alford, J. R., Bigg, D. C. H. and Heal, H. G., *J. Inorg. Nucl. Chem.* **29**, 1538 (1967)
5. Bali, A. and Malhotra, K. C., *Aust. J. Chem.* **29**, 1111 (1976)
6. Banister, A. J. and House, J. R., *J. Inorg. Nucl. Chem.* **33**, 3609 (1971)
7. Banister, A. J. and Dainty, P. J., *J. Chem. Soc. A* 2658 (1972)
8. Banister, A. J. and Clarke, H. G., *Inorg. Synth.* **17**, 188 (1977)
9. Banister, A. J., Clarke, H. G., Rayment, I. and Shearer, H. M. M., *Inorg. Nucl. Chem. Lett.* **10**, 647 (1974)
10. Banister, A. J., Durrant, J. A., Rayment, I. and Shearer, H. M. M., *J. Chem. Soc. Dalton Trans.* 928 (1976)
11. Banister, A., *MTP International Review of Science*, Inorganic Chemistry, Series 2, vol. 3, p. 41 (1975)
12. Banister, A. J., private communication

REFERENCES

13. Barker, C. K., Cordes, A. W. and Margrave, J. L., *J. Phys. Chem.* **69**, 334 (1965)
14. Becke-Goehring, M., *Inorg. Synth.* **6**, 128 (1960)
15. Becke-Goehring, M. and Latscha, H. P., *Z. Anorg. Allg. Chem.* **333**, 181 (1964)
16. Becke-Goehring, M., *Inorg. Macromol. Rev.* **1**, 17 (1970)
17. Bertini, V., DeMunno, A. and Marraccini, A., *J. Org. Chem.* **37**, 2587 (1972)
18. Bishop, M. W. and Chatt, J., *J. Chem. Soc. Chem. Commun.* 780 (1975)
19. Bojes, J. and Chivers, T., *Inorg. Nucl. Chem. Lett.* **12**, 551 (1976)
20. Bojes, J. and Chivers, T., *J. Chem. Soc. Dalton Trans.* 1715 (1975)
21. Bojes, J. and Chivers, T., *Inorg. Chem.* **17**, 318 (1978); *J. Chem. Soc. Chem. Commun.* 391 (1978)
22. Bojes, J., Chivers, T., Drummond, I. and MacLean, G., *Inorg. Chem.* **17**, 3668 (1978)
23. Bragin, J. and Evans, M. V., *J. Chem. Phys.* **51**, 268 (1969)
24. Brinkman, R. and Allen, C. W., *J. Am. Chem. Soc.* **94**, 1550 (1972)
25. Chan, C. H. and Olsen, F. P., *Inorg. Chem.* **11**, 2836 (1972)
26. Chivers, T. and Drummond, I., *Inorg. Chem.* **13**, 1222 (1974)
27. Chivers, T. and Fielding, L., *J. Chem. Soc. Chem. Commun.* 212 (1978)
28. Chivers, T. and Proctor, J., *Can. J. Chem.* **57**, 1286 (1979)
29. Cohen, M. J., Garito, A. F., Heeger, A. J., MacDiarmid, A. G., Mikulski, C. M., Saran, M. S. and Kleppinger, J., *J. Am. Chem. Soc.* **98**, 3844 (1976)
30. Dyke, J. M., Morris, A. and Trickle, I. R., *J. Chem. Soc. Faraday Trans. II* **73**, 147 (1977)
31. Englebrecht, A., Mayer, E. and Pupp, C., *Monatsh. Chem.* **95**, 633 (1964)
32. Flues, W., Scherer, O. J., Weiss, J. and Wolmershäuser, G., *Angew. Chem. Int. Ed. Engl.* **15**, 379 (1976)
33. Garcia-Fernandez, H., *Bull. Soc. Chim. France* 1210 (1973)
34. Gieren, A., Dederer, B., Roesky, H. W., Amin, N. and Petersen, O., *Z. Anorg. Allg. Chem.* **440**, 119 (1978)
35. Gillespie, R. J., Ireland, P. R. and Vekres, J. E., *Can. J. Chem.* **53**, 3147 (1975)
36. Gillespie, R. J., Slim, D. R. and Tyrer, J. D., *J. Chem. Soc. Chem. Comm.* 253 (1977)
37. Glemser, O. and Koch, W., *Angew. Chem. Int. Ed. Engl.* **10**, 127 (1971)
38. *Gmelins Handbuch der Anorganischen Chemie*, Part B, Section 3, "Schwefel", Verlag Chemie, Weinheim (1963); see also Adkins, R. and Turner, A., *J. Chromatogr.* **110**, 202 (1975)
39. Goehring, M., *Ergebnisse und Probleme der Chemie der Schwefelstickstoffverbindungen*, Akademie-Verlag, Berlin (1957)
40. Grushkin, B., *Chem. Abstr.* **76**, 40262, 40283 (1972)
41. Hamada, S., Takanashi, A. and Shirai, T., *Bull. Chem. Soc. Japan* **44**, 1433 (1971)
42. Hamada, S., *Bull. Chem. Soc. Japan* **46**, 3598 (1973)
43. Hamada, S., Kudo, Y. and Kawano, M., *Bull. Chem. Soc. Japan* **48**, 719 (1975)
44. Haworth, D. T., Brown, J. D. and Chen, Y., *Inorg. Synth.* **18**, 124 (1978)
45. Heal, H. G., in *Inorganic Sulphur Chemistry*, ed. G. Nickless, Elsevier, Amsterdam (1968)
46. Heal, H. G., *Adv. Inorg. Chem. Radiochem.* **15**, 375 (1972)
47. Heal, H. G. and Ramsay, R. J., *J. Inorg. Nucl. Chem.* **37**, 286 (1975)
48. Holt, E. M. and Holt, S. L., *Chem. Commun.* 1704 (1970)
49. Holt, E. M. and Holt, S. L., *Chem. Commun.* 36 (1973)

50. Holt, E. M., Holt, S. L. and Watson, K. J., *J. Chem. Soc. Dalton Trans.* 1357 (1974)
51. Jolly, W. L. and Becke-Goehring, M., *Inorg. Chem.* **1**, 76 (1962)
52. Jolly, W. L. and Maguire, K. D., *Inorg. Synth.* **9**, 102 (1967)
53. Jolly, W. L. and Lipp, S. A., *Inorg. Chem.* **10**, 33 (1971)
54. Kruss, B. and Ziegler, M. L., *Z. Anorg. Allg. Chem.* **388**, 158 (1972)
55. Labes, M., Love, P. and Nichols, L. F., *Chem. Rev.* **79**, 1 (1979)
56. MacDiarmid, A. G., *J. Am. Chem. Soc.* **78**, 3871 (1956)
57. Maraschin, N. J. and Lugow, R. J., *J. Am. Chem. Soc.* **94**, 8601 (1972)
58. Mayerle, J. J., Kuyper, J. and Street, G. B., *Inorg. Chem.* **17**, 2610 (1978)
59. Mews, R., Wagner, D.-L. and Glemser, O., *Z. Anorg. Allg. Chem.* **412**, 148 (1975)
60. Mews, R., *Adv. Inorg. Chem. Radiochem.* **19**, 185 (1976)
61. Mews, R., *Angew. Chem. Int. Ed. Engl.* **15**, 691 (1976)
62. Mikulski, C. M., Russo, P. J., Saran, M. S., MacDiarmid, A. G., Garito, A. F. and Heeger, A. J., *J. Am. Chem. Soc.* **97**, 6358 (1975)
63. Mock, W. L. and Mehrotra, I., *J. Chem. Soc. Chem. Commun.* 123 (1976)
64. Monteil, Y. and Vincent, H., *Z. Naturforsch. B* **31b**, 537 (1976)
65. Neubauer, D. and Weiss, J. *Z. Anorg. Allg. Chem.* **303**, 28 (1960)
66. Niinisto, L. and Laitinen, R., *Inorg. Nucl. Chem. Lett.* **12**, 191 (1976)
67. Nyburg, S. C., *J. Cryst. Mol. Struct.* **3**, 331 (1973)
68. Olsen, F. P. and Barrick, J. C., *Inorg. Chem.* **12**, 1353 (1973)
69. Padma, D. K., Bhat, V. S. and Murthy, A. R. V., *Inorg. Chim. Acta* **20**, L53 (1976)
70. Padma, D. K. and Murthy, A. R. V., *J. Sci. Ind. Res.* (New Delhi) **35**, 313 (1976)
71. Paul, R. C., Arora, C. L., Kishore, J. and Malhotra, K. C., *Aust. J. Chem.* **24**, 1637 (1971)
72. Paul, R. C., Sharma, R. P. and Verma, R. D., *Indian J. Chem.* **12**, 761 (1974)
73. Paul, R. C., Sharma, R. P. and Verma, R. D., *Indian J. Chem. A* **15A**, 359 (1977)
74. Peake, S. C. and Downs, A. J., *J. Chem. Soc. Dalton Trans.* 859 (1974)
75. Roesky, H. W., private communication (1978)
76. Roesky, H. W., *Angew. Chem. Int. Ed. Engl.* **18**, 91 (1979)
77. Roesky, H. and Dietl, M., *Angew. Chem. Int. Ed. Engl.* **12**, 424 (1973)
78. Roesky, H. W. and Dietl, M., *Chem. Ber.* **106**, 3101 (1973)
79. Roesky, H. and Wiezer, H., *Angew. Chem. Int. Ed. Engl.* **12**, 674 (1973)
80. Roesky, H. W. and Hamza, A., *Angew. Chem. Int. Ed. Engl.* **15**, 226 (1976)
81. Salaneck, W. R., Lin, J. W-p, Paton, A., Duke, C. B. and Ceasar, G. P., *Phys. Rev. B* **13**, 4517 (1976)
82. Salaneck, W. R., Lin, J. W-p and Epstein, A. J., *Phys. Rev. B* **13**, 5574 (1976)
83. Sasaki, Y. and Olsen, F. P., *Can. J. Chem.* **49**, 271 (1971); see also *Chem. Commun.* 1043 (1971)
84. Scherer, O. J. and Wolmershäuser, G., *Chem. Ber.* **110**, 3241 (1977)
85. Smith, R. D., *Chem. Phys. Lett.* **55**, 590 (1978)
86. Steudel, R., Rose, F., Reinhardt, R. and Bradaczek, H., *Z. Naturforsch. B* **32b**, 488 (1977)
87. Street, G. B., Bingham, R. L., Crowley, J. I. and Kuyper, J., *J. Chem. Soc. Chem. Commun.* 464 (1977)
88. Thewalt, U. and Schlingmann, M., *Z. Anorg. Allg. Chem.* **406**, 319 (1974)

89. Turner, A. G. and Mortimer, F. S., *Inorg. Chem.* **5**, 906 (1966)
90. Villena-Blanco, M. and Jolly, W. L., *Inorg. Synth.* **9**, 98 (1967)
91. Weiss, J. and Piechaczek, H., *Z. Naturforsch.* B **18b**, 1139 (1963)
92. Weiss, J., *Fortschr. Chem. Forsch.* **5**, 635 (1966)
93. Williford, J. D., Van Reet, R. E., Eastman, M. P. and Prater, K. B., *J. Electrochem. Soc.* **120**, 1498 (1973)
94. Wolmershäuser, G. and Street, G. B., *Inorg. Chem.* **17**, 2685 (1978)
95. Wynne, K. J. and Jolly, W. L., *Inorg. Chem.* **6**, 107 (1967)
96. Wynne, K. J. and Jolly, W. L., *J. Inorg. Nucl. Chem.* **30**, 2851 (1968)
97. Zborilova, L., Touzin, J., Navratilova, D. and Mrkosova, J., *Z. Chem.* **12**, 27 (1972)
98. Zborilova, L. and Chybova, J., *Z. Chem.* **17**, 103 (1977)

Notes added in proof
(p. 110) The reaction of triphenylphosphine with S_4N_4 also gives the new product $(Ph_3P=N)_2S_4N_4$, with a 1,5-disubstituted S_4N_4 ring (Bojes, J., Chivers, T., MacLean, G., Oakley, R. T. and Cordes, A. W., *Can. J. Chem.* **57**, 3171 (1979)).

(p. 125) S_5N_6 has been prepared in high yield from $S_4N_4Cl_2$ (Chapter 7, Sec. II.B) and $(Me_3SiN)_2S$ (Sheldrick, W. S., Sudheendra Rao, M. N. and Roesky, H. W., *Inorg. Chem.* **19**, 538 (1980)).

(p. 126) By X-ray crystallography, the NS_4^- ion has been shown to have the formula $[S=S-N=S=S]^-$, with a sickle shape (Chivers, T. and Oakley, R. T., *J. Chem. Soc. Chem. Commun.* 752 (1979)).

7
Unsaturated cyclic sulfur nitride *S*-halides and their *S*-derivatives

I. Introduction and history

The parent compounds discussed in this chapter contain only sulfur, nitrogen, and a halogen covalently bonded to sulfur. Compounds containing these elements and oxygen are excluded because already covered in Chapter 5, as are also ionic S–N–halogen compounds in which the halogen is present merely as one of various possible anions associated with an S–N cation containing no halogen (e.g. S_4N_3Cl; Chapter 6, Sec. III.C). Most of the compounds to be described are cyclic, but this account includes a few non-cyclic compounds, such as the thiazyl chloride monomer NSCl, on the grounds that their chemistry is closely bound up with that of the cyclic compounds. Very many non-cyclic N–S–F compounds (see ref. 10) are, however, excluded because their chemistry has few points of contact with that of S–N heterocycles.

The compounds to be described in this chapter are formally unsaturated, the nitrogen atoms being 2-coordinate; in all, the halogen or ligand replacing it is linked to sulfur. The cyclic thiazyl halides $(NSX)_{3,4}$ have as yet very few substitution products; such as exist will be mentioned here. There is, in contrast, a considerable derivative chemistry of the 5-membered S_3N_2 ring; the compounds concerned may be regarded as derived from the parent compound $[S_3N_2Cl]^+Cl^-$, and some of them have actually been made this way. The present chapter seems the most appropriate place to cover these compounds, and they are dealt with in Section IV.

The important parent compounds $(NSCl)_3$ and $[S_3N_2Cl]^+Cl^-$ were first properly characterized by E. Demarçay in 1880–1, though impure S–N–Cl compounds had been observed earlier. Subsequently, sulfur–nitrogen–halogen compounds received only a little attention until about the middle of the present century, when the modern period of systematic study commenced. Notable advances have been made in recent times by the schools of M. Becke-Goehring, O. Glemser, and W. L. Jolly. In particular, Glemser and collaborators have developed almost the entire subject of sulfur–nitrogen–fluorine compounds, now a large branch of chemistry.

II. Thiazyl halides $(NSX)_n$, $(NS)_nX_2$, and $(NS)_3X$, and their substitution products

A. Thiazyl halides: general

Just as several sulfur nitrides can be regarded as polymers of thiazyl, NS (Chapter 6, Sec. II.D), so there exists a class of sulfur–nitrogen–halogen compounds comprising the monomeric thiazyl halides, NSX, their cyclic trimers, and a tetramer, the chemistry of which will compose the first and largest part of this chapter. These compounds have been well characterized and their structures determined. In all of them the halogen is linked to sulfur, which here has an oxidation number of 4; the molecules are composed of NSX units. Mixtures of NSX and NS units occur in the recently described $(NS)_4F_2$ and $(NS)_4Cl_2$, and may be present in the little known compounds $(NS)_3X$ (X = Cl, Br, I).

B. Cyclo-$(NSF)_4$ and cyclo-$(NSCl)_4$; cyclo-$(NS)_4F_2$ and cyclo-$(NS)_4Cl_2$

The tetrameric fluoride has been isolated in the crystalline state. The corresponding chloride has not been isolated, though it has been postulated as an intermediate. It is obviously very unstable, if it exists at all.

The tetramer $(NSF)_4$ [Fig.7.1(c)] appears to be thermodynamically unstable with respect to the trimer, and so can only be made from the preformed S_4N_4 ring, not from $(NSF)_3$ or NSF.[11] S_4N_4 may be treated with diluted fluorine at $-78°$, giving up to 12% of $(NSF)_4$ (Chapter 6, Sec. II.A.4). The preferred and more convenient method, however, is to treat S_4N_4 with AgF_2 in dry carbon tetrachloride; the mixture is warmed slowly to boiling, filtered, and cooled. White needles of $(NSF)_4$ separate in 14% yield.[9] In both methods $(NSF)_3$ is also formed.

$(NSF)_4$ begins to decompose at 128° and melts at 153°. It is sparingly soluble in carbon tetrachloride.[9]

The molecular structure [Fig. 7.1(c)] has been determined by X-ray diffraction.[11] The molecule is a puckered ring. The single ^{19}F n.m.r. signal and zero dipole moment confirm the cyclic structure.[14] In contrast with the trimers $(NSX)_3$ (X = Cl, F) (see Section C below), the N–S bonds in $(NSF)_4$ are unequal, bond distances of 166 and 154 pm alternating round the ring. A simple interpretation would suggest that π bonding is mainly localized in the shorter bonds, though even the longer distances seem consistent with some

CHAPTER 7. SULFUR NITRIDE S-HALIDES

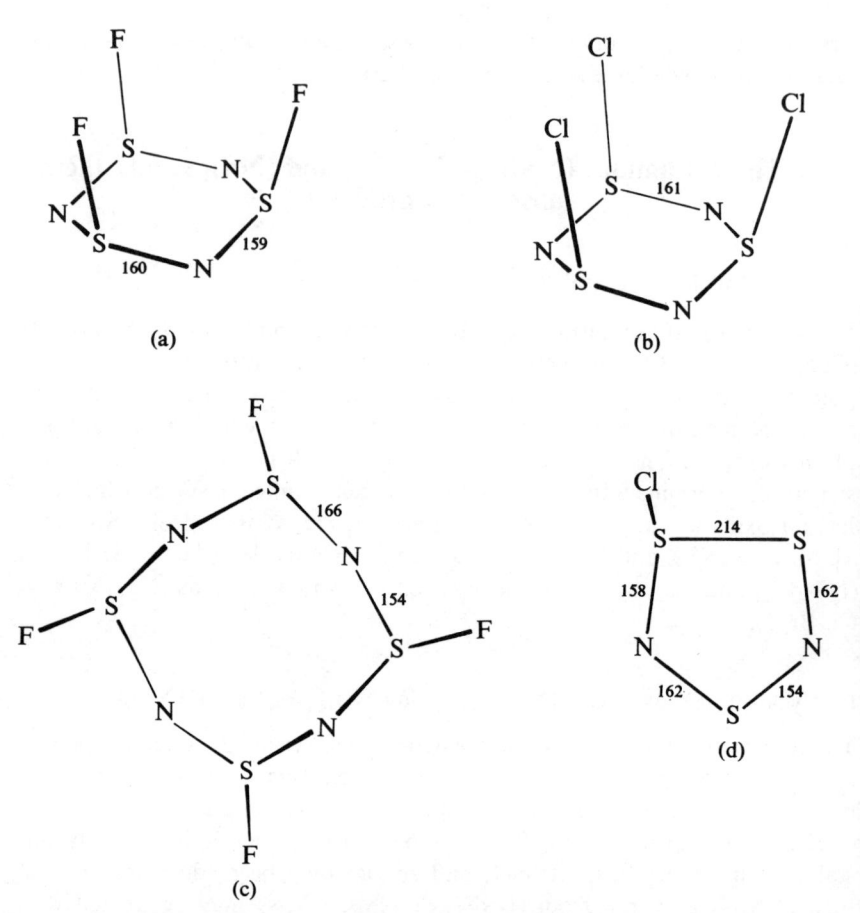

Figure 7.1. Molecular structures of sulfur nitride halides: (a) $(NSF)_3$;[17] (b) $(NSCl)_3$;[11] (c) $(NSF)_4$;[17] (d) $S_3N_2Cl^+$;[7,32] bond distances are in pm.

degree of π bonding (Chapter 9, Sec. II.B). The causes of the alternation of lengths are still being discussed.[7]

The reactions of $(NSF)_4$ have been little studied. It depolymerizes to NSF *in vacuo* at 250°.[10] It forms initially a green solid 1:1 adduct of unknown structure with boron trifluoride; prolonged action of BF_3 at 20° cleaves the ring and gives $NSF \cdot BF_3$.[10] With AsF_5, the salt $[N_4S_4F_3]^+[AsF_6]^-$ may form initially but breaks down to the observed products, $[NS]^+[AsF_6]^-$ and $[N_3S_3F_2]^+[AsF_6]^-$.[18] It is quantitatively hydrolyzed by warm dilute sodium hydroxide solution to ammonia, fluoride, and sulfite.[10] The fluorine atoms can be replaced by chlorine through the action of $SiCl_4$, PCl_5, or $(CH_3)_3$-SiCl, but $(NSCl)_3$ trimer is then formed, not $(NSCl)_4$.[10] With a mixture of oxygen and nitrogen dioxide at 120°, $(NSF)_4$ yields bis(thionylimino)sulfur, $S(NSO)_2$ (Chapter 3, Sec. III.F). Another N–S–O compound, in this case of unknown structure, $S_3N_2O_5$ (Chapter 8, Sec. VI), results from the action of SO_3 on $(NSF)_4$ at 200°.[10] [See note on p. 144 regarding $N_4S_4F_2$.]

$(NSCl)_4$ has not been proved to exist. Early reports of this compound seem to have arisen from equivocal molecular weight determinations on $(NSCl)_3$.[12] The main ultimate product of direct chlorination of S_4N_4 in a solvent is $(NSCl)_3$ (Section C below), but it is clear that a moderately stable intermediate is formed.[1,20,29] Its infrared bands can be observed in the solution during chlorination[20] and in some solid samples[1] of $(NSCl)_3$ prepared by this method. From the relative amounts of S_4N_4 and chlorine consumed in the early stages of chlorination in CCl_4, it has been argued that the intermediate is $N_4S_4Cl_2$, and further evidence[29] suggests two equilibria:

$$S_4N_4 + Cl_2 \rightleftharpoons N_4S_4Cl_2 \qquad (\sim \text{minutes})$$
$$N_4S_4Cl_2 + Cl_2 \rightleftharpoons (NSCl)_3 + NSCl \qquad (\sim \tfrac{1}{2} \text{ h})$$

$(NSCl)_4$ may well be a very short-lived intermediate in the second equilibrium, which is hardly likely to go in one step, but an earlier suggestion[20] that it is the long-lived intermediate observed in chlorination no longer seems tenable. There is a very recent report of the isolation of $N_4S_4Cl_2$.[35]

C. Cyclo-$(NSF)_3$ and cyclo-$(NSCl)_3$

The very similar molecular structures of these two compounds, as determined by X-ray crystallography,[16] are shown in Fig. 7.1 (a) and (b). The rings have a flattened chair form. The halogen atoms are in axial positions, possibly because this enables the lone pairs on sulfur and nitrogen to occupy equatorial positions and so minimize their repulsive mutual interactions,[11] or because the molecule can be stabilized by π_s bonding using equatorial nitrogen lone pairs (cf. Chapter 12, Sec. IX). Both molecules have the halogens in *cis* positions; *trans* isomers are not known. Within each ring the six S–N bond lengths are equal, the length in both fluoride and chloride corresponding to a bond order of about 1.4 [contrast $(NSF)_4$ in Section B above].[11] Accepting the obvious explanation of delocalized π bonding, the question arises whether these are cases of three-center Dewar islands or of more extensive circumannular delocalization. Theoretical studies (Chapter 9, Sec. V.B) favor the former explanation.

$(NSF)_3$ can be made in 90% yield from $(NSCl)_3$ (below) by stirring with AgF_2 in dry carbon tetrachloride at room temperature.[9] After 18–20 h the mixture is filtered, and volatilized and fractionally condensed in a high-vacuum apparatus below room temperature. The $(NSF)_3$ is obtained as colorless strongly refracting crystals, easily sublimed *in vacuo*. $(NSF)_3$ is also formed by allowing NSF (Section F below) to polymerize for 3 days in a sealed glass container.[9]

$(NSF)_3$ melts at 74.2° and boils at 92.5°. It has an appreciable vapor pressure at room temperature. It dissolves readily in carbon tetrachloride or benzene. It is stable in absence of air but turns black in moist air.[10] Its reactions have been little studied.[7] It forms (conveniently in liquid SO_2 solution) 1:1 adducts with BF_3, AsF_5, and SbF_5, low-melting solids, the vibrational spectra of which give evidence for ionic structures such as $[N_3S_3F_2]^+[AsF_6]^-$.[17] $(NSF)_3$ is quantitatively hydrolyzed by cold aqueous NaOH to fluoride, ammonia, and sulfite,[9] but with a little water in a glass vessel it gives some gaseous NSF, HNSO, SO_2, and SiF_4.[10]

$(NSCl)_3$, though prepared by Demarçay nearly a century ago, attracted little attention until the last decade, when its possibilities as an intermediate in synthesis began to be appreciated, thanks largely to Banister. Traditionally, $(NSCl)_3$ has been prepared by treating S_4N_4 with chlorine in carbon tetrachloride at room temperature.[14] The $(NSCl)_3$ seems to form slowly via $N_4S_4Cl_2$ (see Section B above), and overnight standing is required before it crystallizes out. As mentioned above, specimens of $(NSCl)_3$ so prepared may sometimes be contaminated, perhaps with $N_4S_4Cl_2$.[1] However, no method starting from the shock-sensitive and dangerous S_4N_4 (Chapter 6, Sec. II.A) can any longer be recommended, since the following safe and convenient alternative is now available. $[S_3N_2Cl]^+Cl^-$ (Section III.B below) is first prepared, and then chlorinated at room temperature, either by passing chlorine over it:[15]

$$3S_3N_2Cl_2 + 3Cl_2 \rightarrow 2(NSCl)_3 + 3SCl_2$$

or by stirring with sulfuryl chloride:[1]

$$3S_3N_2Cl_2 + 3SO_2Cl_2 \rightarrow 2(NSCl)_3 + 3SCl_2 + 3SO_2$$

The crude $(NSCl)_3$ may be crystallized from dry CCl_4 or SO_2Cl_2. Decomposition occurs if the solutions are heated above 55° or allowed to stand.[1]

$(NSCl)_3$ forms pale yellow moisture-sensitive crystals melting with decomposition at 91°[1] (an earlier report [21] of m.p. 162.5° is wrong), and readily soluble in CCl_4 or benzene.

At temperatures near and above its melting point, $(NSCl)_3$ gives a vapor containing much NSCl,[7] by the equilibrium

$$(NSCl)_3 (g) \rightleftharpoons 3NSCl (g)$$

$\Delta H°$ for the forward reaction has been estimated as 92 ± 12 kJ mol^{-1}, and $\Delta S°$ as 320 ± 40 J mol^{-1} K^{-1}.[21] The dissociation is accompanied by considerable decomposition to S_4N_4, SCl_2, and S_2Cl_2.[29]

A solution of $(NSCl)_3$ in CCl_4 changes color reversibly at around 55° from pale yellow to vivid mint-green.[7,29] NSCl gas is green, and i.r. shows the presence of NSCl in the hot solution, but since the bright green is at once changed to yellow by treatment with chlorine,[7] NSCl is less likely to be the cause of it than some as yet uncharacterized compound formed by partial loss of chlorine from $(NSCl)_3$. Total dechlorination of $(NSCl)_3$ can be achieved with cyclohexene, giving $(SN)_x$ (Chapter 6, Sec. II.C), or by shaking its CCl_4 solution with mercury or tin, giving S_4N_4.[29] S_4N_4 is also formed in high yield when $(NSCl)_3$ is reduced by $S_4(NH)_4$ or S_7NH in CCl_4 in presence of pyridine.[13] Dehalogenation with formation of a non-cyclic product [a sulfur(IV) diimide; see Chapter 3, Sec. IV, and Table 6.2] occurs in the following reaction:[7]

$$(NSCl)_3 + 3(C_6F_5S)_2NH \xrightarrow{py} 3C_6F_5S-N=S=N-SC_6F_5 + 3HCl$$

With its equatorial lone pairs on the ring atoms, $(NSCl)_3$ would be expected to function as a Lewis base, and it does. Adducts with Lewis acids are known as follows: with 1 or 2 $AlCl_3$;[7] with 1 or 2 $FeCl_3$;[7] with 1, 2, or 3 $SbCl_5$;[7,17] with 1 $TiCl_4$;[17] with $\frac{1}{2}$ $SnCl_4$;[17] with 3 or 6 SO_3.[7] All are moisture-sensitive solids. Their structures are not known but some may be ionic, for example $[N_3S_3Cl_2]^+[SbCl_6]^-$;[17] the adducts with $AlCl_3$, $FeCl_3$, and $SbCl_5$ act as sources of NS^+ in the preparation of $S_5N_5^+$ salts (Chapter 6, Sec. III.D). Certain other reactions of $(NSCl)_3$ with Lewis acids lead to ring-fission.[7] Thus BCl_3 gives $[N(SCl)_2]^+[BCl_4]^-$,[7] and the tetrachloroaluminate of the same cation arises from $(NSCl)_3$ and $AlCl_3$ in SCl_2; a hexachloroantimonate is also known.[11] The ion $[N(SCl)_2]^+$ is planar and horseshoe-shaped, with S–N distances corresponding to a bond order of 1.7–1.8, and a wide angle at nitrogen suggesting $p\pi$–$d\pi$ bonding to sulfur (Chapter 9, Sec. III). Warming $(NSCl)_3$ with S_2Cl_2 in absence of a solvent gives $[S_3N_2Cl]^+Cl^-$ (Section III.B below). Interesting but incompletely characterized brown moisture-sensitive solids are formed when $(NSCl)_3$ is heated with the hexacarbonyls of chromium, molybdenum, and tungsten in dichloromethane.[31] The molybdenum compound analyzes fairly well for $Mo(NSCl)_3$; it is insoluble in non-polar solvents, therefore probably polymeric, and reacts violently with water. The chromium and tungsten compounds have variable composition.

$(NSCl)_3$ is not known to form adducts with Lewis bases, but it does undergo a few nucleophilic substitution reactions of its halogen. Its fluorination to $(NSF)_3$ has been mentioned above. The known solvolysis reactions of $(NSCl)_3$ destroy the ring. Thus $(NSCl)_3$ decomposes quickly in damp air, becoming brown. Warm aqueous sodium hydroxide solution hydrolyzes it quantitatively to ammonia and sulfite ion.[12] Ethanol reacts analogously, giving ethyl sulfite, and the reaction of dimethyl sulfoxide is also somewhat analogous:[7]

$$\tfrac{1}{3}(NSCl)_3 + 2Me_2SO \rightarrow [Me_2S=N=SMe_2]^+Cl^- + SO_2$$

(NSCl)$_3$ reacts with nitriles, RCN (R = CCl$_3$, Bu^t, Ph) to give 1,2,3,5-dithiadiazolium chlorides [RCN$_2$S$_2$]$^+$Cl$^-$ containing the new aromatic 5-membered CN$_2$S$_2$ ring.[2]

Esters analyzing as (NSOR)$_n$, probably trimers, are formed as low-melting solids from (NSCl)$_3$ and epichlorohydrin or cyclohexene oxide.[3]

D. The compound (NS)$_3$Br$_2$

D.t.a. and t.g.a. curves attributed to this compound have been reported,[34] but the method of preparation has not been described in the literature. The thermal analysis curves resemble those for the well authenticated compound [S$_4$N$_3$]$^+$Br$_3^-$.[30] It seems very doubtful whether S$_3$N$_3$Br$_2$ really exists.

E. The compounds (NS)$_3$X (X = Cl, Br, I)

These compounds are not well known, but it now seems likely that the chloride, at least, is authentic. (NS)$_3$Cl was reported in the early literature as crystallizing in brick-red or copper-red needles when a mixture of 1 mol (NSCl)$_3$ and 2 mol S$_4$N$_4$ in chloroform is slightly warmed and left to stand.[12] Meuwsen (1932) after some difficulty claimed success with this preparation, as did also Zborilova et al. (1972) who reported a melting point of 165°(dec.) and gave an infrared spectrum and a powder diffraction diagram.[33] On the other hand, Vincent and Monteil (1977) got only [S$_4$N$_3$]$^+$Cl$^-$ from the reaction of (NSCl)$_3$ with S$_4$N$_4$.[29] Further investigation, especially of the absorption spectra of the solutions, is clearly needed.

A report[19] of the formation of (SN)$_3$Br by bromination of S$_4$N$_4$ in CS$_2$ cannot be accepted without further investigation, in view of later work on this reaction (Section III.C below).

(SN)$_3$I precipitates as a dark-red amorphous solid when S$_4$N$_4$ reacts with iodine in boiling inert solvents; the reaction is very slow at room temperature.[19] There is no further information about (SN)$_3$I.

F. Monomeric thiazyl halides

The monomeric thiazyl halides NSX (X = F, Cl, Br) have been characterized. The fluoride and the chloride were first reported by Glemser and collaborators, in 1955 and 1961 respectively, while the bromide has been made by Peake and Downs only recently. These compounds will be briefly described because of their relationship to the cyclic trimers and tetramers.

Several methods are available for the preparation of NSF.[10] It is evolved in good yield from a mixture of S$_4$N$_4$ and excess HgF$_2$ in refluxing carbon tetrachloride, or almost quantitatively from the decomposition

$$\text{Hg(NSF}_2)_2 \xrightarrow[110°]{\text{vacuum}} \text{HgF}_2 + 2\text{NSF}$$

It is a colorless pungent-smelling gas which condenses to a pale-yellow liquid, m.p. $-89°$, b.p. $0.4°$. Liquid NSF trimerizes in a matter of days. The gas at near atmospheric pressure can be kept for days in copper, but decomposes within hours in glass, giving S_4N_4, $S_3N_2F_2$ (Section III.C below), etc.[10] NSF is hydrolyzed by water vapor to thionyl imide (Chapter 3, Sec. III.A). Its other reactions, which have been little studied, include 1:1 addition to BF_3 and SbF_5, and to CsF giving the ionic compound $Cs^+[NSF_2]^-$.[10] The fluorine atom in NSF can readily be replaced by the bis(trifluoromethyl) nitroxyl radical by reaction with $Hg[ON(CF_3)_2]_2$, or by reaction with the radical $(CF_3)_2NO$ itself in presence of mercury.[8]

NSCl is formed as a greenish-yellow gas by heating $(NSCl)_3$ in a vacuum or in an inert gas stream.[21] The conditions required are best determined for specific situations by reference to published data on the $NSCl/(NSCl)_3$ equilibrium.[21] This is probably the best method of preparation for substantial quantities, but other preparations are available, such as the pyrolysis of $[S_4N_3]^+Cl^-$.[22] NSCl has usually been made as a reaction intermediate (Chapter 6, Sec. II.A) or in small amounts for structural studies,[22] so its reactions have received little direct study. The trimerization takes days in CS_2 near room temperature.[20]

NSBr has only been made on a very small scale and trapped in an argon matrix for spectral studies.[22] The method used was pyrolysis of $[S_4N_3]^+Br^-$. NSI is unknown; it did not result from the pyrolysis of $[S_4N_3]^+I^-$.[22]

NSF, NSCl, and NSBr are all bent molecules with sulfur as central atom.[22] The infrared spectra, and the microwave spectrum of NSF, have been reported and the molecular dimensions are known. The N–S bonds are very short (about 145 pm) and presumably triple.[22]

G. Substitution products of the cyclic thiazyl halide oligomers

It might be imagined that the cyclic thiazyl halides $(NSX)_{3,4}$ would serve as the parents of a wide range of substitution products. In fact very few such are known. Inadequate investigation is part of the explanation, but in addition there are two fundamental constraints on substitution in these molecules. First, covalent sulfur($+4$) is only stable in association with at least a few very electronegative ligands, as recent work on the sulfuranes has confirmed.[27] Second, the stability of the unsaturated S–N ring itself depends critically on the nature of the exocyclic atoms; thus all attempts to make cyclo-$(NSBr)_3$ have failed, though the chloride and especially the fluoride are rather stable.

As indicated in Sections B and C above, attempts to replace the halogen in $(NSX)_{3,4}$ directly by nucleophilic substitution have led to breakdown of the rings. The best-characterized compound which can be regarded as a substitution product of these halides is obtained otherwise. It is the bis(trifluoromethyl) nitroxide derivative [NS·ON(CF$_3$)$_2$]$_4$.[8] This tetramer (formula IV in Chapter 6) arises from the action of (CF$_3$)$_2$NO for 16 hours at room temperature on S$_4$N$_4$. It is a colorless crystalline solid, decomposing at 65°, and not attacked by cold water though hydrolyzed by warm aqueous NaOH. A less well characterized compound thought to be the trimer is formed when the monomer NSON(CF$_3$)$_2$ stands at room temperature[8] (for preparation of the monomer, see Section F above). The tetrameric and trimeric formulas have been reasonably well established by mass spectrometry, and i.r. data exist for the tetramer.[8] Addition of epoxides to (NSCl)$_3$ gives products which analyze as triesters of the hypothetical acid (NSOH)$_3$, but these require fuller characterization[3,7] (see also Section C above).

The compound Ph$_3$PNS$_3$N$_3$ (Chapter 6, Sec. II; Fig. 6.4) unquestionably contains an S$_3$N$_3$ ring with a ligand on one sulfur atom, and may perhaps be a derivative of (NS)$_3$Cl (Section E above) the structure of which is, however, unknown.

As indicated above, the fully alkylated derivatives (NSR)$_{3,4}$ are not likely to be stable, but related compounds believed to have structure I have been described; they can be regarded as disulfonium salts, with the constitution of adducts of 2 MeBr with compounds of the type N$_4$S$_4$R$_2$Me$_2$. They result from

$$\begin{bmatrix} & R & \\ N\!\!-\!\!S\!\!-\!\!N & \\ \| & & \| \\ Me_2S & & SMe_2 \\ \| & & \| \\ N\!\!-\!\!S\!\!-\!\!N & \\ & R & \end{bmatrix}^{2+} 2Br^-$$

$$\begin{bmatrix} & Ph_2 & \\ N\!\!-\!\!P\!\!-\!\!N & \\ \| & & \| \\ Me_2S & & SMe_2 \\ \| & & \| \\ N\!\!-\!\!P\!\!-\!\!N & \\ & Ph_2 & \end{bmatrix}^{2+} 2Br^-$$

$$\begin{matrix} & Ph_2 & \\ N\!\!=\!\!P\!\!-\!\!N & \\ | & & \| \\ MeS & & SMe \\ \| & & | \\ N\!\!-\!\!P\!\!=\!\!N & \\ & Ph_2 & \end{matrix}$$

(I) (R = Me, Et, benzyl)　　　　(II)　　　　(III)

the reaction of Me$_2$S(NBr)$_2$ with R–S–S–R, and form colorless crystals soluble in water without decomposition.[4] The P–N–S ring compound II can be made analogously to I, and does lose methyl bromide on gentle heating to give III,[7] a typical "onium salt" reaction. See also Chapter 4, Sec. VI.A, regarding these compounds.

The dimeric fluoride and chloride (NSX)$_2$ are unknown, but a "derivative" which may have the structure IV has been reported.[7] It is a colorless liquid produced by the photodecomposition of N(SCF$_3$)$_3$.

$$CF_3\!-\!S\diagdown^{\displaystyle N}\!\!\diagup\!\!S\!-\!CF_3$$
$$\diagup_{\displaystyle N}\diagdown$$

(IV)

III. The sulfur nitride halides $S_3N_2X_2$

A. Introduction and history

By far the most thoroughly investigated compound in this group is thiodithiazyl dichloride, $S_3N_2Cl_2$, really an ionic compound $[S_3N_2Cl]^+Cl^-$ with a cyclic cation. It was first made in an impure state about 1851 by Fordos and Gelis.[12] For a long time it attracted little attention, receiving only a few lines in Goehring's 1957 book,[13] but of late a strong interest has arisen because of the finding that it can be used in several syntheses of sulfur–nitrogen compounds as a safer intermediate than the explosive S_4N_4.

The fluoride $S_3N_2F_2$ was first reported from Glemser's laboratory in 1959, but very little work has been done on it since. It is almost certainly covalent, i.e. structurally unlike the chloride.

The question of existence of $S_3N_2Br_2$ and $S_3N_2I_2$ is still open.

B. Preparation and properties of $[S_3N_2Cl]^+Cl^-$

The traditional method of preparation is to warm $(NSCl)_3$ (Section II.C above) gently with S_2Cl_2 until dissolved, and leave the solution to crystallize. Red-brown crystals of $[S_3N_2Cl]^+Cl^-$ then separate in good yield. Nothing is known of the mechanism or of the other products.[12]

This method is inconvenient because $(NSCl)_3$ is not sold; indeed today it is usually made from $[S_3N_2Cl]^+Cl^-$. Jolly and Maguire report a simple alternative synthesis for $[S_3N_2Cl]^+Cl^-$ in one step from common materials.[15] S_2Cl_2 is refluxed with ammonium chloride under an air-condenser. The overall equation is

$$4S_2Cl_2 + 2NH_4Cl \rightarrow [S_3N_2Cl]^+Cl^- + 8HCl + 5S$$

Urea can be used in this preparation instead of ammonium chloride.[26] The $[S_3N_2Cl]^+Cl^-$ collects in the air-condenser as well formed rust-colored crystals. It is so easy to make in this way that it serves as a convenient intermediate for the preparation of $[S_4N_3]^+Cl^-$ and $[S_6N_4]^{2+}Cl_2^-$ (Chapter 6, Secs. III.C and E) and for $(NSCl)_3$ (Section II.C above).

The tetrachloroaluminate and tetrachloroferrate can be prepared by stirring the chloride for a day at 20° with a solution of $AlCl_3$ or $FeCl_3$ in thionyl chloride.[7]

X-ray diffraction studies of the chloride[32] and tetrachloroferrate[7] have given the structure of the $S_3N_2Cl^+$ ion [Fig. 7.1(d)]. It is a slightly puckered 5-membered ring with an exocyclic chlorine atom attached to sulfur and lying to one side of the ring. The presence of a covalently bonded chlorine atom is confirmed by n.q.r.[6] The S–N bond lengths are again short and indicate considerable π bonding (see Chapter 9, Sec. V.B).

$[S_3N_2Cl]^+Cl^-$ is a rust-brown crystalline solid, which melts with decomposition at 90° in a sealed capillary. It is insoluble in common organic solvents. It is instantly hydrolyzed by water;[15] controlled hydrolysis gives S_3N_2O [26] (see also Chapter 8, Sec. IV).

When heated in air, the chloride can explode with a flash of light,[13] but it decomposes quietly when carefully heated *in vacuo*:[15]

$$3[S_3N_2Cl]^+Cl^- \xrightarrow{90°} [S_6N_4]^{2+}Cl_2^- + 2NSCl + SCl_2$$

The dark-green product $[S_6N_4]^{2+}Cl_2^-$ (Chapter 6, Sec. III.E) is also formed by reducing $[S_3N_2Cl]^+Cl^-$ with anhydrous formic acid at room temperature.[6] Treatment of $[S_3N_2Cl]^+Cl^-$ with chlorine at room temperature produces $(NSCl)_3$ quantitatively;[15] this is the best preparation of $(NSCl)_3$ (Section II.C above). Its reaction with S_2Cl_2 has been mentioned (Chapter 6, Sec. II.A.1). With phenylmagnesium bromide it gives the linear compound $Ph_2N_2S_3$ (Table 6.2).[5]

Could the chlorine atom in $[S_3N_2Cl]^+$ be replaced by other groups without destroying the ring? Some recent results show that it can. Reaction of $[S_3N_2Cl]^+Cl^-$ with the phosphazene derivative $N_3P_3F_5N(SnMe_3)_2$ gives **V**, the structure of which has been confirmed by X-ray analysis,[23] while a similar

(V)   (VI)

reaction with the cyanuric derivative $N_3C_3F_2N(SnMe_3)_2$ gives a product formulated as **VI**.[23]

Derivatives **V** and **VI** are examples of an increasing class of covalent compounds containing the S_3N_2 ring, bonded from one of its sulfur atoms to an exocyclic group. These compounds are listed in Table 7.1 and further discussed below in Section IV.

C. Other compounds $S_3N_2X_2$ (X = halogen) of unknown structure

The preparation of $S_3N_2Br_2$ has been claimed[19,33] by the action of bromine on S_4N_4 in CS_2 or CCl_4, but recent work on this reaction in CS_2 has shown the products to be $[S_4N_3]^+Br^-$, $[S_4N_3]^+Br_3^-$, and a new compound $CS_3N_2Br_2$ which was presumably mistaken for $S_3N_2Br_2$ in earlier work.[30] Neither $S_3N_2Br_2$ nor $S_3N_2I_2$ is well established.

There is a very inaccessible compound $S_3N_2F_2$ which seems not to have been investigated recently. It is produced in small yield when NSF stands at room temperature and 600 Torr in a glass flask.[9,10] After a week, green crystals will have formed on the walls. These are removed and sublimed in high vacuum, giving (if the operation is done at 40°) yellow-green platelets or (at 65°) bright-green pointed crystals, both apparently polymorphs of $S_3N_2F_2$. The compound melts at 83° and explodes above 100°. With moisture it turns black, and alkaline hydrolysis converts the nitrogen quantitatively into ammonia. The compound is soluble in CCl_4, giving a yellowish-green solution with an absorption maximum at about 375 nm. This solubility rules out the structure $[S_3N_2F]^+F^-$. Covalent structures such as **VII** have been suggested (cf. Table 6.2). The ring structure **VIII** now seems rather likely in view of the recent discovery of its oxygen analog **IX** (Chapter 8, Sec. IV).

$$F-S-N=S=N-S-F$$
(VII)

(VIII)   (IX)

IV. Covalent compounds containing the five-membered S_3N_2 ring

A. Introduction

As stated above in Sections I and III.B, the chloride $[S_3N_2Cl]^+Cl^-$ can be regarded as the parent of a variety of covalent compounds containing the S_3N_2 ring. Compounds in this category known up to 1977 are listed in Table 7.1. The possibility that many such compounds might exist seems first to have been appreciated by Roesky and Janssen[24] about 1975, though the $F_2P(O)N=$ derivative had been made three years earlier (and a 6-membered ring structure erroneously ascribed to it). In 1977 Steudel and coworkers[28] published the characteristic vibrational wavenumbers of the S_3N_2 ring and so provided an easy way of recognizing its presence in new compounds.

B. Preparation of S_3N_2 derivatives

Each of the following methods has been used for more than one compound.

(1) Substitution of the chlorine in $[S_3N_2Cl]^+Cl^-$ has been achieved with several trimethylstannylamino derivatives (cf. Chapter 1, Sec. III.D). A representative reaction of this kind is with the bis(trimethylstannylamino) derivative of cyclodifluorophosphazene trimer (Chapter 12, Sec. VII.D):[23]

$$[S_3N_2Cl]^+Cl^- + N_3P_3F_5N(SnMe_3)_2 \rightarrow 2Me_3SnCl + N_3P_3F_5N=S_3N_2$$

Table 7.1. Derivatives of the $\begin{array}{c} S=N \\ \| \quad \diagdown \\ N-S \end{array} S=$ ring.

Ligand	Reference	Ligand	Reference
\multicolumn{4}{c}{*Derivatives with well or fairly well established structures*}
=Cl$^+$	Chap. 7, Sec. III.B	=NSO$_2$F	23
=O	Chap. 8, Sec. IV	=NSO$_2$CF$_3$	26, 28
=NCOCF$_3$	28	=NSO$_2$C$_4$F$_9$	28
=NCOCCl$_3$	28	=NP$_3$N$_3$F$_5$	23; Chap. 7, Sec. III.B
=NOPF$_2$	28	=NP$_4$N$_4$F$_7$	24
=NC$_3$N$_3$F$_2$	23; Chap. 7, Sec. III.B	=NP$_5$N$_5$F$_9$	24
\multicolumn{4}{c}{*Possible S$_3$N$_2$ derivatives needing further investigation*}
=S*	Chap. 6, Sec. III.B	(–F)$_2$	Chap. 7, Sec. III.C

* Since no crystallographic structure determination has been reported for S$_4$N$_2$, it seems advisable to draw attention here to the possibility of its being a member of the class of S$_3$N$_2$= ring compounds, but as indicated in Chapter 6, Sec. III.B, present evidence is against this; in particular, the i.r. spectrum does not accord with Steudel's list[28] of characteristic S$_3$N$_2$ ring vibrations.

The product has structure **V**. Its homologs containing 8-membered and 10-membered phosphazene rings (Table 7.1) are made similarly,[24] as is the cyanuric derivative **VI**.

The hydrolysis of [S$_3$N$_2$Cl]$^+$Cl$^-$ by formic acid or acetic anhydride[26] gives S$_3$N$_2$O (Chapter 8, Sec. IV).

(2) From the oxide S$_3$N$_2$O (Chapter 8, Sec. IV), by reaction with isocyanate or sulfinylamine groups:[25]

$$S_3N_2O + CF_3SO_2N=C=O \rightarrow CF_3SO_2N=S_3N_2 + CO_2$$
$$S_3N_2O + XSO_2N=S=O \rightarrow XSO_2N=S_3N_2 + SO_2 \qquad (X = F, CF_3)$$

(3) From tetrasulfur tetranitride, by reaction with carboxylic acid anhydrides.[28] This reaction takes place readily in methylene chloride, near room temperature, with (CF$_3$CO)$_2$O and (CCl$_3$CO)$_2$O, giving moderate yields of CX$_3$CON=S$_3$N$_2$ (X = F, Cl) and other products not identified.

C. Properties of S$_3$N$_2$ derivatives

Most of these derivatives form yellow or brown crystals which can be sublimed at low pressure. A few, notably S$_3$N$_2$O (Chapter 8, Sec. IV), are liquids distillable *in vacuo*. Apart from S$_3$N$_2$O, their chemistry has been little studied. The CF$_3$CON= derivative decomposes in moist air.[28]

D. Structures of S_3N_2 derivatives

X-ray crystallographic studies of compound V,[23] of the $FSO_2N=$ derivative,[23] and of the $CF_3CON=$ derivative[28] have been reported. All these molecules contain a puckered S–N ring resembling that in $[S_3N_2Cl]^+Cl^-$, but with longer S–S bonds; longer indeed than most S–S single bonds. The exocyclic groups in all cases lie far out of the mean plane of the ring.

V. Sulfur–nitrogen halides of unknown structure

When sulfur and ammonium chloride are heated together at 280° ± 20° in presence of thiourea, there is formed a brown silky powder with the empirical formula $(S_5N_4Cl)_n$. The cryoscopic molecular weight of this compound and a product obtained from it by pyrolysis at 400–500° correspond to $S_{10}N_8Cl_2$ and $S_6N_2Cl_2$ respectively. I.r. data have been reported but the structures are very speculative.[7]

References

1. Alange, G. G., Banister, A. J. and Bell, B., *J. Chem. Soc. Dalton Trans.* 2399 (1972)
2. Alange, G. G., Banister, A. J., Bell, B. and Millen, P. W., *Inorg. Nucl. Chem. Lett.* **13**, 143 (1977)
3. Alange, G. G., Banister, A. J. and Bell, B., to be published.
4. Appel, R., Hänssgen, D. and Müller, W., *Chem. Ber.* **101**, 2855 (1968)
5. Banister, A. J. and House, J. R., *J. Inorg. Nucl. Chem.* **33**, 4057 (1971)
6. Banister, A. J., Clarke, H. G., Rayment, I. and Shearer, H. M. M., *Inorg. Nucl. Chem. Lett.* **10**, 647 (1974)
7. Banister, A., *MTP International Review of Science*, series 2, vol. 3, p. 41 (1975)
8. Eméleus, H. J. and Poulet, R. J., *J. Fluorine Chem.* **1**, 13 (1971)
9. Glemser, O., *Prep. Inorg. Reactions* **1**, 227 (1964)
10. Glemser, O. and Mews, R. *Adv. Inorg. Chem. Radiochem.* **14**, 333 (1972)
11. Glemser, O., *Z. Naturforsch.* B **31b**, 610 (1976)
12. *Gmelins Handbuch der Anorganischen Chemie*, Teil B, "Schwefel", Verlag Chemie, Weinheim (1963)
13. Goehring, M., *Ergebnisse und Probleme der Chemie der Schwefelstickstoffverbindungen*, Akademie-Verlag, Berlin (1957)
14. Heal, H. G., in *Inorganic Sulphur Chemistry*, ed. G. Nickless, Elsevier Amsterdam (1968)
15. Jolly, W. L. and Maguire, K. D. *Inorg. Synth.* **9**, 102 (1967)
16. Krebs, B., Pohl, S. and Glemser, O., *J. Chem. Soc. Chem. Commun.* 548 (1972)
17. Mews, R., Wagner, D.-L. and Glemser, O., *Z. Anorg. Allg. Chem.* **412**, 148 (1975)

18. Mews, R., *Adv. Inorg. Chem. Radiochem.* **19**, 185 (1976)
19. Monteil, Y. and Vincent, H., *Z. Naturforsch. B* **31b**, 673 (1976)
20. Nelson, J. and Heal, H. G., *Inorg. Nucl. Chem. Lett.* **6**, 429 (1970)
21. Patton, R. C. and Jolly, W. L., *Inorg. Chem* **9**, 1079 (1970)
22. Peake, S. C. and Downs, A. J., *J. Chem. Soc. Dalton Trans.* 859 (1974)
23. Roesky, H. W., *Z. Naturforsch. B*, **31b**, 680 (1976)
24. Roesky, H. W. and Janssen, E., *Chem. Ber.* **108**, 2531 (1975)
25. Roesky, H. W., Holtschneider, G., Wiezer, H. and Krebs, B., *Chem. Ber.* **109**, 1358 (1976)
26. Roesky, H. W., Schaper, W., Petersen, O. and Müller, T., *Chem. Ber.* **110**, 2695 (1977)
27. Schwenzer, G. and Schaefer, H., *J. Am. Chem. Soc.* **97**, 1393 (1975)
28. Steudel, R., Rose, F., Reinhardt, R. and Bradaczek, H., *Z. Naturforsch. B* **32b**, 488 (1977)
29. Vincent, H. and Monteil, Y., *Synth. React. Inorg. Met.-Org. Chem.* **8**, 51 (1978)
30. Wolmershäuser, G., Street, G. B. and Smith, R. D., *Inorg. Chem.* **18**, 383 (1979)
31. Wynne, K. J. and Jolly, W. L., *J. Inorg. Nucl. Chem.* **30**, 2851 (1968)
32. Zalkin, A., Hopkins, T. E. and Templeton, D. H., *Inorg. Chem.* **5**, 1767 (1966)
33. Zborilova, L., Touzin, J., Navratilova, D. and Mrkosova, J., *Z. Chem.* **12**, 27 (1972)
34. Zborilova, L. and Chybova, J., *Z. Chem.* **17**, 103 (1977)
35. Zborilova, L. and Gebauer, P., *Z. Chem.* **19**, 32 (1979)

Note added in proof
(p. 133) The fluoride $N_4S_4F_2$ has recently been reported from fluorination of $N_4S_4Cl_2$ (Zborilova, L. and Gebauer, P., *Z. Anorg. Allg. Chem.* **448**, 5 (1979)).

8
Unsaturated cyclic sulfur nitride *S*-oxides and *S*-oxide ions

I. Introduction and history

The compounds discussed in this chapter are only a few of the known compounds made up exclusively of sulfur, nitrogen, and oxygen, as a glance at *Gmelin* will show. Some S–N–O compounds, such as the nitrosonium sulfates, are outside the scope of this book, while others are mentioned elsewhere in it, namely, the saturated cyclic compound $(S_7N)_2SO$ (Chapter 2), the noncyclic $S(NSO)_2$ (Chapter 3, Sec. III.F), and the SO_3 adduct of S_4N_4 (Table 6.1). Our present concern is with unsaturated S–N rings and cages resembling those described in Chapters 6 and 7 but with exocyclic oxygen ligands. This chemistry is of recent date, beginning with the isolation in 1953, by Meuwsen and Lösel, of a red oil with the formula S_3N_2O and probably cyclic in its molecular structure. Another S–N–O compound reported by Goehring and Heinke in the same year, $S_3N_2O_2$, was at first thought to be cyclic but later shown to have the non-cyclic structure $S(NSO)_2$ (Chapter 3, Sec. III.F). Progress in this field was slow until the late 1960s, since when X-ray crystallography and new methods of synthesis have led to important advances.

The compounds will be classified here according to the size of the S–N ring; some have 8-membered rings and are structurally related to S_4N_4 (Chapter 6), others 6-membered rings related structurally to the sulfanuric compounds (Chapter 5); and there is at least one compound with a 5-membered ring which can be regarded as a structural relative of $[S_3N_2Cl]^+$ (Chapter 7). Finally there are compounds of unknown, but very possibly cyclic, structure.

As with the sulfur nitride halides discussed in Chapter 7, the present compounds invariably have the exocyclic group (in this case oxygen) attached to sulfur, and the ring nitrogen atoms are always 2-coordinate, showing the rings to be unsaturated.

II. Compounds with eight-membered S–N rings

A. $S_4N_4O_4$

This compound, described as "the first oxide of tetrasulfur tetranitride", has been made from the versatile synthetic reagent bis(trimethylsilyl) sulfur diimide (Chapter 3, Sec. IV.C) by the reaction[4]

$$2(Me_3SiN)_2S + 2O(SO_2F)_2 \xrightarrow{CCl_4} S_4N_4O_4 + 2Me_3SiF + 2Me_3SiOSO_2F$$

The yield of purified $S_4N_4O_4$ was only 5%. The preparation did not succeed when sulfuryl chloride was used instead of fluorosulfuric acid anhydride.

(I)

$S_4N_4O_4$ is a yellow solid which sublimes *in vacuo* at 80–90° and decomposes at 115–120°. The structure **I** has been suggested on the evidence of the infrared and mass spectra; the crystal structure has not yet been reported.

In presence of moisture, $S_4N_4O_4$ turns brown and decomposes with loss of SO_2.

B. $S_4N_4O_2$

The synthesis of this compound, reported in 1975, represents yet another synthetic application of bis(trimethylsilyl) sulfur diimide. The diimide (Chapter 3, Sec. IV.C) reacts with $FSO_2N{=}S{=}O$ (Chapter 3, Sec. III.G) in methylene chloride to give the molecular compound $S_4N_4O_2$, and also an ionic compound with the same empirical formula, $[S_5N_5]^+[S_3N_3O_4]^-$ (Section III.A below).[7]

Another and more convenient synthesis of $S_4N_4O_2$ has been described.[9] $[S_3N_2Cl]^+Cl^-$ (Chapter 7, Sec. III.B) is refluxed in carbon tetrachloride with sulfamide, $SO_2(NH_2)_2$. This procedure might have been expected to produce $SO_2(NS_3N_2)_2$ (cf. Chapter 7, Sec. IV), but instead the oxide $S_4N_4O_2$ results in 50% yield.

Molecular $S_4N_4O_2$ occurs as yellow-orange needles, melting with decomposition at 166–168°. The molecular structure, from X-ray crystallography,[8] is shown in Fig. 8.1(a), and consists of an alternant S_4N_4 ring with two oxygen atoms attached to one sulfur. The structure is discussed in Chapter 9, Sec. V.B.

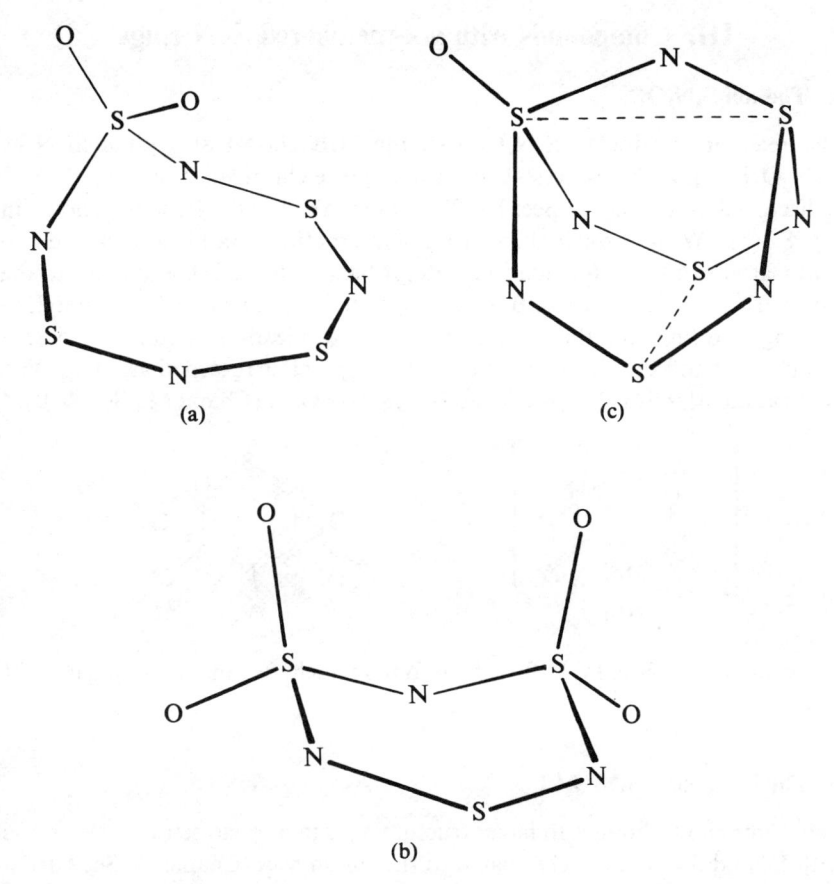

Figure 8.1. Molecular structures of sulfur nitride oxides: (a) $S_4N_4O_2$;[8] (b) $S_3N_3O_4^-$;[8] (c) $S_4N_5O^-$.[11]

The reactions of $S_4N_4O_2$ are now being investigated.[13] With $(R_3Si)_3N$ it undergoes ring contraction, yielding **II** and the sulfur diimide $(R_3SiN)_2S$. The tin homolog of **II** can be made similarly. With excess $(Me_3Sn)_3N$, however, the 8-membered ring **III** is formed.

III. Compounds with six-membered S–N rings

A. The ion $S_3N_3O_4^-$

The reaction producing $S_4N_4O_2$ (Section II.B above) also gives $[S_5N_5]^+$ $[S_3N_3O_4]^-$, violet-black crystals from methylene chloride.[7] The cation $S_5N_5^+$ is discussed in Chapter 6, Sec. III. The anion (IV) has the structure shown in Fig. 8.1(b).[8] With its 6-membered ring of alternating S and N atoms, and two sulfur atoms in the +6 oxidation state, it bears a formal resemblance to the sulfanuric rings (Chapter 5). But two opposite S–N bonds are long enough to be single, ruling out the circumannular delocalization postulated for the sulfanurics; and the ring π bonding is perhaps better regarded as comprising two Dewar islands (–N–S–N–) and (–O$_2$S–N–SO$_2$–) (Chapter 9, Sec. V.B).

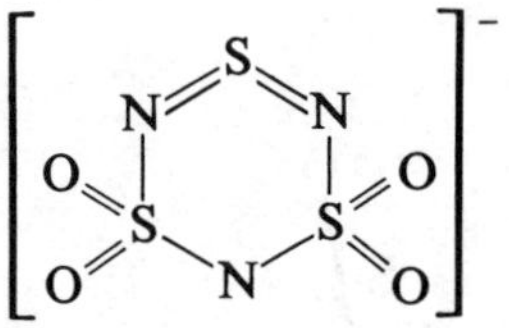

(IV) 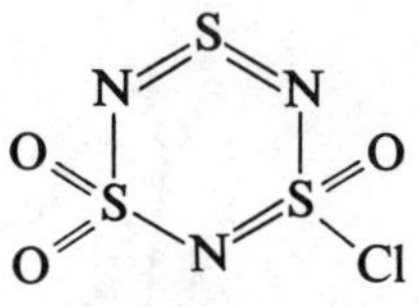   (V)

See Chapter 4, Sec. IV.B, for a compound which may be the conjugate acid of **IV**.

B. The compound $S_3N_3O_2Cl$

This compound, thought to have structure **V**, is in a broad sense isoelectronic with **IV** and likewise closely related to the sulfanurics (Chapter 5, Sec.III.D). It is produced by the photochemical decomposition of ClO$_2$S–N=S=O,[3] which can be made by treating FO$_2$S–N=S=O with BCl$_3$ at 35°.

$S_3N_3O_3Cl$ is a yellow, vacuum-sublimable solid, m.p. 105–108°, soluble in benzene and acetonitrile and readily hydrolyzed by moisture. Structure **V** has been suggested on the evidence of the mass spectrum and infrared spectrum; there are no crystallographic data.

IV. Compounds with five-membered S–N rings: the oxide S_3N_2O

Two sulfur nitride oxides formulated as S_3N_2O have been reported, one by Meuwsen and Lösel[2] in 1953 and the other by Roesky and Wiezer[5] in 1975. It is not clear whether these reports refer to the same compound. Both products are described as red liquids which do not wet glass, but Roesky's compound

IV. FIVE-MEMBERED RINGS

was stable for weeks at room temperature whereas Meuwsen's decomposed in 5 hours at 25°. Perhaps the difference arose from impurities in Meuwsen's sample.

Meuwsen's material resulted from the reaction of thionyl chloride with $Hg_5(NS)_8$ (cf. Chapter 6, Sec. III.B, and Chapter 2, Sec. III.B.5). There is no further report on this reaction.

Roesky's compound, with formula **VII**, is, in contrast, well characterized. It is best prepared by the "hydrolysis" of a suspension of $[S_3N_2Cl]^+Cl^-$ (Chapter 7, Sec. III.B) in methylene chloride by gradual addition of formic acid or acetic anhydride.[9] The product is purified by distillation (b.p. 50°/0.01 Torr). The original method of preparation is far less convenient but worth mentioning for its mechanistic interest: compound **VI**, when treated with

SOF_2, gave **VII**.[5] This substitution of one ring member by another is an unfamiliar type of reaction. A similar reaction of **VI** with COF_2 gives the ketone **VIII**,[6] and no doubt the principle could be further exploited.

The structure **VII** suggested for S_3N_2O is based on its mass spectrum and infrared spectrum,[5,12] and is supported by a crystallographic study of the derivative **IX**[8] which results from the action of $FSO_2N=S=O$ (Chapter 3, Sec. III.G) on S_3N_2O:

$$FSO_2N{=}S{=}O + S_3N_2O \rightarrow IX + SO_2$$

Another similar derivative formulated as **X** can be obtained in either of two ways:[8]

$$CF_3SO_2N{=}S{=}O + S_3N_2O \rightarrow X + SO_2$$
$$CF_3SO_2N{=}C{=}O + S_3N_2O \rightarrow X + CO_2$$

Compounds containing the S_3N_2 ring are more fully described in Chapter 7, Sec. IV.

V. A sulfur nitride oxide ion based on the S_4N_4 cage: $S_4N_5O^-$

The ammonium salt of this ion is one of several products obtained when thionyl chloride dissolved in pyridine is added to excess liquid ammonia at $-78°$.[10] $[NH_4]^+[S_4N_5O]^-$ separates from evaporating aqueous solutions in yellow crystals. It is soluble also in pyridine. It can be heated to 100° without decomposition. The aqueous solution has pH 5 and conducts electricity well. An insoluble yellow silver salt precipitates on adding silver sulfate to an aqueous solution of the ammonium salt, and a sodium salt has been made from the latter by ion exchange. Infrared spectra of the ammonium and silver salts and the electronic spectrum of the aqueous ion have been reported.[10]

X-ray crystallography of the ammonium salt gave the structure shown in Fig. 8.1(c).[11] The ion has a cage structure closely resembling that of S_4N_4 (Fig. 6.2) and related to the structure of the recently discovered $S_4N_5^-$ ion (Chapter 6, Sec. III.G). The $S_4N_5O^-$ structure is discussed in Chapter 9.

Acidification of an aqueous solution of the ammonium salt produces an unstable violet compound which may be the parent acid $H[S_4N_5O]$.[10] It would be desirable to investigate the relationship of this to the presumed NSO^- ion (Chapter 3, Sec. III.G) which is also reported from the $SOCl_2-NH_3$ reaction.

VI. Sulfur nitride oxides of unknown structure

A compound of formula $S_3N_2O_5$, to which structure **XI** was attributed on the evidence of hydrolysis and other reactions, was reported in 1954.[1] It was made

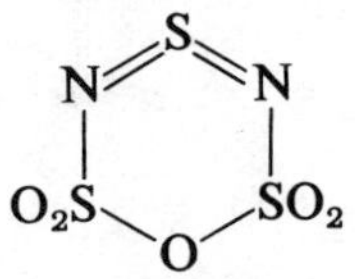

$$\begin{array}{c} S \\ N \diagup \diagdown N \\ | \quad \quad | \\ O_2S \diagdown \quad \diagup SO_2 \\ O \end{array}$$

(XI)

by the action of excess SO_3 on S_4N_4 or $S(NSO)_2$. It was described as forming colorless, strongly refracting crystals which can be sublimed at 70–80°/1 Torr.

There is no recent information on this substance, and further study is needed of the conditions under which sulfur trioxide oxidizes S_4N_4, as opposed to simply adding to it (cf. ref. 34 of Chapter 6; see also p. 133).

References

1. Goehring, M., Hohenschutz, H. and Ebert, J., *Z. Anorg. Allg. Chem.* **276**, 47 (1954)
2. Meuwsen, A. and Lösel, M., *Z. Anorg. Allg. Chem.* **271**, 221 (1953)

3. Roesky, H. W., *Angew. Chem. Int. Ed. Engl.* **10**, 266 (1971)
4. Roesky, H. W. and Petersen, O., *Angew. Chem. Int. Ed. Engl.* **11**, 918 (1972)
5. Roesky, H. W. and Wiezer, H., *Angew. Chem. Int. Ed. Engl.* **14**, 258 (1975)
6. Roesky, H. W. and Wehner, E., *Angew. Chem. Int. Ed. Engl.* **14**, 498 (1975)
7. Roesky, H. W., Böwing, W. Grosse, Rayment, I. and Shearer, H. M. M., *J. Chem. Soc. Chem. Commun.* 735 (1975)
8. Roesky, H. W., *Z. Naturforsch.* B **31b**, 680 (1976)
9. Roesky, H. W., Schaper, W., Petersen, O. and Müller, T., *Chem. Ber.* **110**, 2695 (1977)
10. Steudel. R., *Z. Naturforsch.* B **24b**, 934 (1969)
11. Steudel, R., Luger, P. and Bradaczek, H., *Angew. Chem. Int. Ed. Engl.* **12**, 316 (1973)
12. Steudel, R., Rose, F., Reinhardt, R. and Bradaczek, H., *Z. Naturforsch.* B **32b**, 488 (1977)
13. Witt, M., Aramaki, M. and Roesky, H. W. to be published; see further Roesky, H. W., Witt, M., Krebs, B. and Korte, H.-J., *Angew. Chem. Int. Ed. Engl.* **18**, 415 (1979)

9
Bonding and electron-counting in S–N heterocycles

I. Introduction

The descriptive chemistry of sulfur–nitrogen rings and cages has occupied seven chapters of this book, in the course of which a considerable amount of structural information has been presented and briefly discussed. Many S–N ring and cage structures do not conform to patterns previously established in chemistry, and their existence was not foreseen. However, enough structural data on these compounds now exist to both permit and demand some attempt at rationalization. That is the purpose of the present chapter.

The approach adopted here will be empirical: to look for factual regularities and explain them by means of simple and accepted ideas of bonding. No detailed account will be given of quantum-mechanical calculations, for the following reasons. First, too much space would be required for any presentation of this subject which could do it justice and enable the reader properly to evaluate published work. Secondly, it is a fact that experimental chemists, not theoreticians, are still making the running in this area of chemistry, as they have done for a century and a half. Nevertheless, useful progress has been made with molecular-orbital calculations on S–N heterocycles, especially by the CNDO/2 technique; the interested reader may gain access to this work through references 1 and 12.

The present account is guided by the following questions, to all of which more or less fruitful answers are now possible.

(1) What patterns can be discerned in the bond lengths and angles found in S–N rings and cages, and how can the patterns be rationalized? (Here attention is on individual bonds and the configurations around individual atoms.)

(2) To what extent is it useful to discuss known structures, and their stability, in terms of "aromatic" numbers of π electrons or cluster theory? (Here, in contrast, attention is on the cooperative bonding action of electrons over the whole ring or cage.)

(3) Can predictions be made about the stability of as yet unknown structures which may serve as a guide to research?

The ideas in this chapter have mostly evolved in the last decade, a period which has seen the discovery of most of the necessary facts.

The compounds to be discussed fall into two groups, which will be treated in turn: first, those containing 3-coordinate nitrogen atoms with all bonds to neighboring atoms formally single ("saturated" rings); and secondly, the more diverse and interesting compounds with 2-coordinate nitrogen atoms, in a formal sense doubly bonded to neighbor atoms ("unsaturated" rings).

Diagrams of many of the molecules discussed here, with appropriate references, will be found in earlier chapters. The present chapter is concerned mainly with generalizations.

II. S–N bond parameters in general

A. Length–force constant correlation

Vibrational spectra have been published for nearly all the sulfur–nitrogen heterocycles mentioned in this book (see reviews in refs. 2 and 3, and primary papers for more recent work), but only for S_4N_4, S_2N_2, S_7NH, and $S_4(NH)_4$ do we have complete vibrational analyses.[11,28] To this information can be added data for several simple non-cyclic molecules with S–N bonds, giving Fig. 9.1. This shows, as expected, a fairly smooth inverse relationship between bond length and force constant, except for S_2N_2, which has a low force constant for its S–N distance, perhaps because of severe angular strain.

B. Length–order correlation

Bond lengths and force constants are firmly based experimental quantities, but bond orders cannot be measured. They may be defined in different ways, and hence calculated giving different answers, none of which has any absolute claim to rightness.[1,12,20] Therefore, correlations between S–N bond orders and other parameters of the bond ought not to be taken too seriously. Nevertheless, such correlations have often been found useful in discussion, so they deserve a little space here.

Two recent order–length curves for S–N bonds[15,24] are shown in Fig. 9.2. In both, the orders have been deduced from classical canonical formulas, on the assumption that a nitrogen atom always forms three bonds, and disregarding any bonding action of the nitrogen lone pair. Many non-cyclic compounds, besides the cyclic ones, have provided data for the curves. The relationship shown is quite similar to the well known one for carbon bonds, or that for carbon–nitrogen bonds.[24] There is no such thing as a "true" S–N single-bond distance; the single bond in sulfamic acid, formed by sp^3-hybridized nitrogen,

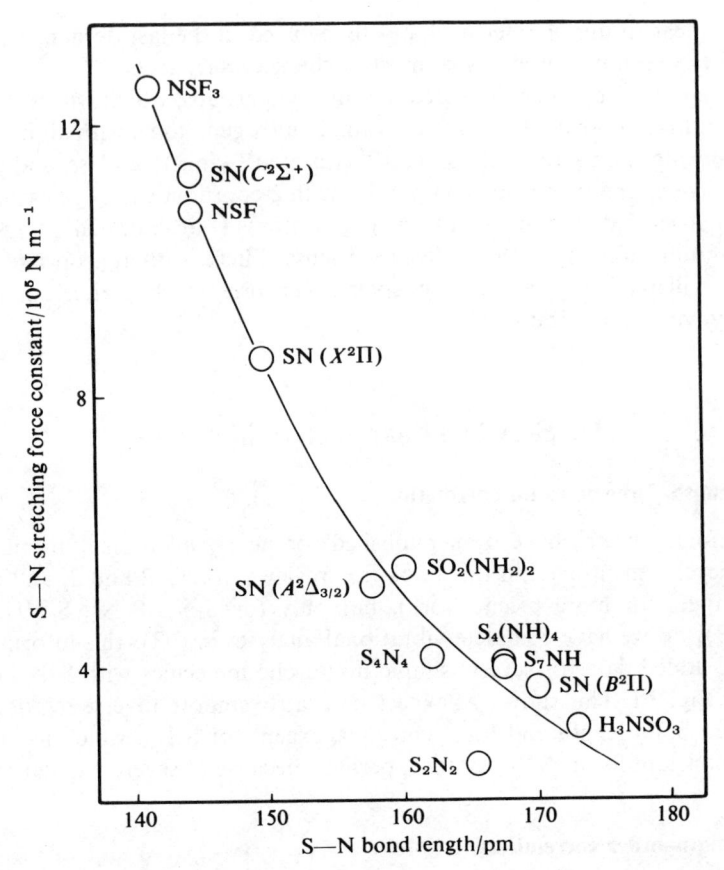

Figure 9.1. Relationship between stretching force constant and length for S–N bonds; data from refs. 11, 17, 22, and 28.

is naturally longer (173–176 pm) than that in $S_4(NH)_4$ where the nitrogens are sp^2-hybridized (167 pm). The short N–S bonds in NSF and NSF_3 are usually assumed to be triple, but it has been argued on the evidence of the sulfur $2p_{3/2}$ electron binding energy that the "real" order in NSF_3 is 1.75, and that further shortening of the bond is caused by charges $-1\frac{1}{4}$ on N and $+2$ on S.[19]

C. Energy–order correlation

Rather few determinations of S–N bond-energy terms or bond-dissociation enthalpies have been reported. The most recent data are summarized in Table 9.1. As expected, the energies correlate fairly well with bond orders. It should be borne in mind, however, that the calculation of the energies from thermo-

II. S–N BONDS 155

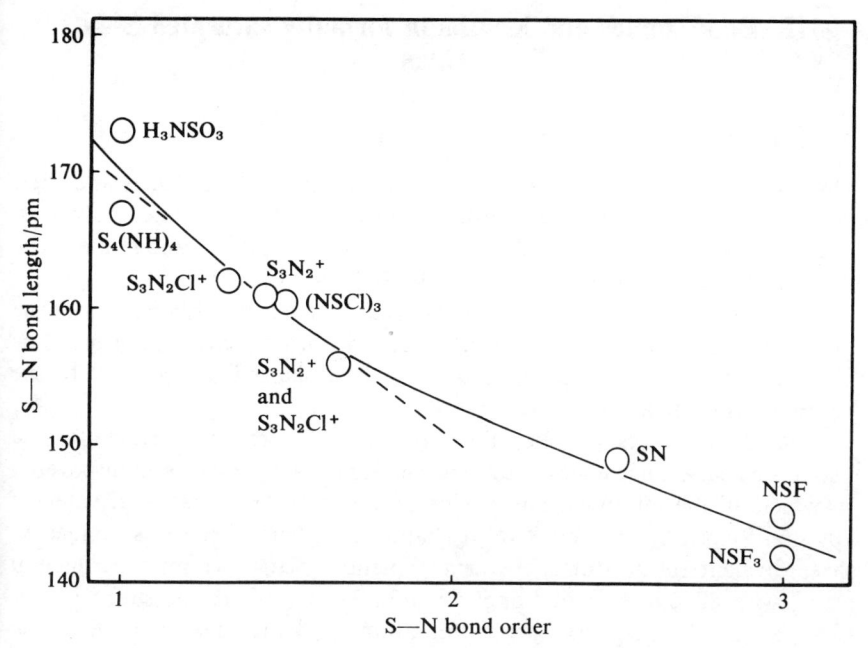

Figure 9.2. Order–length relationship for S–N bonds: full line and points, data from ref. 15; broken line, least-squares line based on 18 organic compounds.[24]

chemical data requires assumptions (for which see the original papers) while the bond orders are subject to the uncertainties just mentioned.

It is interesting that a semi-empirical calculation of the S–N single-bond energy by Sanderson's method[13] gives a result, 243 kJ mol^{-1}, in good agreement with the calorimetric values.

Table 9.1. S–N bond energy terms and bond-dissociation enthalpies.

Compound	Method	Bond-energy term* or Bond-dissociation enthalpy† (kJ mol^{-1})	Canonical bond order	Ref.
$S_4(NH)_4$	combustion in oxygen	*249 ± 10	1	13
$S(NC_5H_{10})_2$		*243 ± 17	1	13
$(SNC_5H_{10})_2$		*247 ± 4	1	13
S_4N_4	thermal decomp.	*301 ± 6	1.5	quoted in 13
$S_3N_2O_2$	combustion in O_2	*335 ± 23	2	13
SN	spectroscopy	†463 ± 24	2.5	25
NSF_3	combustion of NSF in fluorine	†418 ± 8	3	21

III. Bond angles and lengths in formally saturated S–N rings

A. Configurations at nitrogen

Three-coordinate nitrogen atoms occur in the cyclic sulfur imides and their derivatives, in the "saturated" sulfur nitrides $(S_7N)_2S_x$ and related polymers, in the nitride $S_{11}N_2$, in the oxide $(S_7N)_2SO$ (all in Chapter 2), and in the cyclic S(IV) imides (Chapter 3) and S(VI) imides (Chapter 4).

An important question here is whether the arrangement of bonds about the nitrogen atom is pyramidal (with a "stereochemically active" lone pair), or planar, or neither. This question will first be discussed with respect to the compounds of divalent sulfur.

As more structures are determined, it becomes increasingly evident that a planar or near-planar configuration is usual; no example has been discovered in which all the interbond angles closely approach the tetrahedral value of 109°. In $S_{11}N_2$, $S_{16}N_2$, and $S_{17}N_2$ (Chapter 2), where nitrogen is bonded to three divalent sulfur atoms, the arrangement is planar within experimental error, with all the interbond angles within 3° of 120°. In the sulfur imides (Chapter 2) two neighbors are sulfur atoms and the third hydrogen. The nitrogens in $S_4(NH)_4$ are planar, with interbond angles totalling 359°. In 1,3-$S_6(NH)_2$, a planar arrangement is probable though less sure. In crystalline S_7NH, the hydrogen atoms lie about 14° out of the S–N–S plane, a configuration still much nearer planar than tetrahedral. The N–C bonds in $S_4(NMe)_4$ are 11° out of the S–N–S plane.

It appears, then, that the nitrogen lone pairs in all these compounds are not, or are only to a small extent, stereochemically active. Are they involved in $p\pi$–$d\pi$ bonding to sulfur? This explanation has been put forward,[10] and in circumstantial support of it might be adduced the weakness of the basic and nucleophilic character of the nitrogen atom in the sulfur imides, of their alkyl derivatives, and of $S_{11}N_2$ (Chapter 2, Sec. III.B, C, and D). But $p\pi$–$d\pi$ bonding is probably not important, for the following reasons. First, as X-ray photoelectron spectroscopy has shown,[10] the sulfur atoms joined to nitrogen in the sulfur imides are almost electroneutral, having positive charges of magnitude probably less than 0.1 e, very much smaller than on the sulfur atoms in, say, SO_4^{2-} or SF_6; a large charge is usually considered necessary to contract the sulfur d orbitals to a suitable size for $p\pi$–$d\pi$ bonding.[2] Secondly, the S–N bond lengths in the compounds under discussion, 167–171 pm, do not suggest much multiple bonding. It is true that a somewhat larger value has usually been assumed for the S–N single-bond length, namely that in sulfamic acid, which is 173 pm (1951 value) or 176 pm (1960 redetermination). But the "shortening" in the sulfur imides is at least partly caused by the strengthening of the S–N σ bond accompanying the change from sp^3-hybridized nitrogen in sulfamic acid to near-sp^2-hybridized nitrogen in the sulfur imides. Nyburg[24] too, in a recent study of S–N bond lengths, has postulated essentially no contribution to the bonding from nitrogen lone pairs. The picture which emerges of S–N bonding in the sulfur imides and related compounds is, then, as follows. The bonds are much nearer single than double. The nitrogen is planar, sp^2-hybridized, or near it; its lone pair is in a p$_z$ orbital perpendicular to the bonding plane, and, since it is not strongly localized on one side of the nitrogen atom, confers little basic character. The sp^2 configuration is probably preferred to sp^3 simply because of the resulting extra strength in the σ bonds.

No crystallographic data have been published on ring compounds with 3-coordinate nitrogen linked to sulfur(IV) (Chapter 3), but there is one structure for a similar compound of sulfur(VI) (Chapter 4, Sec. III.C). The S–N bonds in trimeric methyl sulfimide, (MeNSO$_2$)$_3$, are 166.9 pm long, a little shorter than in the sulfur imides, as would be expected from the higher oxidation state of sulfur and consequent contraction of its σ-bonding orbitals. The N–C bonds are 28.1° out of the S–N–S plane, halfway to the tetrahedral configuration (54.7°). Neither of these values can be construed as evidence for pπ–dπ N→S bonding, despite the relatively favorable oxidation state of the sulfur.

B. Configurations at sulfur

All the S–S–S and N–S–S and N–S–N bond angles in the compounds of divalent sulfur just mentioned are within 2° of the mean value for S$_8$, 108.1°, and all the dihedral angles are, like those of S$_8$, near 90°. In other words, replacement of a sulfur atom or atoms in S$_8$ by NH or NMe has little effect on the hybridization at sulfur, which is perhaps not surprising since the S–N bonds are so nearly non-polar (see Section A above).

The N–S–N angles in (MeNSO$_2$)$_3$ average 104.5°, a little less than the tetrahedral value, but the configuration around sulfur in this compound is roughly tetrahedral.

IV. Bond lengths and angles in formally unsaturated S–N rings

A. Configurations at nitrogen

With 2-coordinate nitrogen, the question of planarity does not arise. The parameters to be discussed are the bond angle at nitrogen and the S–N bond length. Are they related?

Banister and Durrant[7,8] have systematically reviewed all the structural data on S-N compounds with 2-coordinate nitrogen, cyclic and non-cyclic, published to 1977. For non-cyclic compounds and rings large enough to be unstrained (6, 7, 8, or 10 members), with the sulfur in an average oxidation state of 4 ± 1, they find that the S-N bond length d_{SN} is related to the bond angle at nitrogen N by the equation

$$d_{SN} = 180 - 0.160N \qquad (1)$$

This resembles the corresponding relationship for the phosphazenes (Fig. 12.7) and is similarly explained: the opening up of the nitrogen bond angle towards 180° implies decreasing stereochemical activity of the nitrogen lone pair, and is thought to reflect its increasing involvement in pπ-dπ bonding to sulfur; also, there is some strengthening of the N-S σ bond associated with the change of nitrogen hybridization from sp^2 to sp. The soundness of this correlation can only be properly judged when account is taken of experimental errors, which vary widely between different structure determinations; in fact, 62 of 63 points plotted lie within 3 estimated standard deviations (in bond distance) of the straight line represented by equation (1), and most are far closer.

Equation (1), then, holds reasonably well when the bond angle at nitrogen is free to adjust. In the planar or nearly planar 5-membered rings of the ions $S_3N_2^+$, $S_6N_4^{2+}$ (Chapter 6, Secs. III.E and F), of the ion $S_3N_2Cl^+$ (Chapter 7, Sec. III.B), and of the neutral S_3N_2 derivatives (Chapter 7, Sec. IV; Chapter 8, Sec. IV) the geometry imposes some constraint on the nitrogen bond angle, which is nearly constant at 118–120°. One might nevertheless expect that this angle would still control the amount of π overlap and that the S-N distance would accordingly adjust to fit equation (1). Actually the S-N distances fall many standard deviations below the line of equation (1). It is tempting to attribute this shortening to circumannular π delocalization (Section V.B below), in contrast to the structures fitting equation (1), where delocalization may be restricted to "Dewar islands" (Chapter 12, Sec. IX). Perversely, however, S_2N_2 and $S_3N_3^-$, which are planar and by Banister's reasoning ought to be Hückel species (Section V.B below), fit equation (1) almost perfectly. Another group of molecules in which the S-N distances fall well below the values predicted by equation (1) is the sulfanurics (Chapter 5). Here the higher oxidation state of sulfur (+6) and higher charge on the sulfur atoms is probably the cause, strengthening the pπ-dπ bonding by contraction of the sulfur d orbitals. Roesky similarly has pointed out the shortening of S-N bond lengths associated with increasing coordination number of sulfur.[27]

To summarize: as more structures are determined, patterns are becoming discernible in the S-N bond distances and nitrogen bond angles of unsaturated S-N rings; but current explanations of the patterns have still an *ad hoc* flavor.

B. Configurations at sulfur

Banister and Durrant have also demonstrated inverse empirical correlations between S–N bond distances and N–S–N angles in unsaturated heterocycles and non-cyclic compounds.[8] Because of the varying coordination numbers and oxidation states of sulfur in these molecules, a proper discussion of their conclusions would need an unwarranted amount of space, but one result of the study will be mentioned here. By using the fact that angles of minimum strain at nitrogen and sulfur are both correlated with S–N bond length and hence with each other, one can sometimes decide which of several alternative structures for a compound or ion is the least strained. Banister and Durrant in this way successfully predicted the shape of $S_4N_4^{2+}$ (Chapter 6, Sec. III.F) and argued in support of the recent azulene structure for $S_5N_5^+$ (Fig. 6.5) against the earlier belief in a heart-shaped structure.

V. Synoptic treatments of bonding in unsaturated S–N heterocycles

A. Introduction

As has been shown, S–N bond lengths in S–N heterocycles cover an almost continuous range of values from about 154 pm to 173 pm, and do not obviously group themselves into the categories "single", "double", and "triple". So attempts to assign canonical bonding formulas to these molecules are often unprofitable and sometimes misleading. Efforts are therefore being made to rationalize these structures by a different approach, in which attention is directed at the ring or cage as a whole, and its bonding electrons as a whole, rather than at individual atoms, bonds, and electron pairs. This general methodology may be called "synoptic". Two synoptic viewpoints, applying to different groups of compounds, will be discussed here. One is based on the Hückel theory of aromaticity and the other on cage or cluster theory.

B. Aromaticity in near-planar rings

Following earlier molecular-orbital calculations on $S_4N_3^+$,[14] Banister[4] and Jolly[18] independently suggested in 1972 that the Hückel theory of aromaticity might apply as a general principle to near-planar unsaturated S–N rings. A recent discussion by Roesky[27] confirms the usefulness of this idea.

In addition to the σ bonding of the ring, π bonding similar to that in benzene is postulated. This would be based essentially on atomic p orbitals, though some involvement of sulfur d orbitals by hybridization is not ruled out. The electrons of each 2-coordinate sulfur atom are assumed to be used as follows: two for σ bonding to neighboring atoms; two in a lone pair; and two contributed to delocalized ring π orbitals. Each nitrogen atom uses two electrons in σ bonding, has two in a lone pair, and contributes one to delocalized π orbitals. A molecule or ion with $4n + 2$ delocalized π electrons ($n =$ an integer) is described as a Hückel species and postulated to have appreciable "aromatic stabilization". Electron-counts on this principle show that the ions and molecules in Table 9.2 may be Hückel aromatic species. This table is not exhaustive; the omissions include compounds with a large excess of nitrogen, which Banister has suggested would be explosive.[4]

Of the neutral species listed, S_2N_2 (known to be planar) and S_4N_2 (perhaps near-planar) exist (Chapter 6, Secs. II.B and III.B). The other neutral species have not been found despite searches.[26] Of the cations, $S_4N_3^+$ and $S_5N_5^+$ were known when this theory was published, while $S_4N_4^{2+}$ (near-planar) and $S_3N_2^{2+}$ (almost certainly planar by analogy with $S_3N_2Cl^+$ and because of geometrical constraint) have been discovered since (Chapter 6, Secs. III.F and G). The u.v. spectrum of $S_4N_3^+$ shows the $\pi \rightarrow \pi^*$ transitions to be expected in an aromatic system.[14] Though Banister originally thought that a negative charge on the ring might destabilize it, the anion $S_3N_3^-$ has recently been discovered (Chapter 6, Sec. III.H) and has D_{3h} symmetry; it is a planar 6-membered ring. $S_4N_5^-$ is an interesting case. An ion of this formula is known, but it is a cage (Chapter 6, Sec. IV.B; Section C below), not a flat ring. A cyclic NS_2^- seems unlikely because of ring strain.

Certain other S–N heterocycles may usefully be discussed from the standpoint of aromaticity. $S_3N_2Cl^+$ and the related compounds with S_3N_2 rings

Table 9.2. Planar or near-planar S–N rings expected to have $4n + 2$ delocalized π electrons (Hückel species); known species are in heavy type.

	Anions	Molecules	Cations 1+	Cations 2+
6π	NS_2^-	**S_2N_2**	**$S_2N_3^+$**	**$S_3N_2^{2+}$**
10π	**$S_3N_3^-$**	S_3N_4	**$S_3N_5^+$**	**$S_4N_4^{2+}$**
		S_4N_2	**$S_4N_3^+$**	$S_5N_2^{2+}$
14π	$S_4N_5^-$*	S_4N_6	**$S_5N_5^+$**	$S_5N_6^{2+}$
		S_5N_4	$S_6N_3^+$	$S_6N_4^{2+}$†
18π	$S_6N_5^-$	S_6N_6	$S_6N_7^+$	$S_7N_6^{2+}$

* An ion of this formula is known (Chapter 6, Sec. III.G) but has a cage structure.
† An ion $S_6N_4^{2+}$ exists (Chapter 6, Sec. III.E) but consists of three fused rings.

(Chapter 7, Sec. IV; Chapter 8, Sec. IV) are Hückel 6π species if the 3-coordinate sulfur atom is, reasonably, assumed to contribute one electron to the π system. $S_4N_4O_2$ (Chapter 8, Sec. II.B) is a 10π molecule and $S_3N_3O_4^-$ (Chapter 8, Sec. III.A) is a 6π molecule, if the sulfur atoms with oxygen ligands contribute no electrons to the π system. However, the postulated structure of $S_4N_4O_4$ (Chapter 8, Sec. II.A) would make this compound an 8π species by similar reasoning. In the chair-form rings of $(NSCl)_3$ and $(NSF)_3$ (Chapter 7, Sec. II.B) each sulfur atom has one electron available for ring π bonding, which makes the rings in a formal sense 6π species. However, the sulfur orbitals which would be required for benzene-like π delocalization are occupied by the axial halogen atoms, and CNDO/2 calculations show alternating positive and negative charges (on sulfur and nitrogen atoms respectively) so large that aromatic ring currents are out of the question.[12] The equal bond lengths in these two rings are probably best explained in terms of three-center "Dewar islands" (compare Chapter 12, Sec. IX) rather than by an "aromatic" model.[12]

An important point in support of Hückel-type aromaticity in S–N rings is that only one small ring is known, the radical-cation $S_3N_2^+$, that does not have $(4n + 2)$ π electrons. Its dimer $S_6N_4^{2+}$ (Fig. 6.4) can, by a reasonable method of electron counting, be construed as a Hückel species.[6]

One should not try to characterize the aromaticity theory of S–N rings as either "right" or "wrong". Hückel-type π bonding almost certainly does stabilize these flat or nearly flat rings. What we do not know is how often it is decisive for their structure and stability. A judgment on this will be easier after further successes and failures in the synthesis of new S–N rings.

The Hückel formalism cannot be applied to the commonest of all S–N molecules, S_4N_4, which is neither a planar ring nor a "$4n + 2$" species. It and related structures can be discussed under the cage or cluster formalism now to be described.

C. The cage or cluster viewpoint

Polyhedra of atoms feature in many inorganic molecules, and in the molecules of some carbon compounds. Wade[29] has shown that the bonding in such molecules can often be rationalized most satisfactorily by means of electron-counts based on the whole polyhedron rather than on individual atoms and bonds. The first and most notable success of this approach came in Williams and Wade's interpretation of the structures of the boranes, borane anions, and carboranes. The skeletons of these have far too few bonding electrons to supply one pair for each polyhedron edge (they are "electron-deficient"). There is consequently no alternative to the cluster treatment, apart from an older three-center bond treatment which became increasingly cumbersome and unrealistic as the size of the polyhedron increased. The Williams–Wade treatment of the boranes and other electron-deficient polyhedra starts from a previous finding of molecular-orbital theory, viz. that a polyhedron with n vertices requires $n + 1$ skeletal bonding electron pairs to completely fill its bonding molecular orbitals. Williams and Wade showed that if $n + 2$ pairs are available the structure adopted is a polyhedron minus one vertex ("nido" structure), if $n + 3$ pairs are available it is a polyhedron minus two vertices ("arachno" structure), and so on. In other words, the effect of adding extra skeletal bonding electrons is progressively to "open up" the polyhedron.

Some polyhedra occurring in molecules are not electron-deficient. If they have just enough skeletal bonding electrons to provide one pair for each polyhedron edge, they may be called "electron-precise"; the P_4 tetrahedron in white phosphorus is an example,[23] or the C_8 polyhedron in the hydrocarbon cuneane [Fig. 9.3(b)]. Here again, however, the effect of adding extra electron pairs will be to open up the polyhedron. One edge bond is broken (being replaced by two lone pairs) for every electron pair added. In this respect the formalism resembles that for the electron-deficient case, and the resulting structures can similarly be described as nido or arachno.

Some S–N structures are based on the S_4N_4 polyhedron and lend themselves well to this kind of treatment, viz. S_4N_4, $S_4N_5^-$, $S_4N_5^+$, S_5N_6, and $S_4N_5O^-$. It is not, however, obvious which of the atoms of S_4N_4 are best regarded as forming the polyhedron and which, if any, are "exo" ligands. Consequently two alternative cluster descriptions of S_4N_4 have been suggested as follows.

(1) Banister[5] regards the S_4N_4 molecule as a tetrahedron of sulfur atoms bridged along four of its six edges by nitrogens [Fig. 9.3(a)]. This is justified by the shortness of the six S–S distances forming the tetrahedron edges

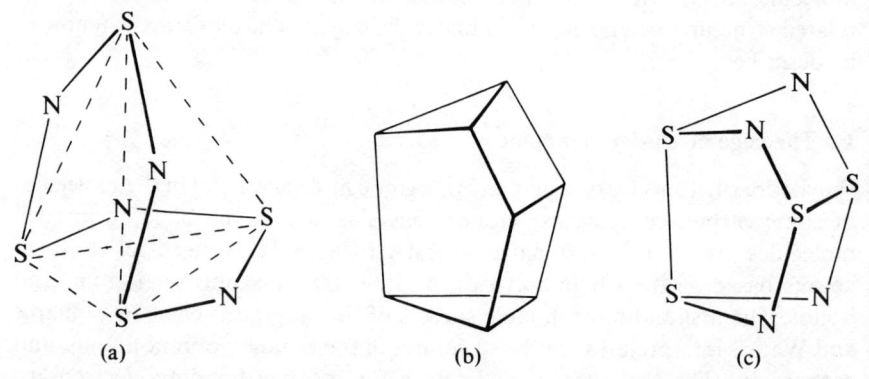

Figure 9.3. "Cage" or "cluster" viewpoints on the bonding of tetrasulfur tetranitride: (a) based on a tetrahedron of sulfur atoms (broken lines, Banister[5,6]); (b) the skeleton of cuneane, C_8H_8; (c) S_4N_4 depicted as a cuneane-like skeleton with two edge bonds missing (Mingos[23]).

(2 × 258 and 4 × 269 pm), which are much less than the van der Waals contact distance of 369 pm, though longer than typical S–S single bonds (204 pm in S_8). The following electron count on S_4N_4 shows the S_4 tetrahedron as electron-precise, that is, with one pair of electrons available per edge bond:

total valence electrons, 4 × 5 + 4 × 6 = 44
subtract electrons required for σ framework (8 S–N bonds), 8 × 2 = 16
subtract lone-pair electrons (one lone pair on each S or N), 8 × 2 = 16
electrons remaining = 12

The 12 electrons just suffice for six edge bonds. It will be noted that electrons are divided between lone pairs and bond pairs in the same way as for "aromatic" S–N rings (Section B above), and that any involvement of lone pairs in supplementary pπ–dπ bonding is disregarded. On this picture the S–N bonds are single. Why are they shorter (162 pm) than the presumed S–N single-bond length of around 173 pm? This can be explained by the reasonable suggestion that the S–S bridge across each S–N–S group "compresses" the S–N bonds; which adds emphasis to the point already made in Section II.B above about the speculative nature of bond length–order correlations.

There is another difficulty about the electron count above. It leaves each nitrogen atom 2 electrons short of an octet. In an earlier theoretical study, Gleiter[16] had argued, like Banister, that the S–N bonds are single, and cited the ^{14}N n.m.r. chemical shift in support; however, he supposed that each N atom had two lone pairs and hence an octet, and did not postulate the S–S bonds bridging S–N–S groups that are required for Banister's tetrahedron. Perhaps future theoretical studies will show that the difference between the two descriptions is more formal than real.

The ion $S_4N_5^-$ (Chapter 6, formula **XI**) is essentially an S_4N_4 molecule with an additional nitrogen bridge, spanning the fifth edge of the S_4 tetrahedron. A count of electrons on the same lines as for S_4N_4 again gives 12 electrons for the electron-precise S_4 tetrahedron. The ion $S_4N_5O^-$ is simply $S_4N_5^-$ with an "exo" oxygen atom attached through one of the sulfur lone pairs (Fig. 8.1), and the electron count yet again gives 12 for the S_4 tetrahedron. In each ion, the edges of the S_4 tetrahedron are of about the same length as in S_4N_4, and more nearly equalized than in S_4N_4. $S_4N_5^+$ (Chapter 6, formula **XII**) lacks one electron pair relative to $S_4N_5^-$, i.e. is one pair short of the number required for an electron-precise S_4 tetrahedron. Accordingly, it is pleasing to find that one of its tetrahedron edges (the dotted one in the formula) is much longer (401 pm) than the others (278–281 pm).

An electron count for S_5N_6 (Chapter 6, formula **XIII**), on the same principles, shows that there are seven electron pairs not involved in the σ framework or in lone pairs. Six of these can be assigned to S---S bridges across

S–N–S groups, like those just postulated for S_4N_4 and the cage ions derived from it. There is only one other very short S–S distance (dashed line in the formula), which satisfactorily disposes of the remaining pair.

(2) Mingos[23] regarded the S_4N_4 polyhedron as made up of all eight atoms of the molecule, not merely the sulfur atoms; it resembles the carbon polyhedron in cuneane, C_8H_{18} [Fig. 9.3 (b) and (c)]. The electron count reads as follows:

total valence electrons, $4 \times 5 + 4 \times 6$ $= 44$
subtract lone-pair electrons (one lone pair on each S or N), $8 \times 2 = 16$
electrons remaining $= 28$

The 28 electrons, or 14 pairs, available for bonding the polyhedron are two pairs too many for the twelve edges. Following the formalism mentioned above, each "surplus" pair leads to one broken edge bond, with the result shown in Fig. 9.3(c). No extension of this treatment to $S_4N_5^-$ or $S_4N_5O^-$ has been suggested; if such were attempted on the basis of a nine-atom polyhedron, or a ten-atom polyhedron with a vertex missing, it would lack the attractive simplicity of Banister's treatment.

D. Transannular bonding in general

If the S_4N_4 molecule is thought of as a bent ring, then the six S–S bonds forming the S_4 tetrahedron in Banister's model [Fig. 9.3(a)] can be described as transannular (cross-ring) bonds. As early as 1966, Turner and Mortimer's molecular-orbital calculations on S_4N_4 had pointed to the presence of two fairly strong transannular S–S bonds; this conclusion held for all seven alternative parametrizations they tested, and was soon confirmed by Gleiter in an alternative approach to the problem.[16] So the Banister model has, at least in part, theoretical support. More recent calculations by the CNDO/2 method suggest that transannular bonding (and also antibonding) in S–N rings may not be confined to S_4N_4 and related structures, but may be significant in flatter rings also. Transannular S–S bonds in $(NSCl)_3$, $(NSF)_3$, and $(NSOCl)_3$ are found each to have 7–10% of the energy of a ring S–N bond.[12] It seems likely that too little attention has been paid to this type of bonding and that it could sometimes be decisive for the stability and conformation of S–N rings. No calculations of transannular bond orders or energies for Hückel-type S–N rings (Section B above) have yet been reported.

E. Conclusion

Throughout Section IV and the present Section V of this chapter we have been discussing formally unsaturated S–N rings and cages, all of which contain more valence electrons than are required for the σ bonds of their framework.

To use a common description, they are "electron-rich". The Banister–Jolly theory of aromaticity in flat rings, the Banister cage theory of S_4N_4, and the recent findings of Cassoux, Glemser, and Labarre about transannular bonding, all show ways of disposing of these surplus electrons. The whole subject is at an early stage of development, and it remains to be seen what combination of these and possibly other ideas will eventually be found the most relevant and useful.

References

1. Adams, D. B., Banister, A. J., Clarke, D. T. and Kilcast, D., *Int. J. Sulfur Chem. A* **1**, 143 (1971)
2. Banister, A. J., Moore, L. F. and Padley, J. S., *Spectrochim. Acta* **23A**, 2705 (1967)
3. Banister, A. J., Moore, L. F. and Padley, J. S., "Structural Studies on Sulphur Species", in *Inorganic Sulphur Chemistry*, ed. G. Nickless, p. 137, Elsevier, Amsterdam (1968)
4. Banister, A. J., *Nature (Phys. Sci.)* **237**, 92 (1972)
5. Banister, A. J., *Nature (Phys. Sci.)* **239**, 69 (1972)
6. Banister, A. J., *MTP International Review of Science*, Inorganic Chemistry, Series 2, vol. 3, p. 41, Butterworths, London (1975)
7. Banister, A. J. and Durrant, J. A., *J. Chem. Res. S* 150 (1978)
8. Banister, A. J. and Durrant, J. A., *J. Chem. Res. S* 152 (1978)
9. Barker, C. K., *J. Phys. Chem.* **69**, 334 (1965)
10. Barrie, A., Garcia-Fernandez, H., Heal, H. G. and Ramsay, R. J., *J. Inorg. Nucl. Chem.* **37**, 313 (1975)
11. Bragin, J. and Evans, M., *J. Chem. Phys.* **51**, 268 (1969)
12. Cassoux, P., Glemser, O. and Labarre, J.-F., *Z. Naturforsch. B* **32b**, 41 (1977)
13. Fleig, H. and Becke-Goehring, M., *Z. Anorg. Allg. Chem.* **375**, 8 (1970)
14. Friedman, P., *Inorg. Chem.* **8**, 692 (1969)
15. Gillespie, R. J., Ireland, P. R. and Vekris, J. E., *Can. J. Chem.* **53**, 3147 (1975)
16. Gleiter, R., *J. Chem. Soc. A* 3174 (1970)
17. Glemser, O., Müller, A. and Krebs, B., *Z. Anorg. Allg. Chem.* **357**, 184 (1968)
18. Jolly, W. L., "Sulfur Research Trends", *Adv. Chem. Ser.* **110**, 92 (1972)
19. Jolly, W. L., Lazarus, M. S. and Glemser, O., *Z. Anorg. Allg. Chem.* **406**, 209 (1974)
20. Jug, K., *J. Am. Chem. Soc.* **99**, 7800 (1977)
21. Larson, J. W., Johnson, G. K., O'Hare, P. A. G., Hubbard, W. N. and Glemser, O., *J. Chem. Thermodynamics* **5**, 689 (1973)
22. Mikulski, C. M., Russo, P. J., Saran, M. S., MacDiarmid, A. G., Garito, A. F. and Heeger, A. J., *J. Am. Chem. Soc.* **97**, 6358 (1975)
23. Mingos, D. P., *Nature (Phys. Sci.)* **236**, 99, **239**, 16 (1972)
24. Nyburg, S. C., *J. Cryst. Mol. Struct* **3**, 331 (1973)
25. O'Hare, P. A. G., *J. Chem. Phys.* **52**, 2992 (1970)
26. Ramsay, R. J., Heal, H. G. and Garcia-Fernandez, H., *J. Chem. Soc. Dalton Trans.* 234 (1976)
27. Roesky, H. W., *Angew. Chem. Int. Ed. Engl.* **18**, 91 (1979)
28. Steudel, R., *J. Phys. Chem.* **81**, 343 (1977)
29. Wade, K., *Adv. Inorg. Chem. Radiochem.* **18**, 1 (1976)

10
Phosphorus–sulfur rings and cages

I. Introduction and history

Compared with P–N or S–N compounds, there is rather little chemistry of cyclic P–S compounds, so far as is known. The most fundamental reason for this appears on present evidence to be that P–S rings and cages can only be of the saturated type (i.e. without double bonds), which greatly limits the possible types of structure. However, there are many indications that interesting new and unexpected structures and reactions may soon come to light; the subject seems to be on the verge of a breakthrough stimulated by modern experimental methods. The known P–S cyclic compounds consist of the phosphorus sulfides, which mostly have cage molecules, and some derivatives of them, and some 4-, 5-, and 6-membered P–S rings with organic or halogen exocyclic groups.

The phosphorus sulfides were investigated in the early decades of the nineteenth century by a number of scientists including Berzelius and Faraday, but were not then properly characterized. The field remained in some confusion until the work of Alfred Stock and his coworkers, and of H. Giran, who in the early years of this century produced phase diagrams of the phosphorus–sulfur system. Since the Second World War the structures of most of the sulfides have been determined by X-ray diffraction. Work on the P–S rings proper dates only from 1965.

P_4S_3, discovered by G. Lemoine in 1864, has been used in the manufacture of matches since 1898. P_4S_{10}, however, is now manufactured on a far greater scale, for use in the synthesis of organic sulfur compounds.

Our account will begin with the binary phosphorus sulfides, since they are source materials for some, though not all, of the other compounds to be described. These binary sulfides comprise eight well characterized (and one less well known) small-molecule sulfides, and polymeric sulfides of variable composition which have been very little studied.

II. Small-molecule phosphorus sulfides and derivatives

A. Preparation, characterization, and structure

The best available phase diagram for the phosphorus–sulfur system was reported by H. Vincent in 1972 (Fig. 10.1).[45] It was obtained by differential thermal analysis, combined with X-ray diffraction studies of the solid phases at various temperatures. The diagram shows three congruently melting sulfides, P_4S_{10}, P_4S_7, and P_4S_3, each with different low-temperature and high-temperature crystalline forms, and four incongruently melting compounds, P_4S_9, P_4S_5, P_4S_4, and P_4S_2, the last being stable only below $-30°$. In the right-hand part of the diagram, where elemental phosphorus is involved in the equilibria, there is a stable region and curves corresponding to equilibria with red phosphorus, and a metastable region and curves for white phosphorus.

Present knowledge of the solid phosphorus sulfide phases is summarized in Table 10.1 and Fig. 10.2, which may be studied in conjunction with the phase diagram. Some key points are as follows.

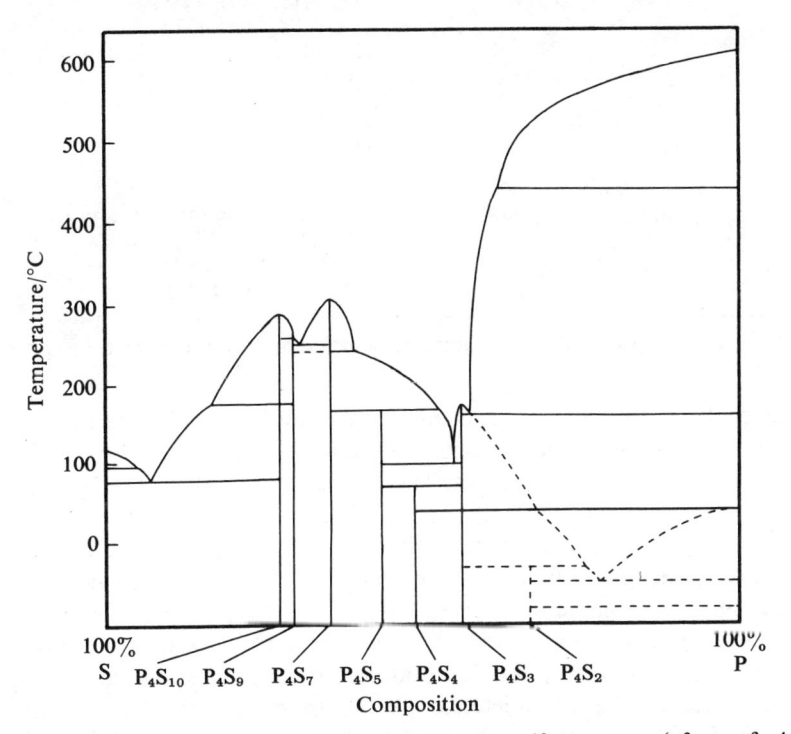

Figure 10.1. Phase diagram of the phosphorus–sulfur system (after ref. 45): full lines, red phosphorus/sulfur; broken lines at right, white phosphorus/sulfur.

Table 10.1. The phosphorus sulfides (excluding poly*

	Crystalline forms, with temperature ranges of stability (°C)	Preparation: recommended methods	M.p. (°C)	B.p. (°C)	Dens* (g cm
P_4S_{10}	α, triclinic, < 177 (14)	purchase (11, 14, 44)			2.09 (44)
	β, cubic, > 177 (14)	(14)	(288) (11, 44, 45)	514 (11, 44)	
P_4S_9	I, rods from CS_2 (34)	(34, 46)	240–270 dec. (34)		2.08 (34)
	II, cubes from α-bromo-naphthalene (34)	(34)	250–259 (34)		
P_4S_7	α, monoclinic, < 242 (45)	(17)			2.19
	β, orthorhombic, > 242 (15)	(11)	308 (45)		
P_4S_5	α, monoclinic (11, 15)	(11, 44)	170–220 dec. (11)		2.17 (11)
	β, monoclinic (22)	(23, 34)			2.26
	amorphous, polymeric (?) (15)	(15)			
P_4S_4	α, monoclinic (21)	(10, 21)			
	β, (21)	(10, 21)			
P_4S_3	α, orthorhombic, < 40 (15)	purchase			2.06
	β, form unknown, > 40 (15)		173 (45)	408 (11)	
P_4S_2	triclinic, < −30 (46)	(46)	dec. above −30 (46)		

* Some of these values are estimates only, and for some, error estimates are available

Like the oxides, the sulfides of phosphorus all have molecules based on a tetrahedron of phosphorus atoms and conceptually derivable from the P_4 tetrahedron of white phosphorus. Their sulfur atoms, like the oxygen atoms in the oxides, either bridge the edges of the tetrahedron or are attached terminally to the phosphorus atoms. In detail, however, the structural parallel between oxides and sulfides is quite limited. P_4S_{10}, it is true, has the same structure as the volatile form of P_4O_{10}. But whereas removal of all terminal oxygens from P_4O_{10} gives a well characterized oxide P_4O_6 (Fig. 11.1) no sulfide P_4S_6 appears in the phase diagram, nor has any crystalline compound of this formula ever been characterized; except in the case of P_4S_9, there are no structural analogies between the lower sulfides and the lower oxides.

II. PHOSPHORUS SULFIDES   169

...ariable stoichiometry mentioned in Section IV).

ΔH_f° (kJ mol^{-1})*	Solubility in CS$_2$ (g per 100 g)	X-ray crystallography (references)		Spectroscopic data (references)		
		complete structure	powder, unit cell	^{31}P n.m.r.	mass. spec.	i.r., Raman
09 (46)	0.222$_{17}$ (44)	(12)	(11)	(2)	(38, 42)	(11, 14, 34)
28 (14)			(11, 14)			(14)
92 (46)	0.53$_{25}$ (46)		(14, 34)	(2)	(42)	(34)
		(34)	(34)			(34)
54 (46)	0.029$_{17}$ (44)	(11, 12)	(11)		(38, 42)	(11, 15, 34)
		(11, 12)	(15)			(15)
11 (46)	0.3–0.5 (11)	(12)		(2, 22)	(38, 42)	(11, 15)
	soluble (22)	(23)		(22)		
35 (14)			(15)			(15)
71 (46)	soluble (21)	(10, 21)	(46)	(10, 21)	(21)	(21, 46)
			(46)	(21, 23)	(21)	(21, 46)
54 (15)	100$_{17}$ (11)	(12, 32)		(2)	(38, 42)	(11, 15, 32, 34)
04 (46)			(46)			

...rences given.

The α designations in Fig. 10.2 and Table 10.1 are applied to forms stable at room temperature, the β designations to less stable forms. The high-temperature β-P$_4$S$_7$ has been shown by X-ray crystallography to have the same molecular structure as the low-temperature α-form,[11] and the same probably applies to the forms of P$_4$S$_{10}$ and P$_4$S$_3$.[11,15,17] Metastable cubic P$_4$S$_9$, designated the II-form, likewise probably has the same molecular structure as the stable I-form.[34] In contrast, the α and β designations of P$_4$S$_4$ and P$_4$S$_5$ refer to different molecular structures (Fig. 10.2). These isomers of P$_4$S$_4$ have been made by directed syntheses; neither has any terminal sulfur atoms. Cooling of a phosphorus–sulfur melt of appropriate composition[46] gives a crystalline P$_4$S$_4$ which has not been fully characterized; since its i.r. spectrum gives

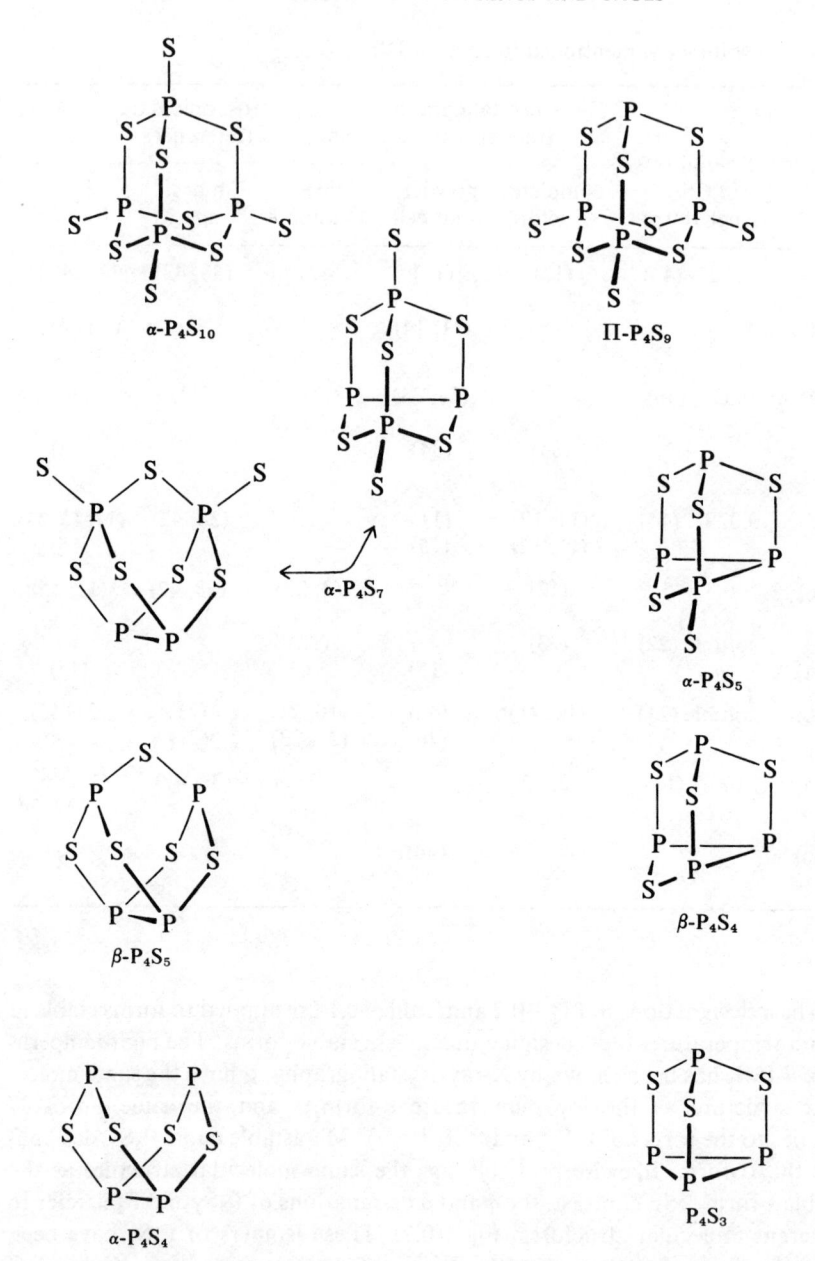

Figure 10.2. Molecular structures of the phosphorus sulfides, shown diagrammatically and not to scale, in order to emphasize relationships; the structure of α-P_4S_7 is shown from two points of view (for sources of structural data, see Table 10.1).

evidence of terminal sulfurs, it may not be identical with either the α- or β-isomer. Obviously, much remains to be learnt about the phosphorus sulfides; this is confirmed by a report of ^{31}P n.m.r. spectra of three previously unknown sulfides in P_4S_3–P_4S_7 mixtures.[21,23]

As Fig. 10.2 shows, the lower phosphorus sulfides can be divided on structural grounds into two groups. P_4S_7 belongs to both. Below it at the left are α-P_4S_4 and the related β-P_4S_5, both, like P_4S_7, having (vertical) two-fold axes of symmetry. To obtain the structures of α-P_4S_5, β-P_4S_4, and P_4S_3 on the right-hand side, however, sulfur atoms must be subtracted from the P_4S_7 structure in ways which destroy the two-fold symmetry.

References to the best methods of preparation for individual phosphorus sulfides are given in Table 10.1. The following is a short commentary on routes to these compounds. In the first place, it is worth noting that P_4S_{10} is widely available from laboratory suppliers, and cheap. P_4S_3 can be bought from specialist firms.[10,28] All the phosphorus sulfides, except P_4S_2, can be prepared by heating red phosphorus with the stoichiometric amount of sulfur, in an inert atmosphere.[24,37] Temperatures above 300° are required, but the mixtures must be heated with care since reaction begins exothermically. Slow cooling is required if one wishes to obtain the incongruently melting sulfides or the low-temperature forms of the congruently melting sulfides.[37] Alternatively, for P_4S_3, P_4S_5, P_4S_7, and P_4S_9, the sulfide P_4S_{10} may be fused with the stoichiometric amount of red phosphorus,[15,24,46] while for P_4S_9, P_4S_3 or P_4S_7 may be fused with sulfur, or a mixture of P_4S_7 and P_4S_{10} heated.[24,46] Obviously many variants are possible within this general approach. In the manufacture of P_4S_3 and P_4S_{10} by direct combination, white phosphorus, not red, is used.[44]

P_4S_2 is not stable at room temperature, and has only been made by cooling a mixture (liquid) of P_4S_3 and white phosphorus to below $-30°$[46] (cf. however, p. 450 of ref. 44).

Alternative methods are more convenient than thermal direct combination for the sulfides P_4S_7, P_4S_5, and P_4S_4. For P_4S_7, white phosphorus and sulfur, dissolved in CS_2 and with a catalytic amount of iodine added, are exposed to daylight for a few days.[17] This produces a deposit of well crystallized α-P_4S_7 in high yield. Alternatively (but using the same principle) a CS_2 solution of $P_2I_4 + S_8$ may be exposed to light.[17] It is interesting that with the ratio $P_4:S_8$ right for P_4S_{10}, the sulfide produced is mainly P_4S_7, probably because P_4S_7 is relatively insoluble and is present in solution as one component in the photochemically generated mixture of products. Similar photochemical procedures starting from P_4S_3 and S_8 give α-P_4S_5.[44]

Some directed syntheses of phosphorus sulfides have recently been developed. Some of these can be understood by reference to Fig. 10.2. Terminal sulfurs (and in one case a bridge sulfur) of the phosphorus sulfides can be progressively removed by phosphines; thus triphenylphosphine effects the following conversions:

$$P_4S_{10} \to P_4S_9 \to P_4S_7 \to [\beta\text{-}P_4S_6] \to \beta\text{-}P_4S_5 \quad \text{(refs. 23, 34)}$$

$$\alpha\text{-}P_4S_5 \to \beta\text{-}P_4S_4 + P_4S_3 \quad \text{(refs. 21, 23)}$$

The only satisfactory route to the separate α- and β-isomers of P_4S_4 is to replace the iodine atoms of α- and β-$P_4S_3I_2$ respectively by sulfur bridges[21] (see also Section E below); a melt of composition P_4S_4 contains both isomers.[21]

All the non-polymeric phosphorus sulfides are yellow crystalline substances. From the practical standpoint, two of their properties call for comment. First, their solubilities: they are dissolved by CS_2 (Table 10.1), and substantial solubility has been reported for some of them in o-dichlorobenzene[11] and benzene.[11] As Table 10.1 shows, the different sulfides have widely different solubilities, a fact often used in their preparation (Table 10.1). Secondly, their susceptibility to hydrolysis by atmospheric moisture: P_4S_3 is unaffected,[44] and reacts only slowly even with boiling water,[13] while at the other extreme P_4S_7 and P_4S_5 react fairly quickly at room temperature, evolving H_2S.[11]

B. Reactions

This account will mainly be confined to reactions of the phosphorus sulfides that clearly show the stability of the P–S cages or their partial and progressive breakdown to new cages or rings. A full account of the reactions in general is available in ref. 24. P_4S_{10} has many uses as a reagent in organic synthesis, but since many of these entail total breakdown of the P–S cage, and have not been mechanistically analyzed, they are not very relevant to inorganic heterocyclic chemistry and will not be discussed here.[11,18] This apart, rather little work has been done on the reactions of P–S cages; for example, little is known of their reduction. Clearly, the lower phosphorus sulfides, containing trivalent phosphorus atoms to which other groups might add, and/or P–P bonds which could easily be opened, are more promising starting materials for the preparation of new rings, cages, and derivatives than is P_4S_{10}; in fact, most of the work along these lines so far reported starts from P_4S_3.

We begin by drawing attention to the reactions already mentioned of P_4S_{10} and P_4S_5, in which a terminal sulfur atom is removed by treatment with triphenylphosphine or other thiophile (Section A above); the cage itself is unaffected. The cage also remains intact in the following reactions of P_4S_3.

Though inactive towards the σ-acceptors BCl_3 and BPh_3 and towards quaternizing reagents,[32] P_4S_3 behaves as a ligand in the reactions below, which yield orange crystalline complexes:

$$\pi\text{-}(C_3H_5)_2Ni + 4P_4S_3 \xrightarrow[-30°]{\text{ether}} Ni(P_4S_3)_4 + C_6H_{10} \quad \text{(ref. 28)}$$

$$C_7H_8M(CO)_4 + 2P_4S_3 \xrightarrow[\text{reflux}]{CS_2} \textit{cis-}(P_4S_3)_2M(CO)_4 + C_7H_8 \quad \text{(ref. 28)}$$
$$(M = Cr, Mo, W) \quad (C_7H_8 = \text{norbornadiene})$$

$$C_7H_8M(CO)_3 + 3P_4S_3 \xrightarrow[\text{reflux}]{CS_2} \textit{cis-}(P_4S_3)_3M(CO)_3 + C_7H_8 \quad \text{(ref. 28)}$$
$$(M = Cr, Mo) \quad (C_7H_8 = \text{norbornadiene})$$

$$Mo(CO)_6 + P_4S_3 \xrightarrow[\text{reflux}]{\text{cyclohexane and diglyme}} P_4S_3Mo(CO)_5 + CO \quad \text{(ref. 13)}$$

^{31}P n.m.r. studies on the products of the second and third reactions, and an X-ray diffraction study on $P_4S_3Mo(CO)_5$, have shown in all cases that P_4S_3 acts as a unidentate ligand through its apical P atom (Fig. 10.2). In this respect it resembles P_4O_6.[11] In contrast, P_4 in the complex $(PPh_3)_2Rh(Cl)P_4$ is probably bonded to rhodium through a face.[13] Presumably the wide S–P–S angle in P_4S_3 (99°) allows the presence of a donor lone pair in an sp^3 hybrid orbital, while sp^3 hybridization is prevented in P_4 by the 60° bond angles. From the fact that P_4S_3 is not displaced from $Ni(P_4S_3)_4$ by PF_3 at 60°,[28] and from the changes in bond lengths accompanying coordination,[13] it seems that P_4S_3 has the same sort of interaction with zerovalent metals as the more common phosphine ligands. Similar complexes P_4S_3CuHal exist.[27]

Some information is available on the breakdown of phosphorus sulfide cages by nucleophilic attack, in hydrolysis, in aminolysis, and under the action of Grignard reagents.

P_4S_3 is fairly quickly hydrolyzed in boiling dilute aqueous NaOH, and the other sulfides hydrolyze more easily.[11] In contrast to the case of P_4O_{10}, which under mild hydrolysis gives largely cyclotetraphosphate, the ring or cage intermediates which must presumably be formed in the hydrolysis of phosphorus sulfides have not been isolated; the products actually reported have been smaller molecules resulting from more extensive breakdown.[11,24] Thus P_4S_{10} in alkaline solution gives almost wholly PO_4^{3-} and sulfide. The lower sulfides give complex mixtures of phosphorus oxoacids in various oxidation states, which have been investigated by paper chromatography but cannot be fully explained because of inadequately understood secondary reactions accompanying hydrolysis. Thus, in cold aqueous alkali plus hydrogen peroxide, hydrolysis of those sulfides containing P–P bonds (P_4S_3, P_4S_5, and P_4S_7) understandably gives substantial amounts of hypophosphate, which contains a P–P bond; however, in addition, the anion $[O_3P\text{-}P(O)_2\text{-}PO_3]^{5-}$, with a three-phosphorus chain, is formed from all these sulfides (including P_4S_7 which does not contain such chains) and also from P_2I_4 (which likewise does not). Thiophosphates are produced in substantial amounts when phosphorus sulfides hydrolyze under acid conditions; thus P_4S_{10} gives monothiophosphate PO_3S^{3-}, and P_4S_7 gives PO_3S^{3-}, HPO_2S^{2-}, H_2POS^-, and $H_2PS_2^-$.

P_4S_3 is not attacked under mild conditions by the tertiary amines Et_3N and Me_2PhN.[7] With primary and secondary amines in benzene at 80°, the cage is largely broken down, giving as one product amine bis-amidodithiophosphates, e.g. (from cyclohexylamine) $[C_6H_{11}NH_3]^+[(C_6H_{11}NH)_2PS_2]^-$. In addition, amorphous yellow substances are produced. These require further characterization but seem to be polymeric salts based on larger fragments of the P_4S_3 molecule containing P–P bonds. The product from cyclohexylamine has the empirical formula $[C_6H_{11}NH_3]^+[P_6S_2H]^-$; its solution in dimethylformamide conducts electricity and precipitates insoluble compounds on treatment with Pb^{2+} or Hg^{2+}.[7,24] In liquid ammonia or liquid methylamine at room temperature, cage breakdown is also extensive, P–P compounds arising from the lower phosphorus sulfides but not from P_4S_{10}.[7,24] Liquid ammonia at −33° reacts with the phosphorus sulfides giving unstable products which analytical data suggest may contain P–S cages or large fragments thereof, but their structures have not been examined by modern methods.[11,24] The ammonolysis of P_4S_{10} as a route to P_3N_5 is mentioned in Chapter 13, Sec. III.C.

Among the most interesting known reactions of the phosphorus sulfides are those within the general category of oxidation or electrophilic attack. In some cases, at least, these give rise to new substances with ring or cage structures, as follows.

Solutions of P_4S_3 in benzene, carbon disulfide, carbon tetrachloride, or o-dichlorobenzene rapidly react with dry air or oxygen, forming a bulky, yellowish-white precipitate with the empirical formula $P_4O_4S_3$ and of unknown, probably polymeric, structure.[11] It deliquesces rapidly in moist air, inflames with a little water, and with much water decomposes and partly dissolves.

With 1 mol iodine in benzene at room temperature, P_4S_3 gives the cage compound β-$P_4S_3I_2$ (Section E below).[26] A large excess of iodine causes the formation of P_4S_7.[11] P_4S_3, P_4S_5, and P_4S_7, but not P_4S_{10}, react with bromine suggesting that, as in the case of iodine, the initial step is the opening of a P–P bond, though the initial stages of bromination have not been fully studied. P_4S_3 consumes up to 4 mol Br_2 in CS_2 at room temperature.[11,23] PBr_3, $PSBr_3$, and α-P_4S_5 are reported bromination products from the lower sulfides. P_4S_7 gives $P_2S_6Br_2$ and $P_2S_5Br_4$ (Section E below).[3]

Chlorine reacts exothermically with P_4S_{10} at room temperature, giving S_2Cl_2, SCl_2, and PCl_5.[48] No ring or cage products have been found, but there is no report of any attempt at mild, controlled chlorination of phosphorus sulfides.

NF_3 is known to behave at high temperatures as a mild and selective fluorinating agent. With P_4S_3 and P_4S_{10} in a nickel trap at 180–215° it gives high yields of the cyclophosphazenes $(NPF_2)_n$, mainly the trimer (Chapter 12);[43] these compounds are similarly obtainable from red phosphorus, phosphides, or P_3N_5; the synthesis owes nothing to the particular molecular structures of P_4S_3 and P_4S_{10}.

The opening up of phosphorus sulfide cages is obviously a rich and little-explored field of investigation. With more use of ^{31}P n.m.r. spectroscopy the facts could now be relatively easily established, and the cluster theory of Wade and Mingos (Chapter 9, Sec. V..C) might provide a new basis for their interpretation.

C. Uses of the phosphorus sulfides

P_4S_3 is used in "strike anywhere" matches. P_4S_{10} is used in tens of thousands of tons annually for the production of additives for lubricants, insecticides, and flotation agents. Details of these and other, minor uses, and references to the extensive patent literature, may be found in ref. 24.

D. The oxosulfide $P_4S_6O_4$

The structure of the compound provisionally assigned this formula has not been determined, no molecular weight has been reported, and the only reported elemental analysis is for sulfur.[1] The method of synthesis:

$$6(Me_3Si)_2S + 4POCl_3 \xrightarrow{heat} 12Me_3SiCl + P_4S_6O_4$$

suggests that it may have structure I, differing from P_4S_{10} (Fig. 10.2) only in the replacement of peripheral sulfur atoms by oxygen.

The presumed $P_4S_6O_4$ forms yellow crystals, m.p. 290–295°.[1] The related compound II, $P_4O_6S_4$, based on a P–O cage, is known and from X-ray analysis definitely has the P_4S_{10} or P_4O_{10} structure;[11] compare also the compound $P_4(NMe)_6S_4$ (Chapter 11, Sec. II.E).

E. Cage and ring compounds from halogenation of phosphorus sulfides

Two isomeric compounds $P_4S_3I_2$, known to have structures **III** and **IV**, have been prepared from P_4S_3, as follows:[26]

$$P_4S_3 + I_2 \xrightarrow[20°]{\text{benzene}} \beta\text{-}P_4S_3I_2$$

$$\beta\text{-}P_4S_3I_2 \xrightarrow[20°, \text{ days}]{\text{benzene soln}} \alpha\text{-}P_4S_3I_2 \downarrow \text{ (slowly crystallizes)}$$

or

$$\beta\text{-}P_4S_3I_2 \xrightarrow[122°]{} \alpha\text{-}P_4S_3I_2$$

Evidently the α-isomer is the more stable under ordinary conditions.

(III) α-$P_4S_3I_2$, m.p. 122–124°

(IV) β-$P_4S_3I_2$, m.p. 111–112°

The $P_4S_3I_2$ isomers form moisture-sensitive orange crystals. Their molecular structures have been determined by X-ray crystallography,[26] and that of α-$P_4S_3I_2$ confirmed by ^{31}P n.m.r. and i.r.[6] Figure 10.3 shows the structural relationship to P_4S_3; iodine apparently breaks open a P–P bond in P_4S_3, giving β-$P_4S_3I_2$, and the α-isomer results from a rearrangement which must entail the breaking of cage bonds.

The behavior of α-$P_4S_3I_2$ as a ligand has been examined.[6] The molecule

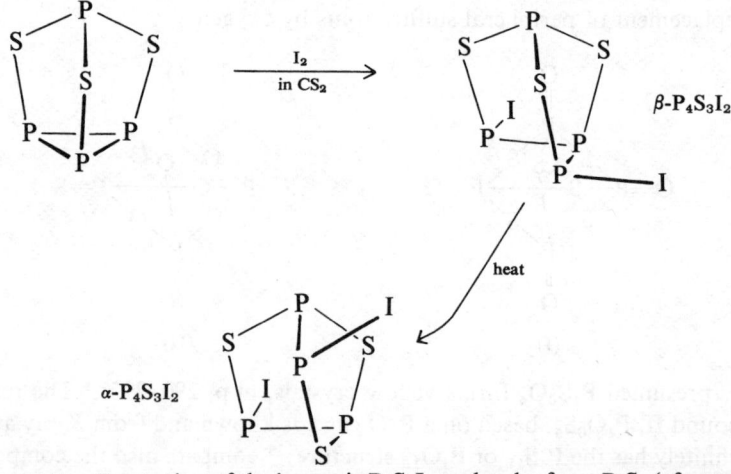

Figure 10.3. Formation of the isomeric $P_4S_3I_2$ molecules from P_4S_3 (after ref. 26).

contains two potential donor P(III) atoms (Fig. 10.3) without attached iodine and therefore relatively free from steric hindrance, but in all its known reactions with metal carbonyls it functions as a unidentate ligand.[6] For example, with $Ni(CO)_4$ in CS_2, or $Fe(CO)_5$ in benzene, the complexes $NiL(CO)_3$, $NiL_2(CO)_2$, $FeL(Co)_4$, and $FeL_2(CO)_3$ are readily obtained as brown air-sensitive solids only slightly soluble in inert solvents. Similar mixed carbonyl complexes of Co, Cr, Mo, and W exist.[6]

It has recently been found that the iodine atoms of α-$P_4S_3I_2$ can be variously substituted without (as ^{31}P n.m.r. shows) altering the molecular skeleton. Refluxing in CS_2 with the appropriate silver halide or pseudohalide gives $P_4S_3Cl_2$, $P_4S_3Br_2$, $P_4S_3(CN)_2$, and $P_4S_3(NCS)_2$ as moisture-sensitive crystals with low melting points.[20] Amines also react with the iodide but the products have not been fully characterized.[20] α- and β-$P_4S_3I_2$ react with $(Me_3Sn)_2S$[21] or with aniline[10] giving respectively the α- and β-isomers of P_4S_4.

With mercury, α-$P_4S_3I_2$ gives P_4S_3; here a skeletal rearrangement must accompany dehalogenation.[20]

The action of bromine on P_4S_3 gives rise to α-P_4S_5, possibly via an unstable intermediate α-$P_4S_3Br_2$.[23]

With about 3 mol Br_2 in CS_2 at 0°, P_4S_7 gives the crystalline compounds $P_2S_6Br_2$ (m.p. 118° dec.) and $P_2S_5Br_4$ (m.p. 90° dec.).[3] The former has, by X-ray diffraction, the molecular structure shown in Fig. 10.4, with a 6-membered P_2S_4 ring in a skew-boat configuration, perhaps imposed by a need for dihedral angles near 90° about the S–S bonds.[16] The structure of $P_2S_5Br_4$ is not known; possibly it is similar but with an exocyclic sulfur replaced by two Br. There is no obvious simple mechanism by which this P_4S_2 ring could be formed from the P_4S_7 cage (Fig. 10.2). Densities and powder diffraction data have been reported for both bromides,[3,16] and $P_2S_5Br_4$ is reported to give PBr_3, $P(S)Br_3$, and unidentified products on decomposition at 200°.[3]

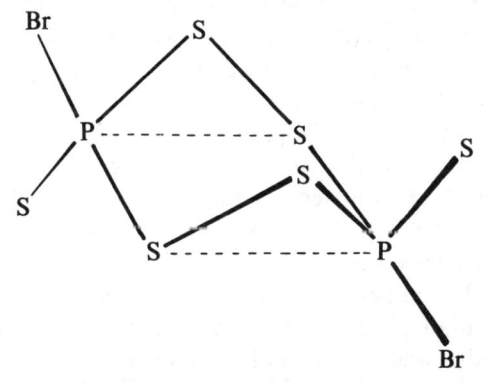

Figure 10.4. Molecular structure of $P_2S_6Br_2$ (after refs. 16 and 24).

III. Organo-substituted P–S rings

A. Six-membered P_3S_3 rings

The existence of compounds **V** has been reported but requires further investigation.[29] They are said to result from the reaction

$$[ROPS_3]^{2-} K_2^+ + 2HCl \xrightarrow[-25°]{CHCl_3} \tfrac{1}{3}(ROPS_2)_3 + 2KCl + H_2S$$

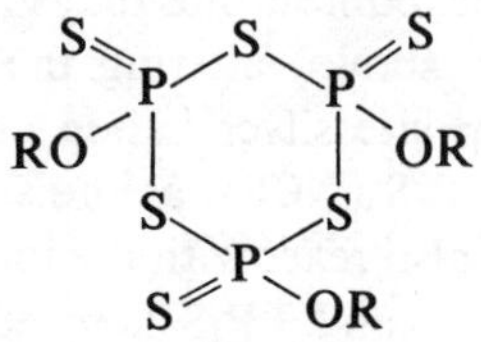

(R = Et, Bu)

(V)

They are dark green liquids, insoluble in petroleum ether but soluble in benzene, acetone, ether, chloroform, and carbon tetrachloride. Rather poor molecular weight values corresponding most nearly to the trimeric formula were reported.[29] Apart from the complicated i.r. spectra, no structural information is available.

B. Five-membered P_3S_2 rings

The compound $Ph_3P_3S_3$ has been obtained by many reactions, of which the following are representative:

$$3(Me_3Si)_2S + 3PhPCl_2 \xrightarrow{heat} 6Me_3SiCl + Ph_3P_3S_3 \quad \text{(ref. 1)}$$

$$3PhPH_2 + 3SCl_2 \xrightarrow[\text{high dilution}]{ether} 6HCl + (PhPS)_3 \quad \text{(ref. 5)}$$

A third method, reaction of a cyclopolyphosphine with sulfur, is known to be a useful route to P–S rings (cf. the end of this section, and Section C below) but the conditions of reaction need further investigation in order to establish its scope and generality. In some cases, e.g. $(PhP)_5$, reaction takes place below 100° and may well involve a nucleophilic ring-opening of S_8 by the phosphine.[5] Failing this, with a too weakly nucleophilic phosphine, reaction may still be possible near 200°,[8] in the temperature range where the S_8 rings open thermally [$(CF_3P)_4$ below]; this method may not have been tested in some reported cases of non-reaction.[33]

Following the determination of its molecular weight in 1964, $Ph_3P_3S_3$ was at first, not unnaturally, imagined to have a symmetrical structure such as **VI**, but recently Baudler and coworkers have shown by means of a difficult and complicated ^{31}P n.m.r. study that the structure is **VII** (R = Ph).[5] The com-

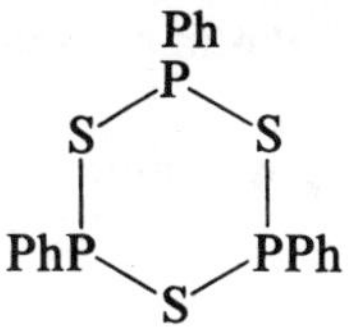

(VI) 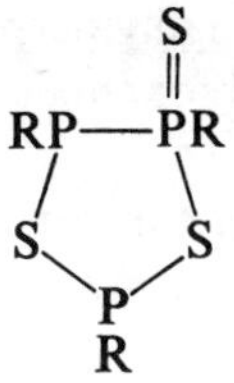 (VII) 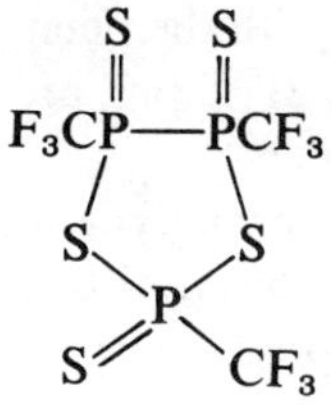 (VIII)

pound does not crystallize well enough for satisfactory X-ray crystallography. ^{19}F and ^{31}P n.m.r., and i.r., on the analogous p-fluoro compound also unequivocally establish structure **VII** (R = p-FC$_6$H$_4$).[31] The indications are, then, that the 5-membered ring of **VII** is more stable than the hypothetical 6-membered ring of **VI**, and that the various syntheses of **VII** are thermodynamically controlled. The ring structure **VII** also occurs in the molecule of α-P_4S_5 (Fig. 10.2).

$Ph_3P_3S_3$ forms cubic colorless crystals melting at 148–149°, sparingly soluble in diethyl ether. It is stable in an inert atmosphere at room temperature but is hydrolyzed by water, producing H_2S.[5]

The compounds just described contain unsymmetrical rings with two P(III) atoms and one P(V) atom. A symmetrical P(V) compound with the same ring atoms, $(CF_3)_3P_3S_5$, shown by ^{19}F n.m.r. to have structure **VIII**, has been made by the reaction[8,33]

$$\tfrac{3}{4}\text{cyclo-}(CF_3P)_4 + \tfrac{5}{8}S_8 \xrightarrow[\text{5 h, 185°}]{\text{sealed tube}} (CF_3)_3P_3S_5$$
(large excess)   (98% yield)

It exists as white sublimable crystals, m.p. 133°, thermally stable to 200°. Attempts to desulfurize it with mercury or Me_3P yielded largely resins, not the analog of **VII**.[8]

C. Five-membered P_4S rings

Compounds of structure **IX** are readily obtained by heating cyclopolyphosphines (RP)$_4$ (R = CF_3 or Ph) with 1 g-atom sulfur per mole and in other

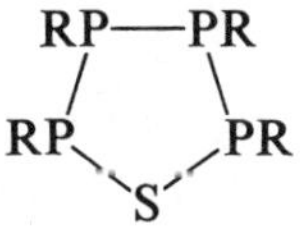

(IX)

ways.[9,32,33] The CF_3 compound is a liquid, the phenyl compound yellow needles. The molecular structure of the latter, determined by X-ray crystallography, is shown in Fig. 10.5(a). The sulfur–phosphorus ring is nearly planar, with the phenyl groups roughly perpendicular to it, and the P–S bond lengths, 211.6 pm, are as expected for single bonds.[9] A ^{31}P n.m.r. spectrum is available.[25]

The thermodynamic stability of this 5-membered ring is demonstrated by the fact that it also results from heating $(CF_3P)_5$ with sulfur; $(CF_3)_5P_5S$ was not so produced.[8]

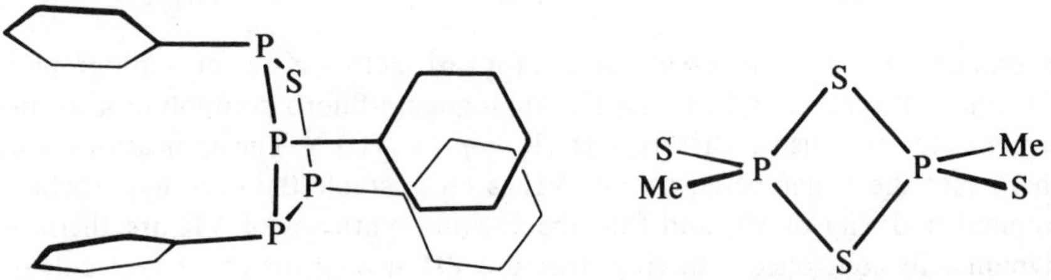

Figure 10.5. Molecular structures of phosphorus–sulfur rings with organic ligands: (a) $(PhP)_4S$ (after ref. 9); (b) $(MePS_2)_2$ (after ref. 47).

D. Four-membered P_2S_2 rings

Compounds $(RPS_2)_2$, often called thiophosphonic acid anhydrides or thionophosphine sulfides, were first clearly described in 1952. The structure of the

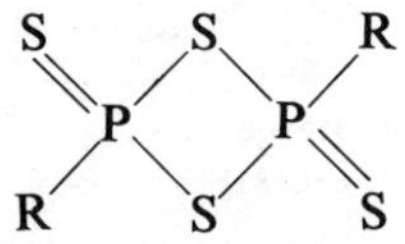

(X)

methyl compound [Fig. 10.5(b)] was reported[47] in 1962 from an X-ray diffraction study; the P–S ring bonds, 214 pm long, are clearly single. The dimeric formula $(RPS_2)_2$ has been established, too, by molecular weight determinations for the phenyl and other derivatives,[4] which, together with i.r. evidence,[4] support the general structural formula X.

These compounds are most conveniently prepared by the reaction[39]

$$2RP(S)Cl_2 + 2H_2S \xrightarrow[160-240°]{reflux} (RPS_2)_2 + 2HCl$$

(R = Me, Et, Pr, Pr^i, Bu, Ph, cyclohexyl)

Alternatively, $RPCl_2$ may be treated with H_2S_2 in ether at room temperature; but H_2S_2 is not readily available.[4] Aromatic $(ArPS_2)_2$ compounds also result from the prolonged heating (160–225°) of aromatic hydrocarbons or ethers with P_4S_{10}.[30] From these syntheses it is clear that the 4-membered ring of X represents the thermodynamically preferred degree of polymerization under ordinary conditions, as in the structurally related phosph(v)azanes (Chapter 11). It is true that crude products from the synthesis of $(RPS_2)_2$ have repeatedly given molecular weight values somewhat too high for dimers, so that less stable trimers or higher polymers may well be present;[4,39] however, these are not formed from purified dimer in the mass spectrometer.[36]

The $(RPS_2)_2$ compounds form high-melting pale yellow crystals, only slightly soluble in most indifferent solvents but recrystallizable from chlorobenzene or o-dichlorobenzene.[30,39] They are readily hydrolyzed by water to phosphonic acids, and therefore sensitive to atmospheric moisture.[30,39] There has been some recent interest in their use as intermediates for the synthesis of insecticides, acaricides, and sludge inhibitors for oils; see the patent literature under "1,3,2,4-dithiadiphosphetane" in *Chemical Abstracts*. The following unusual and interesting reaction falls within the scope of inorganic heterocyclic chemistry:[40]

$$RP(S)(S)PR(S)(S) + Me_3SiN_3 \xrightarrow[20°]{CH_2Cl_2} RP(S)(S)PR(S)(N\text{-}SiMe_3) + \frac{1}{n}S_n + N_2$$

(R = Me or Et) (cf. Chapter 1, Sec. III.F). The cyclic product, which is formed quantitatively, is a colorless sublimable solid melting at 114°. Sodium azide, in contrast, cleaves the ring,[40] as do other nucleophiles, namely alcohols, thiols, amines, and phosphines.[19]

IV. Phosphorus sulfide high polymers

Viscosity and ^{31}P n.m.r. data show that the increasing incorporation of phosphorus into quenched polymeric sulfur melts leads first to branching and crosslinking of the chains, and then to breakdown of the polymers by formation of phosphorus sulfide cage molecules.[35] The high polymers are present in insignificant amounts for S/P atom ratios below 1.8. The phosphorus atoms in them are probably P(v) with terminal sulfurs, i.e. $S=P\lneq$. Other S–P polymers, probably differing from the above in having sulfur chains of a definite rather than random length, can be obtained from chlorosulfanes and perthiophosphoric acid, e.g.[41]

$$2nH_3PS_4 + 3nS_2Cl_2 \rightarrow 2(PS_7)_n + 6nHCl$$

These are light-yellow plastic substances, insoluble in organic solvents, and stable at room temperature.

These P–S high polymers have aroused little interest and might repay further investigation with modern n.m.r. machines and vibrational spectroscopy.

References

1. Abel, E. W., Armitage, D. A. and Bush, R. P., *J. Chem. Soc.* 5584 (1964)
2. Andrew, E. R., Vennart, W., Bonnard, G., Croiset, R. M., Demarcq, M. and Mathieu, E., *Chem. Phys. Lett.* **43**, 317 (1976)
3. Andrews, J. M., Ferguson, J. E. and Wilkins, C. J., *J. Inorg. Nucl. Chem.* **25** 829 (1963)
4. Baudler, M. and Valpertz, H. W., *Z. Naturforsch.* B **22b**, 222 (1967)
5. Baudler, M., Koch, D., Vakratsas, Th., Tolls, E. and Kipfer, K., *Z. Anorg. Allg. Chem.* **413**, 239 (1975)
6. Baudler, M. and Mozaffar-Zanga, H., *Z. Anorg. Allg. Chem.* **423**, 193 (1976)
7. Becke-Goehring, M. and Hoffman, H., *Z. Anorg. Allg. Chem.* **369**, 73 (1969)
8. Burg, A. B. and Parker, D. M., *J. Am. Chem. Soc.* **92**, 1898 (1970)
9. Calhoun, H. P. and Trotter, J., *J. Chem. Soc. Dalton Trans.* 386 (1974)
10. Chang, C.-C., Haltiwanger, R. L. and Norman, A. D., *Inorg. Chem.* **17**, 2056 (1978)
11. Childs, A. F., "The Phosphorus Sulphides", in the Supplement to Mellor's *Comprehensive Treatise on Inorganic and Theoretical Chemistry*, Vol. VIII, Supplement III, Longman (1971)
12. Corbridge, D. E. C., *The Structural Chemistry of Phosphorus*, Elsevier, Amsterdam, London, and New York (1974)
13. Cordes, A. W., Joyner, R. D., Shores, R. D. and Dill, E. D., *Inorg. Chem.* **13**, 132 (1974)
14. Cueilleron, J. and Vincent, H., *Bull. Soc. Chim. France* 1296 (1970)
15. Cueilleron, J. and Vincent, H., *Bull. Soc. Chim. France* 2118 (1970)
16. Einstein, F. W. B., Penfold, B. R. and Tapsell, Q. T., *Inorg. Chem.* **4**, 186 (1965)
17. Falius, H., *Naturwiss.* **50**, 126 (1963)
18. Fieser, L. F. and Fieser, M., *Reagents in Organic Synthesis* (continuing series from 1967), Wiley, New York
19. Fluck, E., Gonzalez, G. and Binder, H., *Z. Anorg. Allg. Chem.* **406**, 161 (1974)
20. Fluck, E., Yutronic, S. N. and Haubold, W., *Z. Anorg. Allg. Chem.* **420**, 247 (1976)
21. Griffin, A. M., Minshall, P. C. and Sheldrick, G. M., *J. Chem. Soc. Chem. Commun.* 809 (1976)
22. Griffin, A. M. and Sheldrick, G. M., *Acta Cryst.* **B31**, 2738 (1975)
23. Griffin, A. M., Minshall, P. C. and Sheldrick, G. M., *Abstracts, 2nd International Symposium on Inorganic Ring Systems*, Akademie der Wissenschaften, Göttingen (1978).
24. Hoffmann, H. and Becke-Goehring, M., *Topics in Phosphorus Chemistry*, **8**, 193 (1976)
25. Hoffman, P. R. and Caulton, K. G., *Inorg. Chem.* **14**, 1997 (1975)
26. Hunt, G. W. and Cordes, A. W., *Inorg. Chem.* **10**, 1935 (1971)
27. Ibáñez, W. F., González, M. G. and Clavijo, C. E., *Z. Anorg. Allg. Chem.* **432**, 253 (1977)

28. Jefferson, R., Klein, H. F. and Nixon, J. F., *Chem. Commun.* 536 (1969)
29. Kolesnikova, N. A., *J. Gen. Chem. (USSR)* **44**, 2628 (1974)
30. Lecher, H. Z., Greenwood, R. A., Whitehouse, K. C. and Chao, T. H., *J. Am. Chem. Soc.* **78**, 5018 (1956)
31. LeGeyt, M. R. and Paddock, N. L., *J. Chem. Soc. Chem. Commun.* 20 (1975)
32. LeGeyt, M. R., *Diss. Abs. Int. B* **35**, 5309 (1975)
33. Maier, L., "Organophosphines and Polyphosphines", in Kosolapoff, G. M. and Maier, L., *Organic Phosphorus Chemistry*, vol. 1, Wiley-Interscience, New York, London, and Sydney (1972)
34. Meisel, M. and Grunze, H., *Z. Anorg. Allg. Chem.* **373**, 265 (1970)
35. Moedritzer, K. and van Wazer, J. R., *J. Inorg. Nucl. Chem.* **25**, 683 (1963)
36. Moedritzer, K., *Phosphorus* **3**, 219 (1974)
37. Monteil, Y. and Vincent, H., *Z. Naturforsch. B* **31b**, 668 (1976)
38. Muenow, D. W. and Margrave, J. L., *J. Inorg. Nucl. Chem.* **34**, 89 (1972)
39. Newallis, P. E., Chupp, J. P. and Groeneweghe, L. C. D., *J. Org. Chem.* **27**, 3829 (1962)
40. Roesky, H. W. and Dietl, M. *Angew. Chem. Int. Ed. Engl.* **12**, 425 (1973)
41. Schmidt, M., *Inorg. Macromol. Rev.* **1**, 101 (1970)
42. Sheldrick, G. M. and Penney, G. J., *J. Chem. Soc. A* 243 (1971)
43. Tasaka, A. and Glemser, O., *Z. Anorg. Allg. Chem.* **409**, 163 (1974)
44. Toy, A. D. F., "Phosphorus", in vol. 2 of *Comprehensive Inorganic Chemistry*, ed. J. C. Bailar *et al.*, Pergamon, Oxford (1973)
45. Vincent, H., *Bull. Soc. Chim. France* 4517 (1972)
46. Vincent, H. and Vincent-Forat, C., *Bull. Soc. Chim. France* 499 (1973)
47. Wheatley, P. J., *J. Chem. Soc.* 300 (1962); Daly, J. J., *ibid.* 4065 (1964)
48. Zur, Z. and Dykman, E., *Chem. and Ind.* 436 (1975)

11
Saturated phosphorus–nitrogen heterocycles: the cyclo- and closo-phosphazanes

I. Introduction

A. Scope, status, and treatment

Following the pattern adopted for sulfur–nitrogen rings, the phosphorus–nitrogen heterocycles will be treated under the following heads: formally saturated rings and cages (phosphazanes), Chapter 11; formally unsaturated rings (phosphazenes), Chapter 12; and finally, the phosphorus nitrides and related compounds, which are polymeric and probably to some degree unsaturated, Chapter 13.

The order of treatment within the present chapter is as follows. First, phosphazanes (rings and cages) based on phosphorus(III); secondly, phosphazanes based on phosphorus(V). The fairly numerous compounds belonging to phosphazane structural types and containing N–N (hydrazido) groups in the ring are dealt with under phosphorus(III) or phosphorus(V) as appropriate.

Despite an extensive literature, the subject of cyclophosphazanes is still at a relatively primitive stage of development compared with that of the cyclophosphazenes. The first well authenticated simple phosph(III)azane rings were described as recently as 1969; there were earlier attempts and claims, but research on this particular subject has suffered to an unusual degree from poor reproducibility of results between different experimenters. Cyclophosph(V)-azanes have been known longer and are more numerous and more tractable. To a remarkable extent, 4-membered P_2N_2 rings dominate the chemistry of both classes of compound; few cyclophosphazanes with rings of other sizes have been reported. A surprising discovery was that attempts to make cyclophosph(III)azanes with MeN< as a ring component by a standard method give instead cage molecules, a result apparently unique to the methyl compounds. It is a notable weakness in the understanding of the phosphazanes that the preferences for these unusual structures have not been rationalized.

In order to keep this account reasonably short and readable, not all known

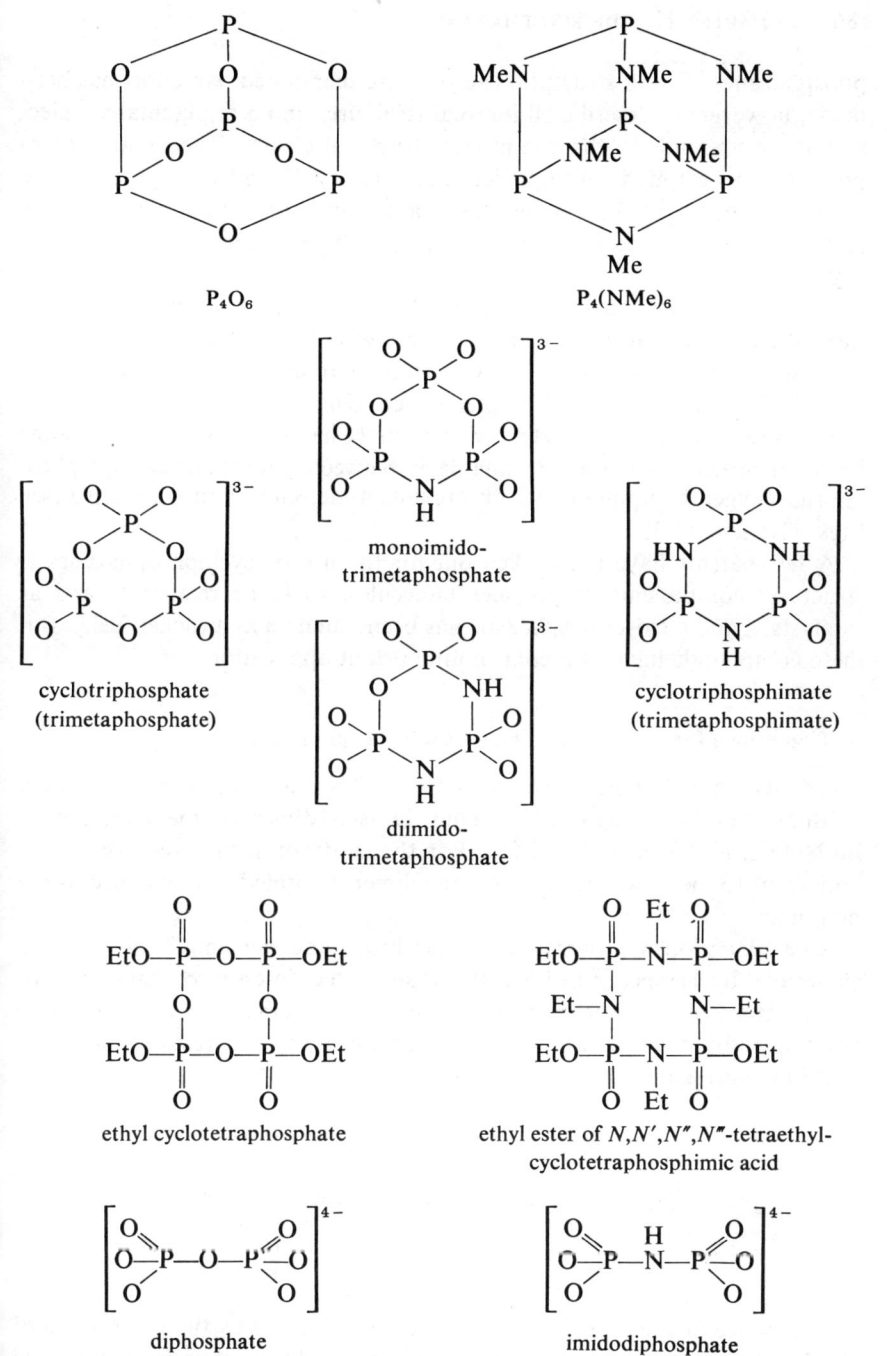

Figure 11.1. A few of the analogies between P–N and P–O compounds brought out by the Franklin "nitrogen system" viewpoint.

phosphazanes in each structural category are mentioned. An effort has been made, however, to describe all unusual structures and a representative selection of compounds in the commoner structural classes. The omitted compounds can be found through Bermann's review,[12] and through the more recent references at the end of this chapter and their bibliographies. The phosphazanes are reviewed from time to time[2,11,12,28,29,35,73] but not regularly.

The relatively few names in this chapter are based on the system in commonest current use, introduced by R. A. Shaw and collaborators in 1962, and endorsed in 1970 by I. Haiduc. A different naming system is used by the Chemical Abstracts Service (Chapter 1, Sec. II).

The Franklin nitrogen-system viewpoint (Chapter 3, Sec. II) has some value in drawing attention to analogies between phosphazanes and phosphorus–oxygen compounds, which are sometimes close and instructive (see Figs. 11.1 and 11.3).

A few patents have been taken out for the use of cyclophosphazanes as structural components of polymer molecules, as flame retardants and as catalysts,[11] and a closophosphazane has been patented as a rocket fuel.[80] But these compounds have at present no important application.

B. Conditions for stability of dimeric cyclophosphazanes

Most cyclophosphazanes have 4-membered P_2N_2 rings and can be regarded as dimers of monomeric units. Thus **Ia** is a dimer of the hypothetical $Bu^tN=PCl$, and **IXa** of $RN=PCl_3$. For the study of such heterocycles it is important to know when, and why, the dimer is formed in preference to the monomer.

An explanation was suggested by Zhmurova and Kirsanov[86] in 1960 and elaborated by Trippett[77] in 1962. It was suggested, in essence, that dimers of the phosph(v)azanes might in appropriate circumstances be stabilized by π bonding within the ring, using the nitrogen lone pairs. The resulting structure could be represented as

The presence of such π bonding seems to be supported by the u.v. spectra of the dimers.[35,77] It was argued that π bonding, and hence dimerization, should be favored when the ligands R on nitrogen are electron-releasing groups, and those on phosphorus are strongly electron-withdrawing or capable of stabilizing a negative charge. The weight of evidence available today certainly seems to support this view. Thus the phosphorus ylides RN=PR$_3$, with no very electronegative groups on phosphorus, are always monomeric, while the compounds (RN·PF$_3$)$_2$ are always dimeric.[11,12] The dimers (ArN·PCl$_3$)$_2$ become increasingly stable relative to the monomers as the base ArNH$_2$ becomes stronger (Section III.C below), though exceptions arise for aryl groups with unusual steric requirements.[12,39,86] Similarly, the alkyl compounds (RN·PCl$_3$)$_2$ are dimeric, as expected from the base strengths of RNH$_2$, except in the case of sterically hindered alkyl groups.[11] Further, the relatively few reported lengths of P–N ring bonds seem consistent with the idea that more electronegative groups on phosphorus strengthen the ring π bonding (references cited in ref. 39). In a few borderline cases, such as

$$[Me_2C(CN)N=PCl_3]_{1,2},$$

both monomer and dimer result from the same preparation[12] and have been separately characterized. In others, (ArN=PCl$_3$)$_{1,2}$, there is some indication of monomer–dimer equilibria in solution,[86] but such equilibria have not been studied in detail.

When Zhmurova, Kirsanov, and Trippett first discussed this question, little was known about the cyclophosph(III)azanes, to which similar considerations might be expected to apply. It does now appear that compounds of this class too are stabilized by electron-releasing groups on nitrogen and electron-withdrawing groups on phosphorus (Section II.C below) but more evidence is needed. The corresponding monomers RN=PR' are almost unknown.[58]

II. Saturated phosphorus(III)–nitrogen heterocycles

A. Introduction

In order to synthesize P(III)–N heterocycles it is necessary to form P(III)–N bonds. Hitherto this has usually been done by the reaction of a phosphorus-(III) halide with ammonia, an amine, or a metal derivative of an amine. Such reactions sometimes yield heterocycles as their final product, but more typically the products are non-cyclic; broadly speaking, the factors discussed above in Section B will explain the results in particular cases. In principle, at least, one might also in some cases expect to obtain linear and other polymers based on the repeating unit –NR–PR–. Almost nothing is known of these, though uncharacterized products which may contain them are mentioned in some published papers.

For purposes of orientation, we begin with a short survey of the reactions of phosphorus trihalides with amines and ammonia.

B. Reactions of phosphorus trihalides with amines and ammonia

PF_3, PCl_3, and PBr_3 are very weak Lewis acids and at low temperatures form very easily dissociated 1:1 adducts with tertiary amines.[16,28] The interaction energy of PCl_3 with Me_3N, for example, is only 27 kJ mol^{-1}.

With secondary alkylamines under mild conditions, PF_3 and PCl_3 probably first form adducts,[37] but hydrogen halide is quickly eliminated and dialkylaminophosphines result. R_2NPX_2, $(R_2N)_2PX$, and $(R_2N)_3P$ can all be obtained under appropriate conditions.[28,45] For example, $(Me_2N)_3P$, a water-soluble, easily oxidized liquid, results from the reaction

$$6Me_2NH + PCl_3 \xrightarrow[20°]{\text{ether}} (Me_2N)_3P + 3(Me_2NH_2)^+Cl^-$$

Certain R_2NPCl_2 compounds may be used for the synthesis of 4-membered ring phosph(III)azanes, via phosphatriazenes or phosphatetrazenes which dimerize (Section C below).

Primary amines react with PCl_3, initially replacing chlorine by RNH-. With excess amine the simple derivatives $(RNH)_3P$ are produced in some cases (e.g. R = benzyl),[28] but with methylamine the main product is the interesting adamantane-like cage $P_4(NMe)_6$ (Section E below). With PCl_3 in excess a major product is $RN(PCl_2)_2$ when R = Me, Et, or Ph,[34,49] but under similar conditions isopropylamine and t-butylamine give the cyclic dimers $(RNPCl)_2$ (Section C below). PF_3 shows interesting differences from PCl_3. $RNHPF_2$ is formed in good yield, but with more amine the phosphorus is oxidized and large quantities of the phosphorane $(RNH)_2PF_2H$ are produced.[37]

PF_3 is known from old observations[32] to react with ammonia, but there are no detailed reports on the products. Rankin recently found that the reaction

$$PBrF_2 + 2NH_3 \xrightarrow[3 \text{ min, } 20°]{\text{gas}} NH_4Br + H_2NPF_2$$

gives the gas aminodifluorophosphine in good yield.[63] Presumably the reaction of PCl_3 with ammonia also involves the substitution of halogen by $-NH_2$, but the course has not been fully elucidated. Becke-Goehring and Schulze[7] found that in chloroform at $-78°$ a white unstable solid was precipitated with the correct analytical composition for the products of the reaction

$$PCl_3 + 6NH_3 \rightarrow (H_2N)_3P + 3NH_4Cl$$

The triamide $(H_2N)_3P$ which may be present has not been isolated, though an adduct of it with BF_3 is well characterized. In ether at $-20°$ the same reaction yields phosphorus(III) amide-imide, $(H_2N)(HN)P_x$, a deammonation product of $(H_2N)_3P$. The amide-imide is a brownish hygroscopic powder, m.p. 115–120°(dec.) and unstable even at 20°. It is only slightly soluble in liquid ammonia and so may well be a polymer. When PCl_3 is added directly to liquid ammonia without a solvent, phosphorus nitride, $(PN)_x$, is obtained in high yield (Chapter 13, Sec. III.B). It is a pale yellow solid amorphous to X-rays. Under these conditions of reaction the P(III) amino derivatives first formed are probably deammonated to the nitride by local heating.[7]

C. Cyclodiphosph(III)azanes: preparation, structure, and properties

These compounds are sometimes called diazadiphosphetidines, in accordance with the IUPAC nomenclature.

Several cyclodiphosphazanes of P(III), with 4-membered rings of alternating N and P atoms, are well established, but attempts to make such rings with certain combinations of ligands on nitrogen and phosphorus have not been successful.

$$\begin{array}{cccc}
\text{Bu}^t & \text{Bu}^t & \text{Pr}^i & \text{Me} \\
\text{ClP}\diagup^{\text{N}}\diagdown\text{PCl} & \text{FP}\diagup^{\text{N}}\diagdown\text{PF} & \text{ClP}\diagup^{\text{N}}\diagdown\text{PCl} & \text{ClP}\diagup^{\text{N}}\diagdown\text{PCl} \\
\text{Bu}^t & \text{Bu}^t & \text{Pr}^i & \text{Me} \\
\text{(Ia)} & \text{(Ib)} & \text{(Ic)} & \text{(Id)}
\end{array}$$

Notwithstanding some earlier relevant observations, the systematic chemistry of cyclophosph(III)azane rings may be considered to date from Scherer and Klusmann's 1969 report of the preparation of **Ia**.[69] It is most easily prepared by the reaction (Jefferson et al., 1973)

$$3\text{Bu}^t\text{NH}_2 + \text{PCl}_3 \xrightarrow[-78°]{\text{ether}} \tfrac{1}{2}(\text{Bu}^t\text{NPCl})_2 + 2(\text{Bu}^t\text{NH}_3)^+\text{Cl}^- \quad (1)$$
(38% yield)

adding the amine in slight excess to the PCl_3.[49] This reaction goes in two stages; the first, requiring 2 mol of amine, gives $\text{Bu}^t\text{N(H)PCl}_2$, which can (as an alternative preparation of **Ia**) be isolated and subsequently dehydrohalogenated with triethylamine.[69] Another alternative preparation uses the lithium salt of trimethylsilylbutylamine instead of butylamine itself:[69]

$$\text{PCl}_3 + \text{Bu}^t\text{NLiSiMe}_3 \rightarrow \tfrac{1}{2}(\text{Bu}^t\text{NPCl})_2 + \text{LiCl} + \text{Me}_3\text{SiCl}$$

Compound **Ia** is a solid, m.p. 40–42°. The molecular structure, determined by X-ray crystallography, is shown in Fig. 11.2.[56] The ring is slightly puckered; the ring angles at nitrogen are a little more than 90° and those at phosphorus less. The four ring P–N bonds do not differ significantly in length. They are shorter than the 177 pm commonly accepted as the length of a single P–N bond; this suggests some $p\pi$–$d\pi$ bonding N→P, using the nitrogen lone pair, a conclusion consistent with the nearly planar configuration around each nitrogen atom. A related trimethylsilyl derivative has a very similar but exactly planar P_2N_2 ring.[57]

The chlorine atoms in **Ia** are easily replaced. Thus mild fluorination with SbF_3 gives **Ib**, a colorless liquid, b.p. 23.5°/4 Torr, which on the evidence of ^{19}F n.m.r. contains the same ring as **Ia**.[49] If 4 mol of amine are used instead of

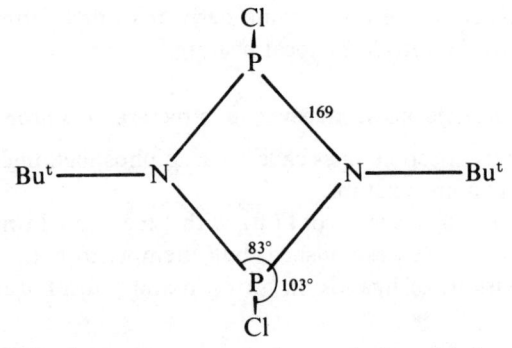

Figure 11.2. Molecular structure of the cyclophosph(III)azane $(Bu^tNPCl)_2$ (after ref. 56); bond distances are in pm.

3 mol in the recommended preparation of **Ia** above, compound **IIa** results[49] while **IIb** results from a large excess of amine.[43]

(IIa)   (IIb)   (IIc)

The phosphorus atoms in **Ia** are easily oxidized without destroying the ring. Thus **IIIa** and **IIIb** result from the action, in 1:1 mol ratio, of dimethyl sulfoxide or sulfur respectively on **Ia**. Compound **IIIc** is formed from dimethyl

(IIIa)   (IIIb)   (IIIc)

sulfoxide and **IIIb**.[49] The isopropyl compound **Ic** resembles **Ia** in its preparation and reactions.

The methyl compound **Id** is unknown. Attempts to make it, e.g. by reaction (1), have failed.[49] The cage compound $P_4(NMe)_6$ (Section E below) seems to be the preferred product. The ethyl analog of **Id** has not been characterized.[49] An 1894 report by Michaelis and Schroeter of the phenyl analog is discredited,[34,49] but a compound thought to have structure **IIc** has been made by a transamination:[81]

$$(Me_3N)_3P + 2PhNH_2 \xrightarrow[\text{vacuum}]{100°} \tfrac{1}{2}(PhNPNHPh)_2 + 3Me_2NH$$

and a benzyl analog of **IIb** has been reported.[5]

The monomers RP=NR' from which the cyclic phosph(III)azanes can be considered to be derived are in most cases unknown, since the reactions from which one might hope to make them (Section B above) have hitherto afforded other products. However, series of compounds in which R is a secondary amino group (phosphatriazenes) or a hydrazino group (phosphatetrazenes) have recently been synthesized and their dimerization to cyclic phosph(III)-azanes has been observed in some cases.[70] Thus compound **IVa** results as a

$$Pr^i_2N-P=NBu^t$$
(IVa)

$$\begin{array}{c} Pr^i_2NP-NBu^t \\ | \quad\quad | \\ Bu^tN-PNPr^i_2 \end{array}$$
(IVb)

$$\begin{array}{c} (Me_3Si)_2N(Me)NP-NSiMe_3 \\ | \quad\quad | \\ Me_3SiN-PN(Me)N(SiMe_3)_2 \end{array}$$
(IVc)

greenish-yellow liquid, capable of distillation at reduced pressure, from the reactions

$$Pr^i_2NLi + PCl_3 \longrightarrow LiCl + Pr^i_2NPCl_2$$
$$Pr^i_2NPCl_2 + (Me_3Si)Bu^tNLi \longrightarrow Pr^i_2N-PClNBu^t(SiMe_3) + LiCl$$
$$Pr^i_2N-PClNBu^t(SiMe_3) \xrightarrow{heat} Me_3SiCl + Pr^i_2N-P=NBu^t$$

On standing for 1–4 months in a closed flask the monomer **IVa** loses its color, and colorless crystals of the dimer **IVb** separate in quantitative yield.

Compound **IVc**, prepared analogously, is particularly interesting as the only cyclic phosph(III)azane to date that has actually been observed in equilibrium with its monomer. In a 20% solution in 1-chloronaphthalene, the equilibrium yields 100% dimer at 100°, whereas at 150° the equilibrium mixture is approximately 35% dimer and 65% monomer.[70]

All the well authenticated 4-membered ring phosphazanes have electron-releasing groups on nitrogen and strongly electronegative atoms attached to phosphorus. One can accordingly guess that π bonding stabilizes the rings, as discussed in Section I.B above, and may indeed compensate for the strain resulting from the near-90° ring angles; the puzzle remains, however, why larger rings with less-strained angles are not preferred.

Since the configuration around phosphorus in this group of compounds is not planar, *cis–trans* isomerism may arise according to the relative positions of the exocyclic groups on the two phosphorus atoms of a ring. Some cyclo-diphosph(III)azanes are known in *cis* form (e.g. Fig. 11.2), others in *trans* form,[70] and some in both. *Cis* or *trans* configurations can be assigned by several criteria, most conveniently by the ^{31}P chemical shift.[50] Slow isomerization of the compounds *cis*-(PhNPOR)$_2$ (R = Me, Et) to mixtures of cis- and trans-isomers has been observed by ^{31}P n.m.r. in chloroform solution at room temperature.[83]

Coordination complexes of platinum and rhodium exist[19] in which cyclodiphosph(III)azanes are ligands, bonded to the metal through one phosphorus atom of the ring.

D. Cyclotri- and cyclotetra-phosph(III)azanes

$$\begin{array}{c} \text{Et} \\ \text{ClP}^{\diagup\text{N}\diagdown}\text{PCl} \\ | \qquad\qquad | \\ \text{EtN}\diagdown\quad\diagup\text{NEt} \\ \text{P} \\ \text{Cl} \\ \text{(Va)} \end{array} \qquad \begin{array}{c} \text{Et} \\ \text{ClP—N—PCl} \\ |\qquad\quad| \\ \text{EtN}\qquad\text{NEt} \\ |\qquad\quad| \\ \text{ClP—N—PCl} \\ \text{Et} \\ \text{(Vb)} \end{array} \qquad \begin{array}{c} \text{Me} \\ \text{MeP—N—PMe} \\ |\qquad\quad| \\ \text{MeN}\qquad\text{NMe} \\ |\qquad\quad| \\ \text{MeP—N—PMe} \\ \text{Me} \\ \text{(Vc)} \end{array}$$

Far less is known about 6- and 8-membered phosphazane rings than about the 4-membered rings.

Compounds believed to be **Va** and **Vb** were prepared by Abel, Armitage, and Willey (1962) in one of the earliest applications of a silylamine to heterocyclic synthesis:[1]

$$\text{PCl}_3 + \text{EtN}(\text{SiMe}_3)_2 \xrightarrow[20°]{\text{toluene}} 2\text{Me}_3\text{SiCl} + \tfrac{1}{3}(\text{EtNPCl})_3 \quad 48\%$$
$$\tfrac{1}{4}(\text{EtNPCl})_4 \quad 14\%$$

and Zeiss and coworkers have recently confirmed the existence of these 6- and 8-membered rings by preparing **Vc** and the methyl analog of **Va** by similar methods.[84,85] There is also mass-spectral evidence of substantial amounts of **Va** in the products of the PCl$_3$–ethylamine reaction (Section C above), though EtN(PCl$_2$)$_2$ is the main product.[49]

Compound **Va** is an easily hydrolyzed and easily oxidized liquid, b.p. 128–131°/0.2 Torr. It is isomeric with trichlorotriethylcyclotriphosphazene (Chapter 12, Sec. V.B and VII.G), but has been characterized as distinct by its i.r. spectrum, from which the high-frequency unsaturated P–N stretch of the phosphazenes is missing. **Vc** forms colorless crystals. X-ray crystallography has shown the 8-membered P–N ring to have a perfect crown shape like that of S$_8$;[84] it is the only P–N ring known with this shape. All the methyl groups are on the same side of the ring, and the configuration around the nitrogen atoms is planar.

E. Cage phosph(III)azanes: preparation, structure, and properties

There is a well known cage phosph(III)azane, with the formula P$_4$(NMe)$_6$. It has been considerably studied and forms a number of derivatives which apparently contain the intact cage. The molecule has almost certainly the adamantane-like structure **VI**. No name for this compound which is both convenient and systematic has been found. The discoverer, R. R. Holmes (1961), called it phosphorus tri-*N*-methylimide. The systematic name* used in *Chemical Abstracts* has, understandably, not caught on, and Holmes's name is still in use. For easy reference, **VI** has been called birdcage compound.

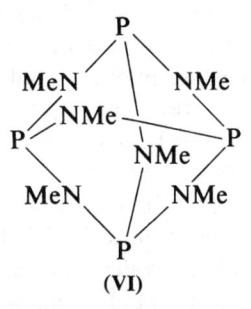

$$
\begin{array}{c}
\text{P} \\
\text{MeN} \quad \text{NMe} \\
\text{NMe} \\
\text{P} \quad \text{NMe} \quad \text{P} \\
\text{MeN} \quad \text{NMe} \\
\text{P}
\end{array}
$$

(VI)

$P_4(NMe)_6$ is prepared by adding PCl_3 to strongly cooled methylamine in the absence of a solvent:[43]

$$4PCl_3 + 18MeNH_2 \rightarrow P_4(NMe)_6 + 12(MeNH_3)^+Cl^-$$

It is extracted with petroleum ether from the product mixture. It is the only heterocyclic compound known to be formed in the PCl_3–methylamine reaction (cf. Sections B and C above). $P_4(NMe)_6$ forms colorless sublimable crystals, m.p. 123°. The liquid boils (under nitrogen) at 303–304°/737 Torr. The compound is readily soluble in indifferent organic solvents and insoluble in water.

The molecular weight of $P_4(NMe)_6$ has been established both cryoscopically and by vapor density. Its molecular structure has not yet been found by X-ray crystallography, though the space group and unit-cell dimensions are known.[47] However, n.m.r. and i.r. data prove structure **VI** beyond reasonable doubt.[28] The ^{31}P n.m.r. spectrum consists of one peak broadened by coupling with protons, showing the four P atoms to be equivalent. The ^1H n.m.r. shows equivalent protons, each split by two equivalent P atoms, and with the chemical shift expected for NMe groups.

The presumed structure of $P_4(NMe)_6$ closely resembles that of P_4O_6. $P_4(NMe)_6$ is, in a loose sense, isoelectronic with, and a nitrogen-system analog of, P_4O_6 (Fig. 11.1). Chemically too they are alike, in that both have stable frameworks and coordinatively unsaturated phosphorus atoms prone to oxidation and Lewis base activity. The reactions of $P_4(NMe)_6$ now to be described, with oxygen, sulfur, diborane, and nickel tetracarbonyl, are all closely paralleled by those of P_4O_6 (Fig. 11.3).[76]

$P_4(NMe)_6$ is inert towards oxygen at room temperature, but at 170° absorbs

---

* 2,4,6,8,9,10-Hexamethyl-2,4,6,8,9,10-hexaaza-1,3,5,7-tetraphosphatricyclo[3.3.1.1^{3,7}]-decane.

oxygen up to an amount corresponding roughly to the formula $P_4(NMe)_6O_4$. This oxide is best prepared from $P_4(NMe)_6$ and Me_3NO.[20] In solution, $P_4(NMe)_6$ takes up sulfur,[20] forming four well defined compounds, $P_4(NMe)_6S_n$ ($n = 1, 2, 3, 4$), which have been shown by ^{31}P n.m.r. spectroscopy to exist in equilibrium in solution.[68] A crystal-structure determination has shown $P_4(NMe)_6S_4$ to have, as expected, the cage of formula VI with a terminal sulfur atom on each phosphorus.[47] Sulfur can be abstracted from these compounds with Ph_3P, restoring $P_4(NMe)_6$. Mixed sulfide-oxides $P_4(NMe)_6S_nO_{4-n}$ ($n = 1$ to 3) can be made from the sulfides and Me_3NO.[20] Similar series of complexes are formed with 1, 2, 3, or 4 mol BH_3 by reaction with diborane,[67] with 1, 2, 3, or 4 PhN by reaction with phenyl azide,[13] and with 1, 2, 3, or 4 $Ni(CO)_3$ by reaction with nickel tetracarbonyl.[66] Evidently each phosphorus atom can add a divalent radical or a molecule of a neutral Lewis acid. The equilibrium constants for successive additions of S or BH_3 differ considerably from values calculated by purely statistical reasoning.[67,68]

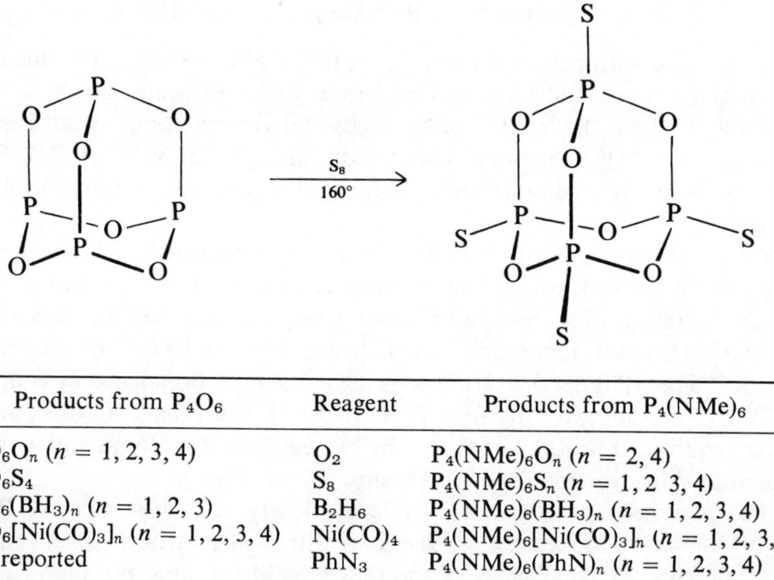

Products from P_4O_6	Reagent	Products from $P_4(NMe)_6$
$P_4O_6O_n$ ($n = 1, 2, 3, 4$)	O_2	$P_4(NMe)_6O_n$ ($n = 2, 4$)
$P_4O_6S_4$	S_8	$P_4(NMe)_6S_n$ ($n = 1, 2\ 3, 4$)
$P_4O_6(BH_3)_n$ ($n = 1, 2, 3$)	B_2H_6	$P_4(NMe)_6(BH_3)_n$ ($n = 1, 2, 3, 4$)
$P_4O_6[Ni(CO)_3]_n$ ($n = 1, 2, 3, 4$)	$Ni(CO)_4$	$P_4(NMe)_6[Ni(CO)_3]_n$ ($n = 1, 2, 3, 4$)
not reported	PhN_3	$P_4(NMe)_6(PhN)_n$ ($n = 1, 2, 3, 4$)

For literature on P_4O_6, see ref. 76; for literature on $P_4(NMe)_6$, see Section II.E of this chapter

Figure 11.3. Structures of P_4O_6 and the sulfide $P_4O_6S_4$ formed from it by the action of sulfur (see ref. 12 in Chapter 10), and a comparison of the chemical behavior of P_4O_6 and $P_4(NMe)_6$ with various reagents. $P_4(NMe)_6$ is believed to be analogous in structure to P_4O_6, and to add the listed ligands one at a time to its phosphorus atoms, giving products $P_4(NMe)_6L_4$ structurally analogous to $P_4O_6S_4$.

$P_4(NMe)_6$ readily adds 1 mol methyl iodide at room temperature, giving a water-soluble phosphonium salt.[46]

At $-78°$, $P_4(NMe)_6$ is completely broken down by HCl according to the equation[43]

$$P_4(NMe)_6 + 18HCl \rightarrow 4PCl_3 + 6(MeNH_3)^+Cl^-$$

A P–N bond-energy term for $P_4(NMe)_6$ has been deduced from its heat of combustion to be 288 kJ mol^{-1}, a typical figure for P–N single bonds.[27]

There are patents concerning the use of $P_4(NMe)_6$ as a rocket fuel.[80]

F. Heterocyclic hydrazido derivatives of phosphorus(III)

A few hydrazido derivatives of P(III) contain heterocyclic rings. In contrast to the phosphazanes described in Sections C to E above, which have alternating P and N atoms in the rings, the hydrazido compounds are characterized by pairs of directly linked nitrogen atoms.

<pre>
          P
       / | \
     MeN NMe NMe
      | | |
     MeN NMe NMe
       \ | /
          P
        (VII)
</pre>

$P_2(NMe)_6$, now known to have structure **VII**, was discovered by Payne, Nöth, and Henniger in 1965. The equation

$$2(Me_2N)_3P + 3MeNH-NHMe \xrightarrow[\text{reflux 12 h}]{\text{benzene}} P_2(NMe)_6 + 6Me_2NH$$

shows the most convenient reaction for its preparation.[59] This is really a transamination, recalling a reaction already described (Section C above) between $(MeN)_3P$ and aniline.

$P_2(NMe)_6$ forms needles, m.p. 116–117°, subliming at 70–80°/1 Torr. Structure **VII**, which is suggested by the mode of formation, molecular weight, and n.m.r. spectra,[28] was confirmed recently by single-crystal X-ray diffraction.[79] Each P–N–N–P group is planar. The P–N bond lengths, averaging 168 pm, are shorter than the accepted single-bond value of 177 pm, but this cannot be ascribed to $p\pi$–$d\pi$ bonding since the methyl groups are considerably out of the P–N–N–P planes. In the crystal there are two kinds of molecules (with enantiomeric arrangements of their methyl groups) randomly distributed over the lattice sites.

The phosphorus atoms in **VII** are in the same oxidation state as, and in a similar environment to, those in **VI**, so it is not surprising that $P_2(NMe)_6$ resembles $P_4(NMe)_6$ in its reactions, adding various groups to phosphorus with preservation of the cage. Thus $P_2(NMe)_6$ forms $P_2(NMe)_6O_2$, m.p. 320–325°, when treated with ethanolic H_2O_2. With sulfur or selenium it gives $P_2(NMe)_6S_2$ and $P_2(NMe)_6Se_2$ respectively. With diborane in tetrahydrofuran, $P_2(NMe)_6\cdot(BH_3)_2$, m.p. 250°, is formed.[33] Triethylaluminum in toluene gives the adduct $P_2(NMe)_6(AlEt_3)_2$, m.p. 153–155°.[74] Chlorine and bromine are taken up readily, giving, it is believed, $P_2(NMe)_6\cdot 2Hal_2$. 1 mol of methyl iodide is added with difficulty; it is likely that the adduct is a phosphonium salt.[33] $P_2(NMe)_6$ reacts with phenyl azide and with diphenylphosphinyl azide in benzene on heating, giving the crystalline compounds $P_2(NMe)_6(PhN)_2$ and $P_2(NMe)_6[Ph_2P(O)N]_2$ respectively.[14]

Goetze, Nöth, and Payne have pointed out that $P_2(NMe)_6$ is generally less reactive towards reagents adding to the phosphorus atoms than is $P_4(NMe)_6$; in explanation they adduce different degrees of cage rigidity.[33]

The controlled degradation of the cage structure **VII** is of interest as the source of the new heterocycles **VIIIa** and **VIIIb**.

(VIIIa)   (VIIIb)   (VIIIc)   (VIIId)

When **VII** is refluxed in benzene with 1 mol of PCl_3 or PBr_3, $ClP(MeN-NMe)_2PCl$ and $BrP(MeN-NMe)_2PBr$ result in high yield. They are crystalline solids, melting at 95° and 96° respectively. Their structures are believed to be **VIIIa** and **VIIIb** respectively, on the evidence of n.m.r. and mass spectra.[59] These compounds, through their halogen function on phosphorus, open up possibilities, now beginning to be investigated, for the synthesis of new derivatives and new heterocycles. Thus **VIIIc** results from the action of 4 mol of dimethylamine on **VIIIa**.[59] Compound **VIIId** comes from **VIIIa** with 1 mol of $MeN(SiMe_3)_2$:[59]

$$P_2Cl_2(NMe)_4 + MeN(SiMe_3)_2 \rightarrow 2Me_3SiCl + P_2(NMe)_5$$

G. Phosph(III)azane polymers

The phosphine $CHF_2CF_2PH_2$ reacts with liquid ammonia at $-78°$ to give a solid polymer with empirical formula $C_6H_9F_6N_2P_3$ and unusually stable to heat, acids, and alkalis. The structure is quite speculative, but linked and crosslinked phosphazane rings may be present.[35]

III. Saturated phosphorus(v)–nitrogen heterocycles

A. Introduction

The phosph(v)azane heterocycles resemble their phosphorus(III) analogs in the stability of 4-membered rings and in the existence of methyl phosph(v)-azane cages. In addition, several compounds have been reported with spiro molecules centered on phosphorus(v) atoms. There are, finally, a few cyclophosph(v)azanes with 6-membered and larger rings, which are isomeric with, and result from the rearrangement of, cyclophosphazenes (Chapter 12).

The reaction of phosphorus(v) halides with amines or ammonia is the most important way of forming P(v)–N bonds, and such reactions sometimes give rise to heterocycles. As with the analogous P(III) reactions (Section II.B above) they will therefore be surveyed briefly by way of introduction.

B. Reactions of phosphorus(v) halides with amines and ammonia

From the existence of PF_6^- and PCl_6^- it is evident that PF_5 and PCl_5 have acceptor properties. PF_5 does, indeed, form a stable crystalline 1:1 adduct with Me_3N.[55] PCl_5 forms a 1:1 adduct with pyridine but it oxidizes Me_3N, giving HCl and hence trimethylammonium chloride.[6]

With secondary amines, PF_5 reacts in well defined stages. When the components are mixed in toluene, adducts $R_2HN{\rightarrow}PF_5$ are precipitated. At 100° the adducts with R = Me, Et, or Pr lose HF and give R_2NPF_4. A second fluorine atom can be similarly substituted by Me_2N-. Little is known of the behavior of PCl_5 with secondary amines.[28]

The reactions just mentioned have not yet any direct application in heterocyclic chemistry. Those of P(v) halides with *primary* amines, however, are very important in heterocyclic synthesis. They too probably always begin with adduct formation; with PF_5, in fact, the adduct $RNH_2 \cdot PF_5$ and the first substitution product $RNHPF_4$ can both be isolated in some cases.[38,39] In the presence of triethylamine, HF is readily eliminated from these rather unstable compounds, giving, it may be supposed, $RN{=}PF_3$. This invariably dimerizes to the cyclodiphosph(v)azane $(RNPF_3)_2$ which can be isolated, usually as a crystalline solid.[38,39] When a primary amine reacts with PCl_5, hydrogen halide is so readily eliminated that the adduct and simple amino-chlorophosphorane are not detected; the product is again, usually, a cyclodiphosph(v)azane $(RNPCl_3)_2$, perhaps formed via the monomer $RN{=}PCl_3$ (see below). These reactions constitute the most important route to P(v)–N heterocyclics.

As a historical point it may be mentioned that these primary amine–PF_5 reactions have only recently been examined (since 1968), but the corresponding reactions with PCl_5 are a special case of the Kirsanov reaction (A. V.

Kirsanov, 1952) which has been the subject of many investigations over the last two decades. Kirsanov reactions take place between PCl_5 and a wide variety of $-NH_2$ compounds, including carboxylic acid amides, sulfonamides, and amines.[2,11,86] Except for the dimeric cyclodiphosph(v)azanes just mentioned, Kirsanov reactions generally give rise to monomeric $RN=PCl_3$ compounds, called monophosphazenes, or alternatively trichlorophosphazo compounds. They can also be performed with organochlorophosphoranes and with chlorofluorophosphoranes, giving rise respectively to products such as $RN=PClPh_2$ and $RN=PF_3$. The use of a Kirsanov reaction for the preparation of sulfanuric chlorides has been mentioned in Chapter 5, Sec. II.A. Several syntheses of cyclophosph(v)azanes have been carried out by reactions of phosphorus(v) halides with bis(silyl)amines in place of primary amines (Chapter 1, Sec. III.D).

When ammonia reacts with a phosphorus pentahalide, the sequence of reactions may well be like that for the amines, namely, addition followed by elimination of hydrogen halide. Even with PF_5, however, the simple adduct has not been isolated, and the first identifiable product is $PF_3(NH_2)_2$, an unstable solid, m.p. $41°$:[53]

$$3PF_5 + 4NH_3 \xrightarrow[20°]{\text{gas phase}} PF_3(NH_2)_2 + 2NH_4{}^+PF_6{}^-$$

With PCl_5, no adduct or primary substitution product has been isolated; the first isolable product (from PCl_5 and ammonium chloride in a solvent at 80–140°) is $[Cl_3P=N-PCl_3]^+[PCl_6]^-$ (Chapter 12, Sec. V.B; cf. ref. 8). This can be supposed to arise from a Kirsanov-type reaction on NH_3, giving the unstable intermediate $Cl_3P=NH$, followed by another substitution of hydrogen. From $[Cl_3P=N-PCl_3]^+[PCl_6]^-$, cyclic phosphazenes may then, in appropriate conditions, be formed by a sequence of reactions (Chapter 12, Sec. V.B). Hence the ammonia–PCl_5 reaction is extremely important in the chemistry of (unsaturated) phosphazenes but is not a route to (saturated) phosphazanes.

The reactions of phosphorus(v) oxohalides with amines and ammonia are also of some interest in heterocyclic synthesis. Adducts $OPCl_3 \cdot 3NR_3$ are formed from the components. Their constitution is not known but, since they are soluble in polar solvents and insoluble in benzene, they may be phosphorylammonium salts.[64] With secondary amines, halogen atoms in $OPCl_3$, $OPR'Cl_2$, or OPF_3 are successively replaced under mild conditions by R_2N- groups, giving rise to numerous well characterized aminooxophosphoranes.[28,45] Thermolytic condensation of these compounds,[28] accompanied by elimination of amine, is an important route to cyclodiphosph(v)azanes (Section C below). The same heterocycles can often be made by reacting $OPCl_3$ or $OPR'Cl_2$ with the amine hydrochloride at a high temperature, so telescoping the two steps into one. The R_2N- derivatives (one of these is the well known solvent hexamethylphosphoramide) have so far been little investigated with a view to heterocyclic synthesis, but there are indications that they may give interesting high polymers by pyrolysis.[17]

With ammonia in excess, $OPCl_3$ gives phosphorus oxotriamide, $OP(NH_2)_3$,[52] a compound resembling sulfamide (Chapter 4, Sec. II.B) in constitution and properties, and like sulfamide, undergoing condensation polymerization when heated. Cyclic compounds have not so far been identified in the products[28] or in those resulting from condensation by HCl,[65] but heterocycles are obtainable by thermal condensation of some of the organo-amides (Section C below), and by reacting $OPCl_3$ with ammonium chloride instead of ammonia.[23]

The reactions of phosphorus thiotrichloride, $SPCl_3$, with ammonia and amines closely resemble those of $OPCl_3$.[28] Heterocycles and polymers are similarly derived from the products.[41] The $SPCl_3$–ammonia reaction also provides one route to the very interesting cation $P(NH_2)_4{}^+$ which may well find applications in heterocyclic synthesis.[71]

C. Cyclodiphosph(v)azanes: preparation, structure, and properties

Our knowledge of cyclodiphosph(v)azane derivatives starts from the work of E. Gilpin and A. Michaelis in the early years of this century.[54]

The first alkyl compound with formula **IXa** (R = Me) was reported in 1961 by Chapman, Holmes, Paddock, and Searle,[11] by heating PCl_5 with methyl-

$$Cl_3P\underset{\underset{R}{N}}{\overset{\overset{R}{N}}{\diamond}}PCl_3 \qquad F_3P\underset{\underset{R}{N}}{\overset{\overset{R}{N}}{\diamond}}PF_3$$

(IXa)   (IXb)

ammonium chloride in s-tetrachloroethane. This is a variant of the standard phosphazene synthesis (Chapter 12, Sec. V.B), with the difference that here the methyl group on nitrogen prevents the elimination of HCl which would lead to an unsaturated phosphazene ring. Methyl-**IXa** forms hygroscopic monoclinic needles. The dimeric formula was early established by molecular weight determinations, and the vibrational spectra showed a centrosymmetric structure with a strongly bonded 4-membered ring. The fluorine analog, Me-**IXb**, a liquid boiling at 88°, was soon afterwards prepared by the reaction[28]

$$PF_5 + MeN(SiMe_3)_2 \rightarrow Me_3SiF + \tfrac{1}{2}(MeNPF_3)_2$$

and can also be made from PF_5 and methylamine.[38] The structures of both compounds have been determined by diffraction methods (Fig. 11.4). The rings are planar. Coordination about the nitrogen atoms is planar. About

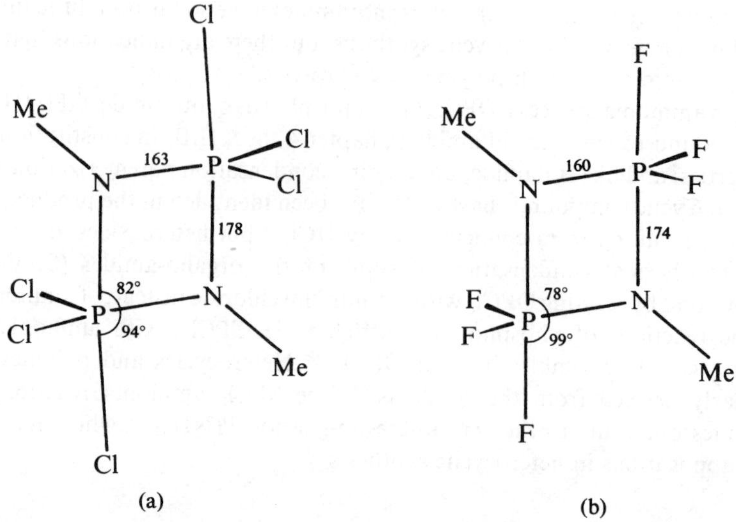

Figure 11.4. Molecular structures of cyclophosph(v)azanes: (a) (MeNPCl$_3$)$_2$ (after ref. 42); (b) (MeNPF$_3$)$_2$ (after ref. 3); bond distances are in pm.

the phosphorus atoms there is trigonal bipyramidal coordination: two of the bonds in the ring are equatorial to the P atoms and short, while the other two are axial to P and longer. Axial and equatorial P–Cl bonds in (Cl$_3$PNR)$_2$ are readily distinguished by ^{35}Cl n.q.r. spectroscopy.[22] The structure is consistent with involvement of the nitrogen lone pairs in *localized* pπ–dπ bonding in the ring. In conformity with the hypothesis mentioned above in Section I.B, the ring bonds in Me-**IXb** are shorter than in Me-**IXa**.

Me-**IXa** and Me-**IXb**, when heated together at 110°, give all the possible mixed chlorofluorides in a redistribution equilibrium.[15]

From the heat of hydrolysis of Me-**IXa**, its average P–N bond energy has been deduced to be at least 311 kJ mol^{-1},[31] about 31 kJ more than in (EtN)$_3$P. Thus the ring is very stable, and dissociation to monomer is probably significant only at high temperatures (cf. Section E below). It is interesting to see how far this stability is affected by changing the ligands on nitrogen or phosphorus. The fluorides **IXb** appear to be quite stable almost irrespective of the nature of R, even when R has steric requirements; thus the methyl, n-propyl, and t-butyl derivatives, and the phenyl and other aryl derivatives, are easily obtained.[38] On the other hand, as Zhmurova and Kirsanov showed, the chlorides **IXa** are only stable as cyclic dimers if the amine RNH$_2$ is a relatively strong base, i.e. if $K_b > 10^{-10}$ in aqueous solution.[11,45] Thus Me-**IXa** is dimeric; Ph-**IXa** is dimeric in benzene but monomeric in dioxan; halogeno-Ph-**IXa** compounds are monomeric in dioxan and dimeric in fresh benzene solutions, but dissociate to the monomer in benzene when the solution is refluxed.[2,11,45] These facts are in general accord with the hypothesis described above in Section I.B.

We next review compounds related to **IXa** and **IXb** but with other ligands on phosphorus. A substantial number of such compounds is known (for some, see refs. 29 and 40). They have been made for the most part in three ways, as follows.

(1) Chlorine atoms in **IXa** may be directly substituted. Thus SO_2 replaces two chlorines by oxygen:[8]

$$(RNPCl_3)_2 + 2SO_2 \xrightarrow[20°]{\text{liquid } SO_2} 2SOCl_2 + (RNPOCl)_2 \quad (R = Me, Et, Pr^n, Ph)$$

giving **Xa**, and **Xa** is readily converted into **Xb** or **Xc** by reaction with Me_2NSiMe_3 or $MeSSiMe_3$ respectively.[36]

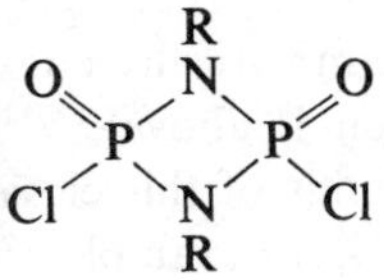

[Structures Xa, Xb, Xc shown]

(Xa) 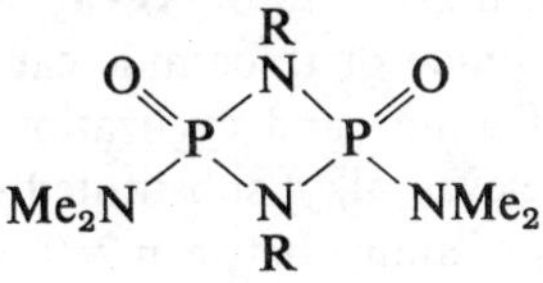 (Xb) 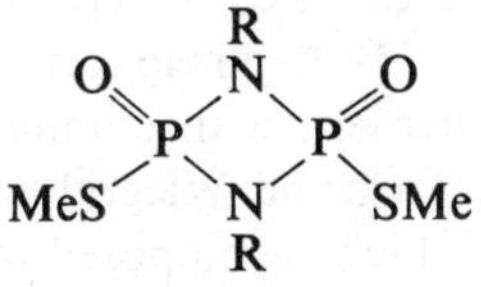 (Xc)

With H_2S two chlorine atoms in Me-**IXa** are replaced by sulfur:[2,8]

$$(MeNPCl_3)_2 + 2H_2S \xrightarrow[20°]{\text{benzene, pyridine}} 4HCl + (MeNPSCl)_2$$

giving **XIa**. The remaining chlorine atoms in **XIa** react readily with diethylamine to give **XIb**, with aniline to give **XIc**, and with ethanol to give **XId**.[11]

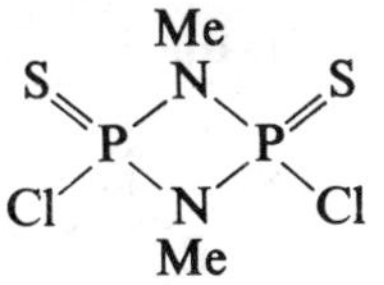

[Structures XIa, XIb, XIc, XId shown]

(XIa) 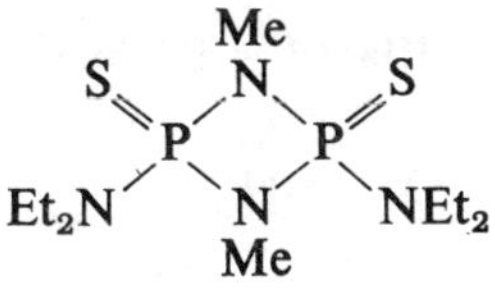 (XIb) 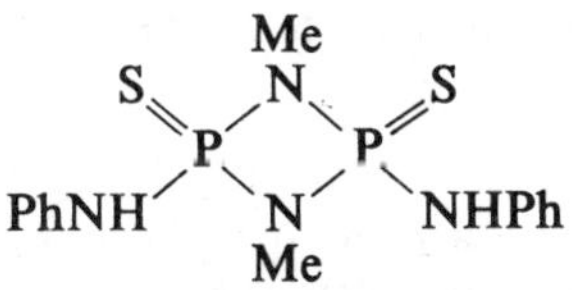 (XIc) 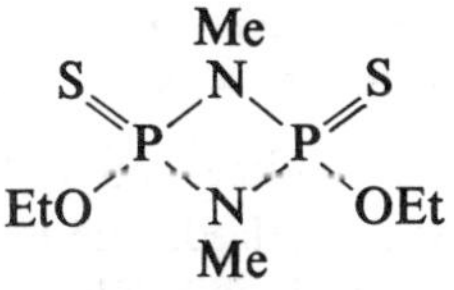 (XId)

(2) The standard preparation, involving the reaction of a phosphorus pentahalide with an amine (Section B above, and the beginning of the present

section) may be adapted by using an organophosphorane.[11] Thus **XII** (R = Ph, Et, CH_2Cl) can be obtained by the reaction[2,72] exemplified by

$$2RPF_4 + 2MeN(SiMe_3)_2 \rightarrow (MeNPRF_2)_2 + 4Me_3SiF$$

$$\begin{array}{c} \text{Me} \\ \text{F} \diagdown \text{N} \diagup \text{F} \\ \text{F—P} \qquad \text{P—F} \\ \text{R} \diagup \text{N} \diagdown \text{R} \\ \text{Me} \end{array}$$

(XII)

Both a monomer and a dimer have been reported from the corresponding reaction with Ph_2PF_3,[39] where the presumed condition for ring stability is only marginally satisfied (Section I.B above).

Another adaptation is to heat with the amine hydrochloride the phosphoranes $OPCl_3$ or $OPRCl_2$, which gives **Xa** or **XIVa** respectively.[54]

(3) A phosphorus(v) oxoamide or thioamide can be thermolytically condensed, with elimination of amine and cyclization (Section B above).[17,41] Bock and Wiegräbe found eleven alkyl-substituted compounds of this class which decomposed with loss of amine between 230° and 300°, for example:[17]

$$O{=}P(NHBu^i)_3 \xrightarrow[280-295°]{N_2} Bu^iNH_2\uparrow + \tfrac{1}{2}[Bu^iNP(O)NHBu^i]_2$$

In four cases they isolated cyclic dimers, **XIIIa** and **XIIIb**, from the products.

$$\begin{array}{c} \text{R} \\ \text{O} \diagdown \text{N} \diagup \text{O} \\ \diagup\text{P} \qquad \text{P}\diagdown \\ \text{RNH} \quad \text{N} \quad \text{NHR} \\ \text{R} \end{array} \qquad \begin{array}{c} \text{R} \\ \text{S} \diagdown \text{N} \diagup \text{S} \\ \diagup\text{P} \qquad \text{P}\diagdown \\ \text{RNH} \quad \text{N} \quad \text{NHR} \\ \text{R} \end{array}$$

(XIIIa) (R = Pr^i, Bu^i) (XIIIb) (R = Pr, Bu^i)

$$\begin{array}{c} \text{Ar} \\ \text{O} \diagdown \text{N} \diagup \text{O} \\ \diagup\text{P} \qquad \text{P}\diagdown \\ \text{Ar'} \quad \text{N} \quad \text{Ar'} \\ \text{Ar} \end{array} \qquad \begin{array}{c} \text{R} \\ \text{S} \diagdown \text{N} \diagup \text{S} \\ \diagup\text{P} \qquad \text{P}\diagdown \\ \text{Ph} \quad \text{N} \quad \text{Ph} \\ \text{R} \end{array}$$

(XIVa) (Ar and Ar' various including Ar = Ar' = Ph) (XIVb) (R = Me, Et, Pr, Bu, Ph)

Yields were sometimes poor, higher polymers evidently being formed. In variants of this method, one may start with a *P*-organo-phosphorus(v) oxoamide or thioamide.[28,41] The phenyl compound **XIVa** (Ar = Ar' = Ph) was thus prepared by Michaelis and his students:

$$2OP(Ph)(NHPh)_2 \xrightarrow[\text{20 Torr}]{\text{dry-distil}} 2PhNH_2\uparrow + [PhNP(O)Ph]_2$$

Compound **XIVb** is similarly made:[61,77]

$$2SP(Ph)(NHR)_2 \xrightarrow{250-270°} 2RNH_2 \uparrow + [PhNP(S)R]_2$$

For evidence as to the structures of these various cyclophosph(v)azanes, **X**, **XI**, **XII**, **XIII**, and **XIV**, reference should be made to the original papers. In the older work, elemental analyses and colligative molecular weight values are usually given, while the more recent publications often also cite extensive n.m.r. data. A special structural point arises when there are two different ligands on each phosphorus atom.[21] If one ligand is divalent (e.g. O= or S=) and the other monovalent, so that the hybridization of phosphorus is roughly tetrahedral, *cis* and *trans* isomers are to be expected. In fact, the ^{31}P n.m.r. spectra of several compounds **Xa** and **Xb** in methylene chloride showed the presence of two isomers with slightly different chemical shifts.[36,50] In the similar case of ethyl-**XIVb**, the isomers have been separated, and the presumed *trans* isomer shown by X-ray crystallography actually to have the *trans* structure.[18] The crystal structure of *trans*-phenyl-**XIVb** has also been solved.[61] Dipole moments also give evidence of the existence of *cis* and *trans* isomers.[35] The crystallographic data on compounds of the class **XIV**, with 4-coordinate phosphorus, show the P–N ring bonds to be practically equal; here, of course, the distinction between axial and equatorial bonds to phosphorus, observed in **IX**, is absent.

Considering the generality of the Kirsanov reaction, and noting that so many cyclodiphosph(v)azanes are stable in the dimeric as opposed to the monomeric state, many other compounds of this class should be available. Some, resembling in constitution those mentioned above, are known and are covered in the review literature.[11,28] More interesting from the structural standpoint are the derivatives believed to have structures **XV**, **XVa**, and **XVI**, with fused rings. An unstable colorless crystalline substance formed from oxamide and PCl$_5$ in 1,2-dichloroethane at 60° is probably **XV**; it readily loses 2HCl to give, it is thought, **XVa**.[10] The cyclodiphosphazane structure is supported in each case by a dimeric molecular weight and a single ^{31}P n.m.r. signal. Similar evidence has been adduced for the structure **XVI** of three compounds formed from acetohydrazide and trichlorophosphoranes.[24]

(XV)   (XVa)   (XVI) (X = Cl, Me, NMe$_2$)

Polymers containing P_2N_2 rings have been reported from the following reactions.[60] Pyrolysis of $PhNP(O)Ph_2$ at 300–420° causes loss of aniline and formation of a crosslinked resin. Polymers based on the repeating unit **XVII** and other similar units have been prepared by transamidative reactions of aromatic diamines with $PhP(O)Im_2$ (Im = imidazol-1-yl) in quinoline solution at 170–300°. The polymer of **XVII** could be handled in air but began to decompose at 170°.

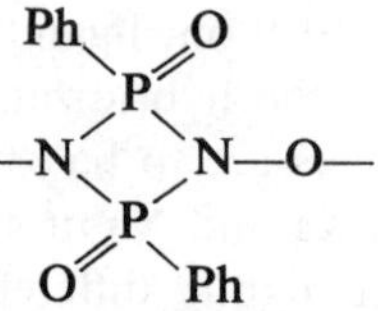

$$\begin{array}{c} Ph \quad O \\ \diagdown \! P \! \diagup \\ -N \quad N-O- \\ \diagup \! P \! \diagdown \\ O \quad Ph \end{array}$$

(**XVII**)

Most of the cyclodiphosph(v)azanes are low-melting crystalline solids, soluble in indifferent, including non-polar, organic solvents. Recent compilations of melting points, i.r. and n.m.r. data, and literature references exist.[2,11,12]

Of the reactions of the cyclophosph(v)azanes,[11] a selection will now be mentioned which show the limits of stability of the rings.

The thermal stability of cyclophosph(v)azane dimers has not been thoroughly investigated, but there are certain indications. Among the least stable thermally are the phenyl and halogenophenyl compounds $(ArNPCl_3)_2$ which, as mentioned above, readily dissociate to monomers in benzene or dioxan. At the other extreme, $(MeNPF_3)_2$ rearranges slowly to a tetramer at 130° (Section F below); this could be a bimolecular reaction and does not necessarily proceed via the monomer. Dissociation to the monomer has been postulated as first step in the addition reaction of $(RNPCl_3)_2$ with isocyanates, CO_2, and CS_2, which set in above 150°, for example:[11]

$$(MeNPCl_3)_2 + 2PhNCO \rightarrow 2Me-N=C=Ph + 2POCl_3$$

At 200°, $(MeNPCl_3)_2$ decomposes to methyl chloride and phosphazene high polymers $(NPCl_2)_n$.[44] Compound **XVIII** can be made by addition of the monomers as indicated by the dashed line. By means of ^{15}N, ^{31}P, and ^1H n.m.r

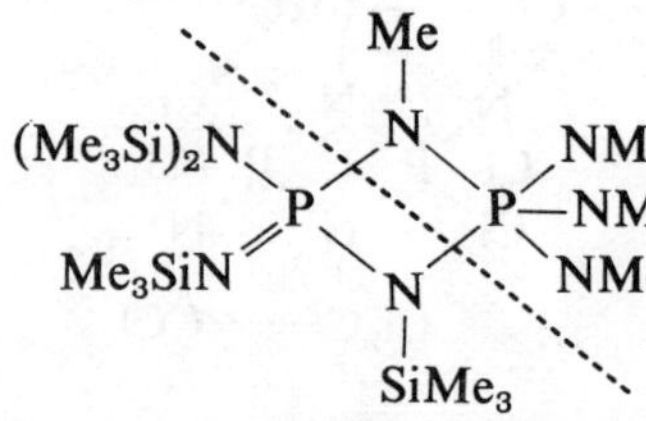

(**XVIII**)

XVIII has been shown to be in equilibrium in deuterobenzene solution with very small amounts of these monomers, at temperatures of 40–50°.[4]

As examples already cited show, chlorine ligands on the phosphorus atoms of cyclophosph(v)azane dimers can frequently be replaced under mild conditions by fluorine, oxygen, sulfur, amino, or ethoxy functions without disrupting the ring. However, the ring is often broken open by nucleophilic reagents. Thus $(ArNPCl_3)_2$ compounds are extensively decomposed by water; hydrolysis under very mild conditions (moist SO_2) gives $ArNH \cdot P(O)Cl \cdot NAr \cdot POCl_2$, evidently formed by the rupture of one ring P–N bond.[11] With excess primary amines at 0°, alkyl compounds $(RNPCl_3)_2$ are ring-opened[11] to give bis-phosphonium salts such as $[(RNH)_3P-NR-P(HNR)_3]^{2+}Cl_2^-$. With liquid ammonia,[11] the alkyl and phenyl compounds $(RNPCl_3)_2$ ring-open to give phosphonium salts of the general formula $[RNH-P(NH_2)_2-N=P(NH_2)_2-NHR]^+Cl^-$. A fuller discussion of aminolysis reactions of the dimers can be found in ref. 11.

D. Cyclophosph(v)azanes: trimers and higher oligomers

Compared with the extensive chemistry of the 4-membered ring phosph(v)-azanes just described, there is surprisingly little to be said about larger rings of this class. There is one report[48] of the formation of a trimer in one of the typical reactions by which the dimers are made, viz.

$$3PhP(S)(NH_2)_2 \xrightarrow{160°} 3NH_3 + [PhP(S)NH]_3$$

but no details were given and there has been no confirmation. Apart from this, all the known 6-membered and larger phosph(v)azane rings are prepared otherwise, by the rearrangement of isomeric phosphazenes.

(XIXa)   (XIXb)

(R = Me, Et, Pr, Pr^i, benzyl; not aryl or CF_3CH_2)

Some phosphazenes **XIXa** (Chapter 12, Sec. VII.E) rearrange at 160–200° to phosphazanes **XIXb**.[2] In the i.r. spectrum, the characteristic P=N stretch of the phosphazene at 1225 cm^{-1} disappears, and a new P=O stretch at 1250 cm^{-1} and a P–N stretch at 700 cm^{-1} appear. The structural alteration is confirmed by different hydrolysis products. The rearrangement is believed to go by an intramolecular attack of a nitrogen atom on the α-carbon atom of an alkoxy group:[2]

$$\text{RO}\diagdown\overset{\text{O---CH}_2\text{R}}{\underset{\text{N}}{\overset{\|}{\text{P}}\diagup\diagdown}}\text{N:} \longrightarrow \text{RO}\diagdown\overset{\text{O}}{\underset{\text{N}}{\overset{\|}{\text{P}}\diagup\diagdown}}\text{N---CH}_2\text{R}$$

When R = CF_3CH_2 the rearrangement does not occur because the nitrogen lone pair is too strongly involved in ring π bonding (Chapter 12, Sec. IX).

Similar rearrangements occur with alkoxycyclotetraphosphazenes, giving for example XX. But pentameric, hexameric, and heptameric methoxycyclo-

$$\begin{array}{c}
\text{O}\quad\quad\text{O}\\
\|\quad\text{Et}\quad\|\\
\text{EtOP---N---POEt}\\
|\quad\quad\quad|\\
\text{EtN}\quad\quad\text{NEt}\\
|\quad\quad\quad|\\
\text{EtOP---N---POEt}\\
\|\quad\text{Et}\quad\|\\
\text{O}\quad\quad\text{O}
\end{array}$$

(XX)

phosphazenes decompose on heating without, as far as is known, forming phosphazanes.[2]

The parent acids of **XIXb** and **XX** result from the hydrolysis, with rearrangement, of cyclodichlorophosphazenes under mild enough conditions to preserve the ring.[2] Thus $(NPCl_2)_3$ in ether solution may be shaken with aqueous sodium acetate; this gives the trisodium salt of **XXI**, which is believed to result from the rearrangement of the primary hydrolysis product **XXII**.[32]

(XXI)   (XXII)   (XXIII)

XXI itself (trimetaphosphimic acid) can be made from the sodium salt by ion-exchange as a dihydrate, plates melting at 105–110°. The structure **XXI** has been established by potentiometric titration of the acid itself and from the i.r. spectra of the salts.[62,76] Three hydrogen atoms (those in OH groups) can be replaced by sodium; all six can be replaced by silver.[32] The tetrameric acid **XXIII** separates as its crystalline dihydrate half an hour after mixing ethereal

($NPCl_2)_4$ with water. The anion has two geometrical isomers.[75] The hydrolysis of $(NPCl_2)_5$ and $(NPCl_2)_6$ gives acids which may well have analogous molecular structures but have so far only been obtained in amorphous form. Sodium and silver salts exist.[32]

Apart from the preparation of salts and structural studies on them, the main chemical investigation so far carried out on the metaphosphimates has been kinetic studies of the hydrolysis of the trimer, **XXI**,[2] aided by chromatographic separation of the products. In aqueous solution **XXI** is degraded stepwise by several routes to the ultimate stable products, orthophosphate and ammonium ion. At 60° and pH 3.6, the half-reaction time for the first step is 1 hour. The most surprising finding is that, in the main hydrolysis sequence, two –NH– bridges are successively replaced by –O– without ring-opening. Replacement of the third –NH– to give cyclotriphosphate (trimetaphosphate) then follows, but is accompanied by a ring-opening reaction. **XXIII** is known to hydrolyze much more slowly than **XXI** but no details have been reported.

E. Spirophosph(v)azanes

Becke-Goehring, Leichner, and Scharf in 1965 found[8] that when $(MeNPCl_3)_2$ (**IXa**) is prepared by heating PCl_5 with methylammonium chloride in tetrachloroethane (Section C above) the mother-liquor after crystallization of **IXa** yields a small quantity (2%) of a benzene-soluble compound $P_4(NMe)_6Cl_8$, m.p. 395°(dec.). In 1967 Weiss and Hartmann showed by X-ray crystallography that this compound has structure **XXIV**.[82] This is a thought-provoking

$$Cl_3P \begin{array}{c} Me \\ N \\ \diagdown \\ \diagup \\ N \\ Me \end{array} \begin{array}{c} Cl \\ | \\ P \\ | \\ N \\ Cl \end{array} \begin{array}{c} Me \\ N \\ \diagdown \\ \diagup \\ N \\ Me \end{array} \begin{array}{c} Me \\ N \\ \diagup \\ \diagdown \\ N \\ Me \end{array} PCl_3$$

(**XXIV**)

discovery, since an isomeric structure of greater symmetry, the adamantane-like molecule **XXVa**, can be imagined and was early discussed by Becke-Goehring as a possibility. **XXVa**, which probably has not been made, would be closely related in structure to $P_4O_6S_4$ and to the various adducts of $P_4(NMe)_6$ which have the structure **XXVb** (see Fig. 11.3 and Section II.E above). However, in **XXVa** (not in **XXVb**) the phosphorus atoms would need to make use of non-equivalent axial and equatorial bonds in the cage; this may be the factor that destabilizes **XXVa** with respect to **XXIV**. Two points arise in this connection.

(1) Compound **XXIV** shows only one ^{31}P n.m.r. signal; i.e. the chemical shifts of the central and terminal P atoms are not distinguishable.[8] This should

be a general reminder that indistinguishable chemical shifts do not necessarily imply identical environments, as has sometimes been assumed with other P–N heterocycles and cages.

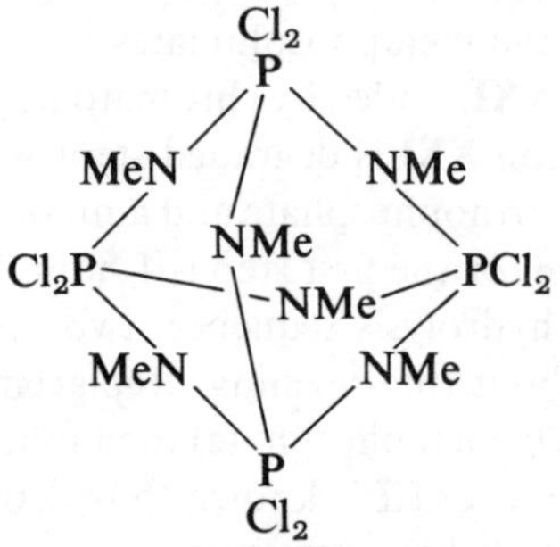

(XXVa)

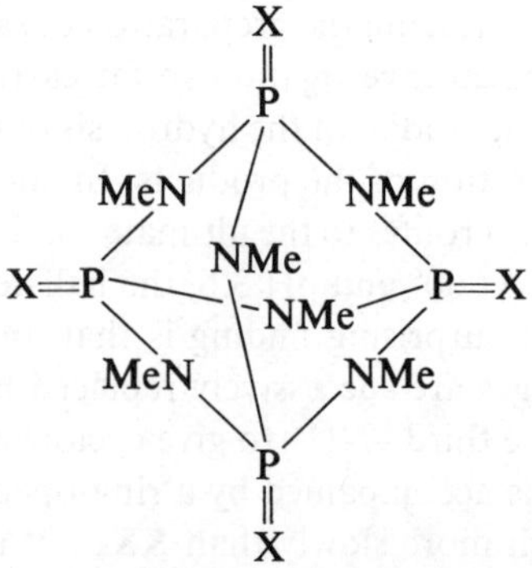

(XXVb) (X = O, S, Se, PhN)

(2) Compound **XXVa** may be capable of existing and perhaps of isomerizing to **XXIV**. This might be tested by the chlorination of $P_4(NMe)_6$ under very mild conditions at low temperatures, so that the cage (**VI**) is not destroyed; this should yield **XXVa**. It is interesting that the analogous reaction of chlorination of P_4O_6 does not seem to have been studied.

A few chemical reactions of **XXIV** have been reported. SO_2 introduces an oxygen atom for two Cl atoms on each terminal phosphorus, and H_2S a sulfur atom similarly, giving $P_4(NMe)_6Cl_4O_2$ and $P_4(NMe)_6Cl_4S_2$ respectively.[8]

The fluoro analog of **XXIV** has recently been described (formula **XXIXb** and Section F below). Unlike **XXIV**, it has an isomer (formula **XXIXa**) but this too has not the adamantane structure. The spiro compound **XXIXc** was obtained in the same investigation.[51]

F. Cage phosph(v)azanes

The compounds of this class with adamantane-like cages, $P_4(NMe)_6S_4$ and $P_4(NMe)_6O_4$, have been mentioned above under the chemistry of $P_4(NMe)_6$ in Section II.E.

Recently Utvary and Czysch[78] found that rearrangement of $(MeNPF_3)_2$ (**IVb**) in a sealed tube at 130° for 7 days gave two new substances, both with the elemental composition of the starting material. One of these is obviously covalent; it forms cubic crystals soluble in CCl_4 and sublimable at 68–70°/ 1 atm. On the evidence of its molecular weight in chloroform, its ^1H and ^{19}F n.m.r. spectra, and its i.r. spectra, it has been provisionally assigned the cubane-like structure **XXVI**. The other is insoluble in CCl_4 but soluble in

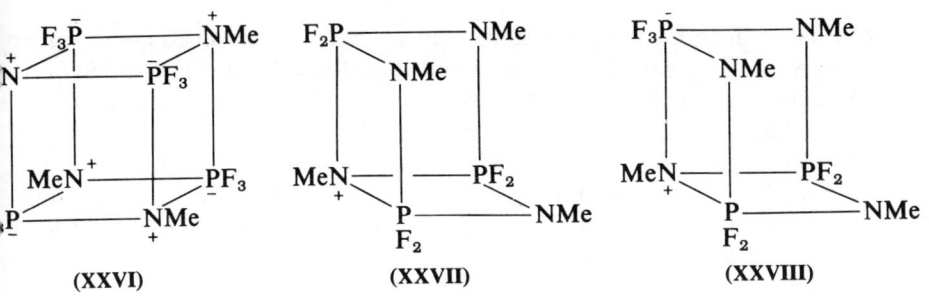

CH$_3$CN and is believed to be a salt of the ion PF$_6^-$, for the presence of which there is n.m.r. evidence. The n.m.r. spectra of the cation are consistent with structure **XXVII**. This second compound loses PF$_5$ when sublimed *in vacuo* at 95°, leaving a CCl$_4$-soluble, presumably covalent, compound (MeN)$_4$P$_3$F$_7$, m.p. 119–121°, for which the structure **XXVIII** (i.e. **XXVII** with an F$^-$ ion incorporated) has been suggested.

In a sealed tube at 130°, compound **XXVIII**, (MeN)$_4$P$_3$F$_7$, decomposes over 117 hours giving as one product the original starting material (MeNPF$_3$)$_2$ (formula **IXb**) and a new substance (MeN)$_6$P$_4$F$_8$, white needles melting at 184–186°.[51] This has been shown by X-ray crystallography to have the structure **XXIXa**. Its isomer **XXIXb**, melting at 190–192°, is one of the major products from a complicated reaction between (MeNPF$_3$)$_2$ and MeN(SiMe$_3$)$_2$ at 110°; it is the fluoro analog of **XXIV** (Section E above).[51] A second important product from the same reaction is **XXIXc**, m.p. 67–69°.[51]

G. Heterocyclic hydrazido derivatives of phosphorus(V)

A few compounds of this type, containing pairs of adjacent nitrogen atoms in the ring, are known. Crystalline compounds, to which structures **XXXa** and **XXXb** are attributed, result from the reactions of 3 mol hydrazine with 1 mol RP(O)Cl$_2$ or RP(S)Cl$_2$ in benzene at 20°.[28] Compound **XXXb** (R = PhO)

210   CHAPTER 11. THE PHOSPHAZANES

and its 1,5-dimethyl derivative have been separated by chromatography into isomers *cis* and *trans* with respect to the phosphorus ligands.[25] In a solid *cis* compound of this type the ring has a twist conformation, while the *trans* isomer has a ring in the chair conformation.[26]

Reference has been made (Section II.F above) to the compounds $P_2(NMe)_6O_2$, $P_2(NMe)_6S_2$, and $P_2(NMe)_6Se_2$ formed from the P(III) cage compound $P_2(NMe)_6$. These are presumably P(v) compounds with structure **XXXI**, related to **XXIX** and **XXX**.

(XXXa) (R = PhO)   (XXXb) (R = Ph, PhO)   (XXXI) (X = O, S, Se)

H. Homologs of the cyclodiphosph(v)azanes with sulfonyl groups as ring components

Sulfonyl, $\rangle S(=O)_2$, is isoelectronic in a broad sense with the groups $\rangle P(=O)$-Hal or $\rangle PHal_3$, and can replace these in cyclophosph(v)azane rings. The resulting compounds, with 4-membered PN_2S rings, have been described in Chapter 4, Sec. V.

IV. The cage phosphazane-phosphazene anion $P_{12}N_{14}S_{12}{}^{6-}$

A salt $K_6{}^+[P_{12}N_{14}S_{12}]^{6-}$, containing the remarkable cage anion shown in Fig. 11.5, results when P_4S_{10} is added to molten KSCN under nitrogen at 180–200°. There is evidence for the complete breakdown of the P_4S_{10} cage, and formation of $SP(NCS)_3$ as an intermediate, in this synthesis. The product is readily crystallized from water as an octahydrate, forming six-sided plates or large octahedra.[30]

The anion, the structure of which has been determined by X-ray diffraction,[30] is a sphere made up of twelve 6-membered P_3N_3 rings. Each ring contains one 2-coordinate nitrogen atom 157 or 160 pm distant from its phosphorus neighbors, and two 3-coordinate nitrogens associated with four longer P–N distances, 167–172 pm. The X-ray photoelectron spectrum of $K_6{}^+[P_{12}N_{14}S_{12}]^{6-}$ distinguishes between the two kinds of nitrogen atoms and points to localization of much of the anionic charge on the 2-coordinate ones. The ^{31}P n.m.r. and vibrational spectra have also been reported.

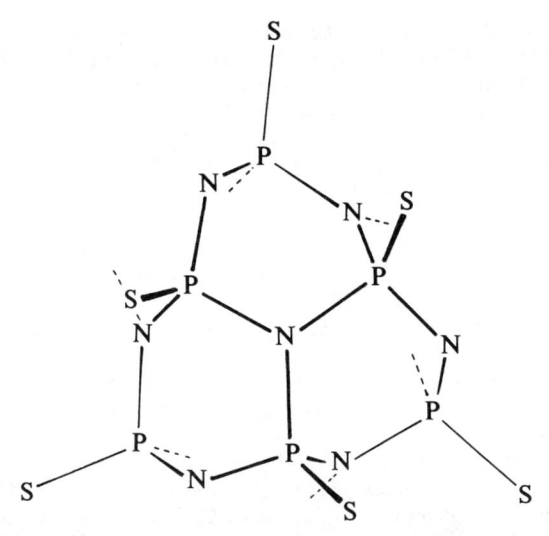

Figure 11.5. Part of the structure of the cage phosphazane–phosphazene anion $P_{12}N_{14}S_{12}^{6-}$ (after ref. 30). The ion is a "sphere" made up of twelve fused P_3N_3 rings, of which for clarity only the three facing the observer are shown. In each ring, all three phosphorus atoms form part of other rings; of the nitrogen atoms, two belong to three rings, and so are 3-coordinate (phosphazane) nitrogens, while the third is a 2-coordinate (phosphazene) nitrogen belonging to two rings only.

Anhydrous $K_6^+[P_{12}N_{14}S_{12}]^{6-}$ does not decompose below 700°, an extraordinary thermal stability recalling the polyhedral borane anions.

References

1. Abel, E. W., Armitage, D. A. and Willey, G. R. *J. Chem. Soc.* 57 (1965)
2. Allcock, H. R., *Phosphorus-Nitrogen Compounds*, Academic Press, New York and London (1972)
3. Almenningen, A., Andersen, B. and Astrup, E. E., *Acta Chem. Scand.* **23**, 2179 (1969)
4. Appel, R., Halstenberg, M. and Knoll, F., *Z. Naturforsch. B* **32b**, 1030 (1977)
5. Barlow, M. G., Green, M., Haszeldine, R. N. and Higson, H. G., *J. Chem. Soc. C* 1592 (1966)
6. Beattie, I. R. and Webster, M., *J. Chem. Soc.* 1730 (1961)
7. Becke-Goehring, M. and Schulze, J., *Chem. Ber.* **91**, 1188 (1958)
8. Becke-Goehring, M., Leichner, L. and Scharf, B., *Z. Anorg. Allg. Chem.* **343**, 154 (1966)
9. Becke-Goehring, M. and Scharf, B., *Z. Anorg. Allg. Chem.* **353**, 320 (1967)
10. Becke-Goehring, M. and Wolf, M. R., *Z. Anorg. Allg. Chem.* **373**, 245 (1970)
11. Bermann, M., *Adv. Inorg. Chem. Radiochem.* **14**, 1 (1972)

12. Bermann, M., *Topics in Phosphorus Chem.* **7**, 311 (1972)
13. Bermann, M. and Van Wazer, J. R., *Inorg. Chem.* **12**, 2186 (1973)
14. Bermann, M. and Van Wazer, J. R., *Inorg. Chem.* **13**, 737 (1974)
15. Binder, H., *Z. Anorg. Allg. Chem.* **384**, 193 (1971)
16. Boal, D. H. and Ozin, G. A., *J. Chem. Soc. Dalton Trans.* 1824 (1972)
17. Bock, H. and Wiegräbe, W., *Chem. Ber.* **99**, 377 (1966)
18. Bullen, G. J., Rutherford, J. S. and Tucker, P. A., *Acta Cryst.* **B29**, 1439 (1973)
19. Burckett St. Laurent, J. C. T. R., Hosseini, H. E., Nixon, J. F. and Sinclair, J., *Abstracts, 2nd International Symposium on Inorganic Ring Systems*, Akademie der Wissenschaften, Göttingen (1978).
20. Casabianca, F., Pinkerton, A. A. and Riess, J. G., *Inorg. Chem.* **16**, 864 (1977)
21. Corbridge, D. E. C., *Topics in Phosphorus Chem.* **3**, 57 (1966)
22. Dalgleish, W. H. and Porter, A. L., *J. Magn. Reson.* **20**, 359 (1975)
23. Derynck, J. M., Puskaric, E., de Jaeger, R. and Heubel, J., *J. Chem. Res. S* 188 (1977)
24. Ebeling, J. and Schmidpeter, A., *Angew. Chem. Int. Ed. Engl.* **8**, 674 (1969)
25. Engelhardt, U. and Merrem, H.-J., *Z. Naturforsch. B* **32b**, 1435 (1977)
26. Engelhardt, U., Merrem, H.-J., Jürgens, G. D. and Bünger, T., in ref. 19.
27. Fleig, H. and Becke-Goehring, M., *Z. Anorg. Allg. Chem.* **376**, 215 (1970)
28. Fluck, E., *Topics in Phosphorus Chem.* **4**, 291 (1967)
29. Fluck, E., Haubold, W., Kosolapoff, G., and Maier, L., Vol. 6, p. 579, in *Organic Phosphorus Compounds*, ed. G. M. Kosolapoff, Wiley-Interscience, New York (1973)
30. Fluck, E., Lang, M., Horn, F., Haedicke, E. and Sheldrick, G. M., *Z. Naturforsch. B* **31b**, 419 (1976)
31. Fowell, P. A. and Mortimer, C. T., *Chem. and Ind.* 444 (1960)
32. *Gmelins Handbuch der Anorganischen Chemie*, Teil, C. "Phosphor", Verlag Chemie, Weinheim (1965)
33. Goetze, R. Nöth, H. and Payne, D. S., *Chem. Ber.* **105**, 2637 (1972)
34. Goldschmidt, S. and Krauss, H.-L., *Ann. Chem.* **595**, 193 (1955)
35. Grapov, A. F., Mel'nikov, N. N. and Razvodovskaya, L. V., *Russ. Chem. Rev.* **39**, 20 (1970)
36. Green, M., Haszeldine, R. N. and Hopkins, G. S. A., *J. Chem. Soc. A* 1766 (1966)
37. Harman, J. S. and Sharp, D. W. A., *J. Chem. Soc. A* 1935 (1970)
38. Harman, J. S. and Sharp, D. W. A., *J. Chem. Soc. A* 1138 (1970)
39. Harris, J. and Rudner, B., *J. Org. Chem.* **33**, 1392 (1968)
40. Harris, R. K., Wazeer, M. I. H., Schlak, O., Schmutzler, R. and Sheldrick, W. S. *J. Chem. Soc. Dalton Trans.* 517 (1977)
41. Healy, J. D., Shaw, R. A. and Woods, W., *Phosphorus and Sulphur* **5**, 239 (1978).
42. Hess, H. and Forst, D., *Z. Anorg. Allg. Chem.* **342**, 240 (1966)
43. Holmes, R. R. and Forstner, J. A., *Inorg. Synth.* **8**, 63 (1966)
44. Horn, H.-G., *Z. Anorg. Allg. Chem.* **406**, 199 (1974)
45. Houben-Weyl, *Methoden der Organischen Chemie*, 4th ed., vol. 12, part 2, Thieme Verlag, Stuttgart (1964)
46. Hunt, G. W., *Diss. Abs. Int. B*, **32**, 5086 (1972)
47. Hunt, G. W. and Cordes, A. W., *Inorg. Nucl. Chem. Lett.* **10**, 637 (1974)
48. Ibrahim, E. H. M. and Shaw, R. A., *Chem. Commun.* 244 (1967)

49. Jefferson, R., Nixon, J. F., Painter, T. M., Keat, R. and Stobbs, L., *J. Chem. Soc. Dalton Trans.* 1414 (1973)
50. Keat, R. and Thompson, D. G., in ref. 19.
51. Kubjacek, M. and Utvary, K., *Monatsh. Chem.* **109**, 587 (1978)
52. Lehmann, H.-A., *Pure Appl. Chem.* **44**, 221 (1975)
53. Lustig, M. and Roesky, H. W., *Inorg. Chem.* **9**, 1289 (1970)
54. Michaelis, A., *Ann. Chem.* **407**, 290 (1915)
55. Muetterties, E. L., Bitner, T. A., Farlow, M. W. and Coffman, D. D., *J. Inorg. Nucl. Chem.* **16**, 52 (1960)
56. Muir, K. W. and Nixon, J. F., *Chem. Commun.* 1405 (1971)
57. Niecke, E., Flick, W. and Pohl, S., *Angew. Chem. Int. Ed. Engl.* **15**, 309 (1976)
58. Niecke, E. and Scherer, O. J., *Nachr. Chem. Techn.* **23**, 395 (1975)
59. Nöth, H. and Ullmann, R., *Chem. Ber.* **107**, 1019 (1974)
60. Parks, L., Nielsen, M. L. and Miller, J. T., *Inorg. Chem.* **3**, 1261 (1964)
61. Peterson, M. B. and Wagner, A. J., *J. Chem. Soc. Dalton Trans.* 106 (1973)
62. Pustinger, J. V., Cave, W. T. and Nielsen, M. L., *Spectrochim. Acta* **11**, 909 (1959)
63. Rankin, D. W. H., *J. Chem. Soc. A* 783 (1971)
64. Revel, M., Navech, J. and Vives, J. P., *Bull. Soc. Chim. France* 2327 (1963)
65. Riesel, L. and Somieski, R., *Z. Anorg. Allg. Chem.* **415**, 1 (1975)
66. Riess, J. G. and Van Wazer, J. R., *Bull. Soc. Chim. France* 1846 (1966)
67. Riess, J. G. and Van Wazer, J. R., *Bull. Soc. Chim. France* 3087 (1968)
68. Riess, J. G. and Wolff, A., *J. Chem. Soc. Chem. Commun.* 1050 (1972); *Bull. Soc. Chim. France* 1587 (1973)
69. Scherer, O. J. and Klusmann, P., *Angew. Chem. Int. Ed. Engl.* **8**, 752 (1969)
70. Scherer, O. J. and Glassel, W., *Chem. Ber.* **110**, 3874 (1977)
71. Schmidpeter, A. and Weingand, C., *Angew. Chem. Int. Ed. Engl.* **7**, 210 (1968); **8**, 615 (1969)
72. Schmutzler, R., *Chem. Commun.* 19 (1965)
73. Shaw, R. A., *Phosphorus and Sulfur* **4**, 101 (1978)
74. Spangenberg, S. F. and Sisler, H. H., *Inorg. Chem.* **8**, 1004 (1969)
75. Steger, E. and Lunkwitz, K., *J. Mol. Struct.* **3**, 67 (1969)
76. Toy, A. D. F., in *Comprehensive Inorganic Chemistry*, ed. J. C. Bailar, H. J. Emeléus, R. S. Nyholm and A. F. Trotman-Dickenson, vol. 2, p. 389, Pergamon, Oxford and New York (1973)
77. Trippett, S., *J. Chem. Soc.* 4731 (1962)
78. Utvary, K. and Czysch, W., *Monatsh. Chem.* **103**, 1048 (1972)
79. Van Doorne, W., Hunt, G. W., Perry, R. W. and Cordes, A. W., *Inorg. Chem.* **10**, 2591 (1971)
80. Verdier, C., *Chem. Abstr.* **73**, 132620 (1970)
81. Vetter, H.-J. and Nöth, H., *Chem. Ber.* **96**, 1308 (1963)
82. Weiss, J. and Hartmann, G., *Z. Anorg. Allg. Chem.* **351**, 152 (1967)
83. Zeiss, W. and Weis, J., *Z. Naturforsch. B* **32b**, 484 (1977)
84. Zeiss, W., Schwarz, W. and Hess, H., *Angew. Chem. Int. Ed. Engl.* **16**, 407 (1977)
85. Zeiss, W. and Barlos, K., *Z. Naturforsch. B* **34b**, 423 (1979)
86. Zhmurova, I. N. and Kirsanov, A. V., *J. Gen. Chem. (USSR)* **30**, 3018 (1960)

12
The phosphazenes

I. General introduction

The cyclic phosphazenes constitute one of the largest and most studied groups of heterocyclics based on inorganic skeletons, being comparable in numbers and diversity to the cyclosiloxanes. Figure 12.1 shows the range of structural types among the cyclophosphazenes. In addition, there exist linear, cyclolinear, and cyclomatrix phosphazene polymers. All these phosphazenes are based on units $-(N=PRR')-$, where R and R' may for example be halogens, pseudohalogens, amino groups, or a wide variety of organic radicals. The phosphorus is in the +5 oxidation state. The structural formulas of phosphazenes are often written with alternating single and double P–N bonds, though in symmetrical phosphazenes the bonds are more often than not equal in length and electronically equivalent. The formal multiple bonding in their skeletons distinguishes the phosphazenes from the phosphazanes based on phosphorus(v) which have been described in Chapter 11.

The cyclodichlorophosphazenes, $(NPCl_2)_n$, are the longest-known phosphazenes and the source of very many derivatives, so they are often regarded as parent compounds.

The chemistry of the cyclic phosphazenes cannot be divorced from that of their non-cyclic precursors and polymers, which will accordingly be described as far as is relevant.

This area of chemistry is much too large for an exhaustive treatment in one chapter of a short book. Moreover, many of the facts of phosphazene chemistry raise no new questions of principle and are easily hunted down in several good recent reviews.[4,7,8,10,11,35,40,41,42,43,52,65,66,77,81,82,83,85] Our aim therefore is to outline the key points, with an adequate selection of illustrations, and to provide a bibliography enabling all the literature on any aspect to be quickly located.

Many chemists would regard the phosphazenes as the most important

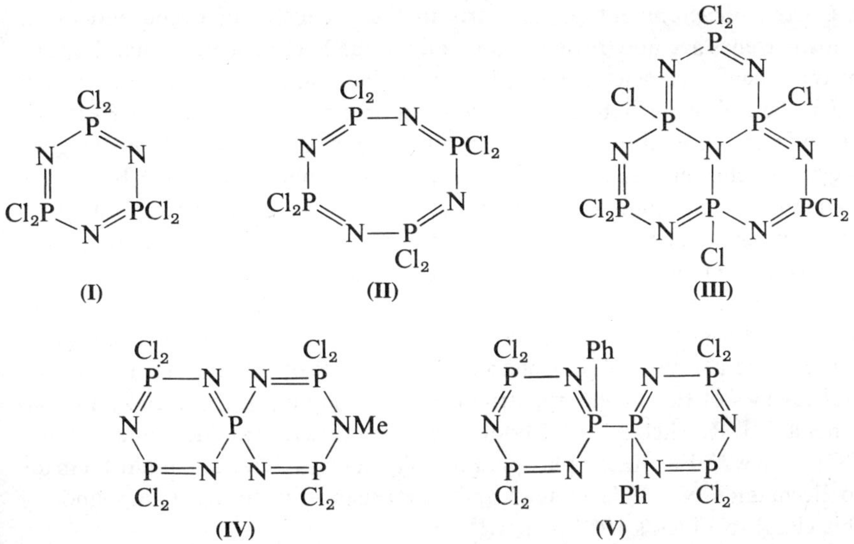

Figure 12.1. Structural types among the phosphazenes, showing some ring types and modes of coupling and fusion of rings.

group of inorganic heterocyclics. Although perhaps less surprising and thought-provoking than the unsaturated sulfur nitrides (Chapter 6), they are indeed unique in modern inorganic chemistry in several ways. They provide the best examples of inorganic covalent homologous series, with many members stable enough to be isolated and studied individually. The cyclophosphazene skeletons are abnormally stable (for inorganic heterocycles), with the result that over a thousand derivatives have been made, and the theory of substitution mechanisms is evolving towards a level of sophistication comparable to organic theories. The structures and bonding of the phosphazenes are of unusual interest with regard to π bonding, delocalization, and aromaticity in the rings. Finally, these are compounds with real possibilities of practical application; a phosphazene synthetic elastomer is now on sale in the United States (Section VI.A below), and certain cyclophosphazenes were reported recently by J. F. Labarre of the University of Toulouse as showing promise in the chemotherapy of cancer. (See also note on p. 249.)

II. Literature

The phosphazene field is so large and complicated that it seems advisable to provide the following short critical guide to the literature in addition to the usual list of references.

Current developments of all sorts in the chemistry of cyclic and linear phosphazenes are now reviewed annually in the Specialist Periodical Reports of the Chemical Society, which began publication in 1970.[41-46] For earlier work the following publications are recommended. An article by Keat and Shaw in the new edition of Kosolapoff's *Organophosphorus Chemistry* gives the most comprehensive listing available of cyclic phosphazenes.[40] It includes over 800 "pure" phosphazenes and also many cyclic compounds in which phosphazene groups are combined with other groups. For each compound are given references (up to 1970) on its physical properties of all kinds, together with actual values of the more easily quoted physical constants. This article is indispensable to any serious worker with phosphazenes. It gives also a good general review of phosphazene chemistry. Probably the best recent general reviews of the chemistry, structure, and bonding of the cyclic phosphazenes are H. R. Allcock's 1972 book and review article[7,8] and Krishnamurthy's 1978 review.[52] For linear phosphazene oligomers, reference should be made to Bermann's two 1972 reviews and (for organo-substituted compounds of this class) to Fluck's 1967 review.[14,15,31]

A number of earlier reviews exist, but are probably less to be recommended because recent advances have made parts of them obsolete and even misleading. But the 1965 *Gmelin* is in a class by itself and will be useful for some time to come, especially for the early literature.[35]

III. History

Cyclodichlorophosphazene trimer, $(NPCl_2)_3$, was made in small amounts in 1834 by Liebig and Wöhler; thus it shares with tetrasulfur tetranitride the distinction of being the longest-known inorganic heterocyclic compound. In 1895-6 the American H. N. Stokes discovered other members of the cyclodichlorophosphazene series, also the linear polymeric material $(NPCl_2)_n$, and suggested structural formulas now known to be correct. In 1924 Schenck and Römer obtained permission to make the polymers in bulk at the Bayer dyestuffs factory in Leverkusen, and there they developed preparative methods still important today. Serious commercial interest in the compounds seems to have started later, with a series of patents from 1949 onwards. A historical landmark was the first X-ray structure determination on a phosphazene, reported in 1939 by Ketelaar and de Vries, which confirmed the long-presumed cyclic structure of $(NPCl_2)_4$. The first industrial polyphosphazene products, elastomers, foams, and films, were marketed in the United States in 1977 by the Firestone Tire and Rubber Company.

IV. General properties of the cyclic phosphazenes

For the detailed physical properties of these compounds reference should be made to the extensive compilations in the review literature.[7,40,52,77] Here we give a short summary of salient features, for the purpose of orientation.

The parent dichlorophosphazenes, $(NPCl_2)_n$, are the best-known. The trimer (m.p. 114°) (formula **I** in Fig. 12.1), tetramer (m.p. 124°) (formula **II**), and pentamer are low-melting solids; the hexamer is a liquid. The lower members have a faint pleasant smell described as organic or camphor-like. Exposure to vapor from the trimer or tetramer in a poorly ventilated room can give rise, after delay, to eye inflammation and difficulty in breathing, but no permanent injury has been reported.

The dipole moments of the dichlorophosphazenes are small (< 0.5 D)[40] so it is not surprising that they are all readily soluble in organic solvents, including non-polar solvents. The tetramer is decidedly less soluble than the trimer, pentamer, and hexamer. All the cyclic chlorophosphazenes have densities near 2 g cm^{-3}. They are stable in moist air at room temperatures, though hydrolyzed under more drastic conditions (Section VII.E below).

The other halogenophosphazenes are generally similar. $(NPF_2)_3$, with b.p. 52° and m.p. 28°, is the most volatile cyclic phosphazene. The fluoro compounds are more easily hydrolyzed than the chloro compounds. The alkyl and aryl phosphazenes melt in the range 90–330° and are quite resistant to hydrolysis, the former being readily soluble in water without decomposition.

V. Primary preparations of the cyclic phosphazenes

A. Introduction

The preparative reactions available for phosphazenes can be divided into primary, in which the basic ring (or chain) structures are built up from simple molecules, and secondary, in which functional groups are altered (usually by nucleophilic substitution) without amendment of the skeleton. Historically the cyclodichlorophosphazenes, $(NPCl_2)_n$, have dominated the field, and they are still the main route of entry into cyclophosphazene chemistry, so this Section V is devoted mainly to them. However, some reference will also be made here to the bromo and other cyclophosphazenes which can be made by variants of the same primary synthesis.

Very many phosphazenes are best prepared by substitution from the dichlorophosphazenes; these secondary syntheses are dealt with below in Section VII.

B. Syntheses from ammonium halides and phosphorus(v) compounds

Liebig and Wöhler first made cyclo-$(NPCl_2)_3$ by the action of gaseous ammonia on phosphorus pentachloride. In essence, this is still the chief route to the phosphazenes, though the principle of it has been extended in scope and the technique greatly improved.

Today the usual procedure is to react ammonium chloride with PCl_5 in a boiling solvent, usually s-tetrachloroethane.[27,44,59,60] Overall:

$$nNH_4Cl + nPCl_5 \rightarrow (NPCl_2)_{3 \text{ to } > 10} + 4nHCl$$

This important reaction has been much investigated, with the following findings. The solvent is not essential, but if the reagents are heated to the temperature of reaction without a solvent, PCl_5 sublimes out of the mixture.[35] This can be prevented by use of an autoclave with provision for venting the HCl, but it is more convenient to reflux in a solvent so that the PCl_5 is continually washed back. s-Tetrachloroethane boils at 146°, a suitable temperature for reaction, and it dissolves PCl_5 readily and is inert towards it. However, it decomposes somewhat, and if re-used gives lower yields.[27] The ammonium chloride does not dissolve, and the rate of reaction is determined largely by its surface area; thus reagent-grade material (small crystals) reacted about three times as fast as analytical grade (larger crystals).[27] The phosphazenes are kinetically stable compounds which do not polymerize or depolymerize much below 250°, so in this synthesis the products are kinetically controlled, and the proportions of rings of different sizes and chains depend upon the dilution.[7] As expected from polymerization theory, the yield of linear products is maximal (about 19%) when the volume of solvent is small, and becomes negligible at higher dilutions.[27] Increasing dilution favors the formation of the cyclic trimer, has relatively little effect on yields of tetramer and hexamer, and reduces the yields of pentamer and heptamer; tentative kinetic explanations have been given.[27] In a typical preparation, from 0.5 mol each of PCl_5 and NH_4Cl in 80 ml solvent, the products were: linear phosphazenes 5%; cyclic phosphazenes, trimer 41%, tetramer 18%, pentamer 22%, hexamer 14%, heptamer 6%.[27]

The separation of pure components from this product mixture is troublesome, and published methods leave much to be desired. The cyclic oligomers, which alone are soluble in petroleum ether, can be separated by extraction with this solvent from the linear material.[7,54] To separate the individual ring compounds, one may to some extent use their different base strengths. Extraction of the mixture with concentrated H_2SO_4 takes out mainly the trimer which can then be recovered by diluting the acid with water and back-extracting into petroleum; this gives trimer pure enough to be cleaned up by crystallization.[54] However, this method is not well adapted to separation of the

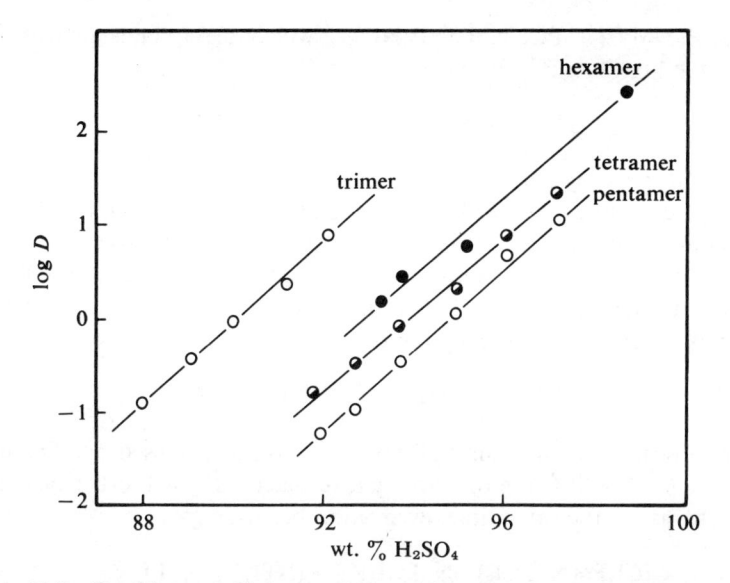

Figure 12.2 Partition coefficients of the chlorophosphazene cyclic oligomers between sulfuric acid and n-hexane; D = molarity in acid/molarity in hexane (after ref. 65).

larger rings, the base strengths of which alter rather little with size (Fig. 12.2).[64,65] These can be separated by fractional distillation under reduced pressure, before or after removal of the trimer.[54,60] Industrial patents describe cumbersome variations of these methods. Probably the best procedure for the higher oligomers is preparative-scale vapor-phase chromatography on the fluoro compounds, made from sodium fluoride and the crude chloro mixture.[12] There are no reports on the use of gel-permeation chromatography, though a polystyrene gel suitable for the range of molecular weights concerned is now available (Bio Beads S-X8, from Bio-Rad Laboratories, Richmond, California).[37]

The composition of phosphazene mixtures is most conveniently and reliably determined by ^{31}P n.m.r spectroscopy.[27,52] I.r. spectra are useful for the trimer and tetramer but do not differentiate well between the larger rings. Paper chromatography[87] and vapor-phase chromatography[18] are satisfactory analytical methods.

The mechanism of the synthesis of phosphazenes from PCl_5 and NH_4Cl has been largely elucidated.[26] Our knowledge comes from the synthesis of some of the intermediates in *ad hoc* experiments, from the electrical conductance and ^{31}P n.m.r. spectra of the mixtures at various stages, and from measurements of HCl evolved. The first intermediate **A** so far isolated is the salt $[Cl_3P=N=PCl_3]^+PCl_6^-$, i.e. P_3NCl_{12}, which precipitates in over 80% yield in the first hour:[26]

$$3PCl_5 + NH_4Cl \rightarrow P_3NCl_{12} + 4HCl$$

It forms colorless moisture-sensitive needles melting at 310°.[11] This first step uses up practically all the free PCl_5, leaving much unreacted NH_4Cl. The next phase consists of successive reactions of chain-lengthening giving rise to the soluble compounds (which have been isolated):[12] $[Cl_3P=(N-PCl_2)_n=N=PCl_3]^+$-$Cl^-$ (intermediate **B** with $n = 1$, **C** with $n = 2$, etc.). The agent in this process is probably PCl_3NH, formed by the reaction

$$PCl_6^- + NH_4^+ \rightarrow PCl_3NH + 3HCl$$

and the process continues until all the PCl_6^- from **A** is used up. The chain cations of **A**, **B**, and **C** are then thought to react with each other producing longer chains; at the same time, cyclization begins, e.g.

$$[Cl_3P=(N-PCl_2)_2=N=PCl_3]^+ \rightarrow (NPCl_2)_3 + PCl_4^+$$

Explanations for the relative yields of the different rings are still quite speculative.

How **A** comes to be formed from the starting materials is not entirely clear, but ^{31}P n.m.r. data show two unstable intermediates.[26] One of these may well be $Cl_3P=NH$, stable organic analogs of which have been made by the Kirsanov reaction (Chapter 11, Sec. III.B):

$$R_3PCl_2 + R'NH_2 \rightarrow R_3P=NR' + 2HCl$$

The following extensions and adaptations of the above phosphazene synthesis have been reported. From ammonium bromide, $PBr_3 + Br_2$, and *s*-tetrabromoethane as solvent, one may obtain $(PNBr_2)_n$ cyclic compounds in good yield; the trimer is more soluble than the tetramer in organic solvents, giving a convenient separation procedure.[7] Mixed bromochlorides result from further obvious adaptations of this synthesis.[7] The analogous use of PF_5 to make fluorophosphazenes would not be convenient (PF_5 is a gas) and has not been reported; the fluorophosphazenes are made by fluorination of the chlorides (Section VII.D below). Organo-chlorophosphoranes $RPCl_4$ with NH_4Cl give cyclo-$[NP(Cl)R]_{3,4}$, i.e. non-geminally substituted chlorophosphazenes (cf. Sections VII.A and G below).[7] Bis-organochlorophosphoranes similarly yield $(NPR_2)_{3,4}$.[63] Alkylamino and perfluoroalkyl derivatives are similarly available.[7]

Improvements in this basic phosphazene synthesis are continually reported.[41–46]

C. Synthesis from aminochlorophosphoranes

The synthesis from bis-organochlorophosphoranes described above in Section B does not work with trifluoroalkyl compounds, but an alternative and allied method is available for these and some other phosphazenes. As a first step, aminophosphoranes $R_2PCl_2NH_2$ are made *either* by chlorination of the P(III) compounds R_2PNH_2 *or* by treating the P(V) compounds $R_2P(O)NH_2$ with PCl_5 (R = CF_3, C_3F_7, for example). In presence of trimethylamine, $R_2PCl_2NH_2$ readily dehydrohalogenates to give a mixture of the cyclophosphazene trimer and tetramer $(R_2PN)_{3,4}$.[7]

D. Oxidative phosphazene synthesis from phosphorus(III) halides

These methods, although not widely used hitherto, could perhaps be further exploited. A phosphorus(III) halide is oxidized by azide or by chloramine, as in the examples:[7]

$$Ph_2PCl + NaN_3 \xrightarrow[\text{no solvent}]{165°} NaCl + Ph_2PN_3 \quad \text{(unstable intermediate)}$$

$$\downarrow -N_2$$

$$(NPPh_2)_4 \xleftarrow[\text{in vacuo}]{268°} (NPPh_2)_n \quad \text{(high polymers)}$$
$$33\%$$

$$Ph_2PCl + NH_2Cl \xrightarrow[\text{diethyl ether}]{\text{cooled}} Ph_2P(Cl)(NH_2)NHP(Cl)_2Ph_2$$

$$\downarrow \begin{array}{c} 300° \\ \text{in vacuo} \end{array} -3HCl$$

$$(NPPh_2)_3 \longleftarrow \quad \longrightarrow (NPPh_2)_4$$
$$20\% \qquad\qquad\qquad\qquad 62\%$$

E. Other primary phosphazene syntheses

Several other syntheses have been described which are not of general importance but can be of value in specific cases.[7,41–46] New syntheses, and minor improvements to existing ones, are continually being reported; they may be found in the Specialist Periodical Reports of the Chemical Society on Organophosphorus Chemistry.

VI. Phosphazene polymers other than simple rings

A. Phosphazene high polymers

Chains of alternating phosphorus and nitrogen atoms are isoelectronic with the silicon–oxygen chains of the siloxanes, and hopes have long been entertained that phosphazene chemistry might generate a range of useful polymers comparable to the silicones. The applied chemistry of the phosphazenes is still in its infancy, but recently an important advance was made; the first phosphazene elastomer, known as PNF, was marketed by the Firestone Tire and Rubber Company of Akron, Ohio. This material is based on the repeating unit

$$\begin{array}{c} OCH_2CF_3 \\ | \\ -N{=}P- \\ | \\ OCH_2-(CF_2)_x-CF_2H \end{array} \qquad (x = 1, 3, 5, 7, \ldots)$$

It can be compounded similarly to conventional synthetic elastomers. It has an exceptional range of service temperatures from $-57°$ to $+177°$, good chemical resistance, and good mechanical properties, and is a promising candidate for mechanical sealing applications under extreme conditions of service. Reference 11 gives the basic chemistry of phosphazene polymers, and technical information may be obtained from Firestone's promotion literature. There is no obvious limitation to the future development of phosphazene polymer products, and Firestone's lead is being followed in other countries, notably Japan.

Ring–chain equilibria occur in several areas of inorganic polymer chemistry and have been most fully investigated for the silicones. Among the phosphazenes they seem to occur, but are harder to study. It has been known since the end of last century that "inorganic rubber" could be made by heating the cyclic dichlorophosphazenes to $350°$.[35] More recent results strongly suggest that at 250–$300°$ there are rapidly established equilibria between the cyclic oligomers and a linear polymer, accompanied by a slower and (under these conditions) irreversible transformation of the latter into a crosslinked polymer.[3,10,11]

$$(NPCl_2)_3 \underset{}{\overset{250-300°}{\rightleftharpoons}} \text{benzene-soluble linear polymer}$$

$$\downarrow 250-300°$$

crosslinked benzene-insoluble "inorganic rubber"

The position of equilibrium between the high polymers and the cyclic oligomers is much affected by pressure. At 70 kbar the high polymers are stable up to $1000°$,[4] while at <0.1 Torr they depolymerize to $(NPCl_2)_{3,4}$ even at $300°$. At moderate pressures and at $600°$ in a sealed tube, the same equilibrium mixture of trimer, tetramer, and higher polymer is formed, irrespective of which of these species was originally put in the tube.[4] The other halogeno- and pseudohalogeno-phosphazene rings will polymerize similarly. Organo-cyclophosphazenes, however, will not, probably for thermodynamic reasons connected with the severity of steric hindrance in the high polymers as compared with the cyclic oligomers.

Polymerization of the halogenocyclophosphazenes is probably often initiated by ionization, e.g.

$$(NPCl_2)_3 \rightleftharpoons N_3P_3Cl_5^+ + Cl^-$$

and is catalyzed by many substances that promote ionization.[11]

The linear polymer $(NPCl_2)_x$ is a colorless transparent elastomer which dissolves slowly but completely in benzene or tetrahydrofuran to give viscous solutions; unlike the cyclic oligomers, it is insoluble in n-heptane and can be precipitated by adding this to a benzene solution. The crosslinked polymer $(NPCl_2)_x$ is a clear colorless transparent rubber which becomes harder and tougher with increasing crosslinking (i.e. longer heat treatment). It is insoluble in organic solvents but absorbs them with swelling. The linear polymer is an important intermediate in the synthesis of organophosphazene polymers, since its chlorine atoms, unlike those of the crosslinked polymer, can be fully substituted by organic nucleophiles.[3,10,11] In spite of their excellent elastomeric properties, the high-polymeric chlorophosphazenes themselves have no practical application because they are slowly hydrolyzed by atmospheric moisture.[10,11] Therefore Allcock's work on the synthesis and nucleophilic substitution of the linear polymer has proved to be the key to successful commercial development of the phosphazenes.

B. Linear phosphazene oligomers

We have mentioned (Section V.B above) that compounds of this series occur as intermediates in the standard preparation of cyclic dichlorophosphazenes. In the early 1960s E. Fluck systematically investigated them.[30,31] He heated cyclic $(NPCl_2)_3$ and $(NPCl_2)_4$ with PCl_5 in a sealed tube at 200–220° for 72 hours. With mol ratios $NPCl_2/PCl_5$ from 1 to 3, all the PCl_5 was used up. The products slowly solidified at room temperature. They were found by ^{31}P n.m.r. spectroscopy to consist of salts with the general formula $[Cl_3P-N(PCl_2=N)_nPCl_3]^+PCl_6^-$ ($n \geqslant 0$), the value of n increasing with the $NPCl_2/PCl_5$ ratio. The first two members of the series ($n = 0, 1$) have been isolated as crystals, also the chlorides of the same two cations.[15,31,32] The ^{31}P n.m.r. studies produced no evidence of chain branching.

Oligomeric phosphazene derivatives containing various other elements are formed when ammonium chloride and phosphorus pentachloride react in the presence of halides of Sb, Al, B, Ti, Zn, Fe, Cu, and Tl. Their structures are not known but may well be related to those just described. They are stable at 500–600° but are easily hydrolyzed.[4]

C. Fused-ring, spiro, and coupled-ring phosphazene compounds

The fused-ring compound $P_6N_7Cl_9$ (formula **III** in Fig. 12.1) was first obtained by Stokes in his pioneer work on the phosphazenes; he reported it in 1897. Recently the compound has been thoroughly investigated by Oakley and Paddock.[63]

It results in small yields (0.5 g from 200 g) when the high polymer ("inorganic rubber") from the heating of cyclic $(PNCl_2)_n$ oligomers at 300° is depolymerized under 0.1 Torr at the same temperature. It then distils, collecting partly in the condenser and partly with the more volatile monocyclic phosphazenes, from which it can be separated by its low solubility in petroleum. It can be recrystallized from benzene as transparent plates, m.p. 235°.

The structure **III** has been fully established by X-ray diffraction, ^{31}P n.m.r., vibrational spectroscopy, and mass spectroscopy.[63] It presents interesting features which are discussed below in Section VIII. The reactions of this compound can be rationalized in terms of its structure, and are discussed below in Sections VII.D and VIII.

The spiro compound (formula **IV** in Fig. 12.1) was prepared in 1969 by building up a second ring on monocyclic $(NPCl_2)_3$, according to the accompanying reaction scheme.[53]

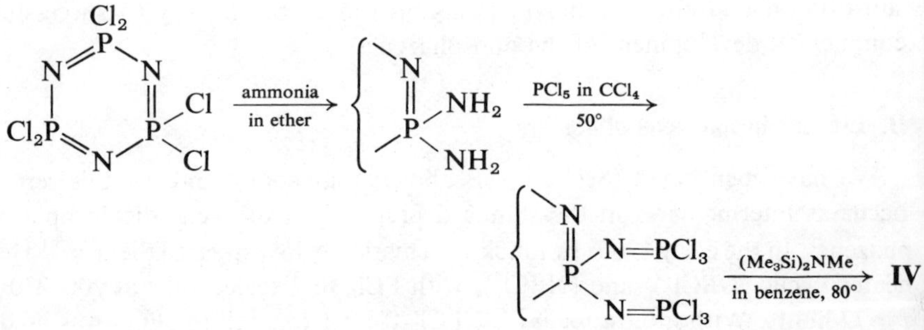

Compound **IV** forms crystals melting at 228° with decomposition. This is another example of the value of silylamines in synthesis; the ring-closure failed with ammonia and with monomethylamine. The structures of **IV** and the intermediates seem fairly well established by ^{31}P n.m.r.[53]

It will be noted that one of the rings in **IV** is not a true phosphazene ring, but rather a phosphazadiene ring containing one saturated nitrogen atom. A neutral, bicyclic spirophosphazene is in fact stoichiometrically impossible but the anion **VIa**, with two true phosphazene rings, has been made.[75,76]

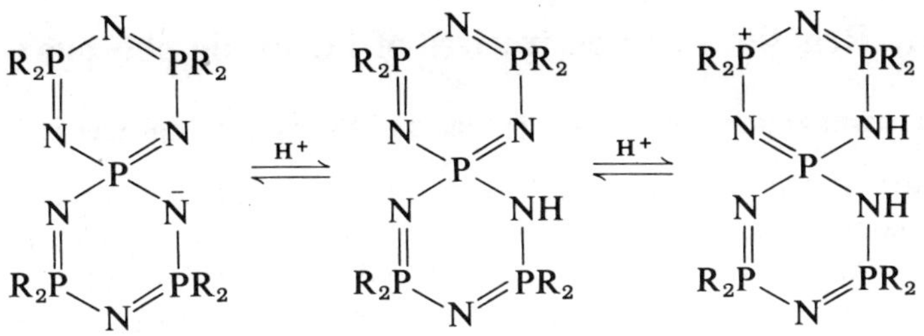

VI. OTHER PHOSPHAZENE POLYMERS

[Structures VIa, VIb, VIc shown with H⁺ equilibria between them]

(VIa)   (VIb)   (VIc)

The hydrochloride of cation **VIc** (containing two phosphazadiene rings) results in over 57% yield when PCl_3 is added to a boiling benzene suspension of $[H_2N-PPh_2=N-PPh_2-NH_2]^+Cl^-$.[75] This rather surprising reaction entails an oxidoreduction of PCl_3 to give some phosphorus(v) (which acts as the central atom of the spiro structure) and an unidentified orange powder containing P(<3). Another route to **VIc** is thermal condensation of the geminal diaminophosphazene $N_3P_3R_4(NH_2)_2$ with the compound

$$[Cl-PR_2=N-PR_2-Cl]^+Cl^-.$$[76]

The action of ammonia on **VIc** in methanol now gives **VIb**, m.p. 235–240°, and further deprotonation of **VIb** with sodium methoxide gives the true spirodiphosphazene anion **VIa**.

The coupled-ring phosphazene (formula **V** in Fig. 12.1), m.p. 250°, is produced in 7% yield by the reaction of $(NPCl_2)_3$ with diphenylmagnesium in refluxing dioxan.[16] Phenylmagnesium bromide does not give this compound (cf. Section VII.G below). The structure shown is consistent with the observed mass spectrum, ^{31}P n.m.r., i.r., and osmometric molecular weight.

Formula **VII** represents a cyclophosphazene with a diimidosulfur bridge. It has been prepared in small yield from the presumed cis-2,6-diamino derivative of $(NPF_2)_4$. Treatment with thionyl chloride affords $N_4P_4F_6(NSO)_2$, which in presence of pyridine loses SO_2 to give **VII** as yellowish crystals melting at 57°.[72]

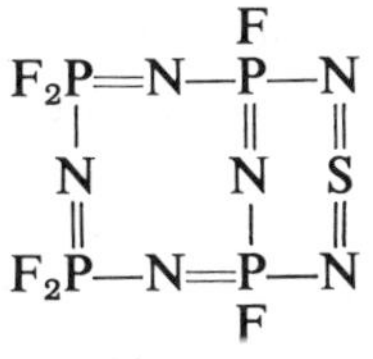

$$\begin{array}{c} F \\ F_2P=N-P-N \\ |\parallel\parallel \\ NNS \\ \parallel|\parallel \\ F_2P-N=P-N \\ F \end{array}$$

(VII)

Notwithstanding the supposed 2,6 structure of the starting material, an X-ray diffraction study of **VII** shows the NSN group to be attached in the 2,4 positions.[33]

VII. Reactions and derivatives of the cyclic phosphazenes

A. Isomerism among substitution products of the cyclophosphazenes

Hitherto, the study of isomerism among phosphazene derivatives has been mainly confined to the trimeric and tetrameric rings. Even with these, the identification of isomers can be difficult and controversial. We shall briefly discuss the naming of isomers and then indicate how mixtures of isomers can be separated and their components identified.

Although seldom exactly planar, phosphazene rings are so flexible that conformational isomerism is only known in the crystalline state (Section VIII below). Isomeric substitution products can be adequately characterized by stating to which phosphorus atoms the substituents are attached, and whether they lie above or below the ring.

There are two ways of stating to which phosphorus atoms the substituents are attached. One way, which cannot be avoided in complicated cases such as the presence of more than two kinds of ligand, is the standard Ring Index procedure of numbering the ring atoms; a nitrogen atom should be numbered 1. The other way, used more in practice, is illustrated in Fig. 12.3; in this method, those isomers with as many ligands as possible paired up on the same phosphorus atom or atoms are called geminal (*gem*), and the other isomers non-geminal. A geminal structure cannot have *cis* and *trans* forms, but a non-geminal one must have them. For the trimer derivatives shown in Fig. 12.3, one prefix, *gem* or *cis* or *trans*, suffices to characterize the isomer completely. For the tetramers, the description vicinal or antipodal must usually be added (Fig. 12.3).[69]

In recent work the separation of isomers has mainly been accomplished by chromatography.[82] Thus the aminocyclophosphazenes can be separated on silica gel by elution with petroleum–benzene mixtures, using columns, or thin layers on the diagnostic or preparative scale. Gas–liquid chromatography has been used for the relatively volatile methylfluorophosphazenes and chlorofluorophosphazenes; for these, fractional distillation is also helpful.[25,69]

In most recent work the identification of isomers has been attempted in the first instance by n.m.r. spectroscopy, making use of ^{31}P and, where applicable, ^{1}H and ^{19}F resonances. ^{31}P chemical shifts are not as a rule highly diagnostic. Recourse must also be had to the analysis of spin structures, which are complicated by long-range couplings, but may be simplified by heteronuclear decoupling. Specific cases will not be discussed here; the general principles have been reviewed,[7,52] while a report on the methyl derivatives of $(NPF_2)_4$ provides a good detailed illustration.[69] Other relatively quick techniques, namely ^{35}Cl n.q.r. spectroscopy,[24,52] infrared spectroscopy, measurements of base strength, and dipole moments have proved useful in particular cases.[7,82]

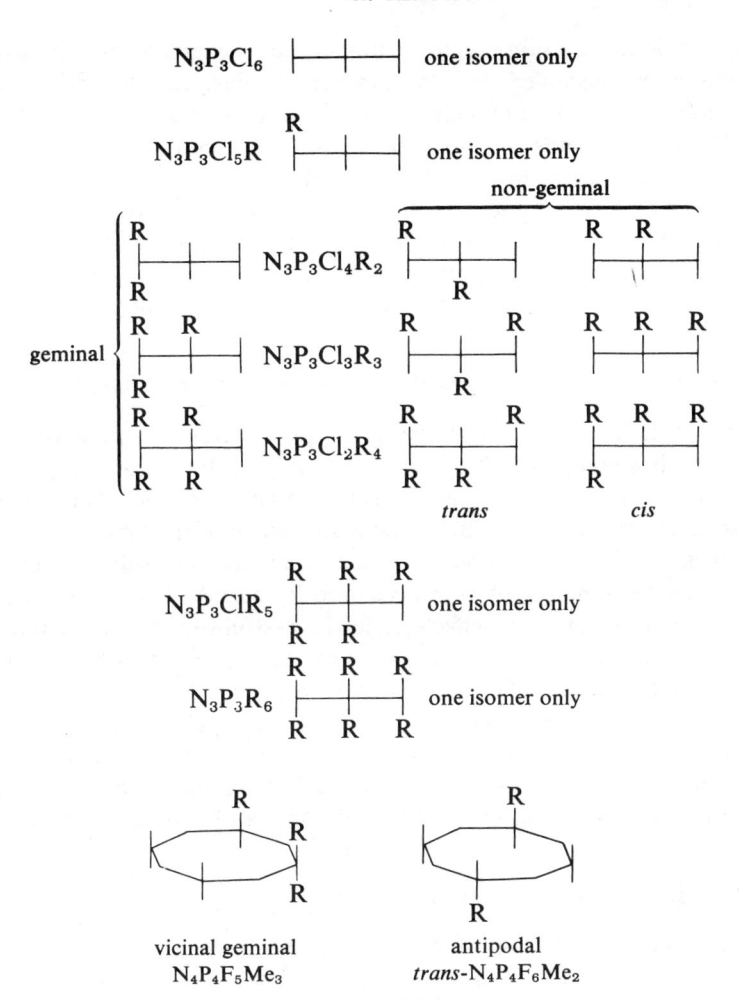

Figure 12.3. The most widely-used way of labelling isomeric substitution products of the cyclophosphazenes, showing all the possible R derivatives of a trimer, followed by two examples of tetramer derivatives.[69,82]

B. Complexes with Lewis bases

One would expect the 4-coordinate, rather electropositive phosphorus atoms of the phosphazenes to function as acceptor centers in the formation of adducts, especially since there is evidence (Sections D and E below) for 5-coordinate phosphorus in the transition states of many nucleophilic substitutions. However, no donor–acceptor complex of a phosphazene with a Lewis base has been really well characterized. $(NPCl_2)_3$ appears to ionize, with solvation, in pyridine solution, but any complexes it forms with pyridine seem to be very unstable.[41,56,77] With trimethylamine, the chlorophosphazenes undergo a slow reaction in which most of the chlorine atoms are replaced by dimethylamino groups; this may begin with formation of an Me_3N adduct, which then eliminates MeCl.[77] When $(NPPh_2)_3$ and s-tetrachloroethane are mixed in benzene, an adduct $(NPPh_2)_3 \cdot 3Cl_2CHCHCl_2$, m.p. 114°, can be crystallized.[84] This has been regarded as a donor–acceptor adduct,[77] but there is no information on its structure. It dissociates to its components *in vacuo*. There is evidence for adducts of $(NPCl_2)_3$ with dimethylformamide.[41]

C. Nucleophilic substitution reactions of the phosphazenes: introduction

The most studied reactions of the cyclic phosphazenes, and also of high-polymer phosphazenes, have been nucleophilic substitutions in which the phosphazene skeleton is preserved while the chlorines, or other ligands on phosphorus, are replaced. This is the main source of the many organophosphazenes described in the literature. Such reactions are easily effected. Their course can to some extent be rationalized by means of the same concepts, such as inductive and steric effects, as are successfully used with organic rings. There remain, it is true, many thought-provoking problems which could well require another decade of experimentation for their full solution[82] but it seems likely that these will eventually be resolved in terms of already known mechanistic concepts. In Sections D to G following, we discuss some important types of nucleophilic substitution, the derivatives obtainable thereby, and current ideas on mechanism. A fuller account can be found in ref. 52.

D. Halogen and pseudohalogen exchange

The isotopic substitution of chlorine in the cyclic $(NPCl_2)_n$ compounds ($n = 3, 4, 5, 6$) by $^{36}Cl^-$ in methyl cyanide is first-order in both reagents.[7] An S_N2 mechanism, involving a transition state with 5-coordinate phosphorus, would account for this. All the chlorines of the substrate are equally reactive. With 10^{-2}M of both reagents, the half-reaction times are of the order of half an hour at 25° for the trimer. The order of rate constants is tetramer > pentamer > hexamer > trimer, for reasons that are not clear.

The cyclic fluorophosphazenes $(NPF_2)_{3,4}$ are readily obtained by refluxing the corresponding chloro compounds with sodium fluoride in methyl cyanide or nitrobenzene respectively.[79] They are low-melting solids easily purified by distillation. Partial fluorination may be effected similarly.[25] In the cyclic trimer, the phosphorus atoms are fluorinated in the order 2, 2', 4, 4', 6, 6', i.e. geminally as far as possible. Fluorination of the tetrameric chloride also proceeds as geminally as possible, the second phosphorus atom to be attacked being the nearest neighbor of the first, and the fluorination sequence in the pentameric chloride, though not fully worked out yet, also seems to be predominantly geminal.[67] These results are consistent with an S_N2 mechanism subject to σ-inductive effects, but for a detailed explanation of the relative rates it is necessary also to invoke π-inductive effects which partly offset the σ effect.[67] Non-geminal $N_2P_3Cl_4F_2$ is not accessible by direct fluorination but can be made as follows:[7]

(making use of Me_2N groups to temporarily block the geminal positions).

Attempts to fluorinate the fused-ring chloride $N_7P_6Cl_9$ (III in Fig. 12.1) led only to breakdown of the skeleton, probably because the electronic effects of fluorination further weaken the already weak central P–N bonds.[62]

Other halogen exchanges[7] and also the replacement of halogen by the pseudohalogens cyano, isocyanato, isothiocyanato, and azido have been reported.[7,70,73]

E. Nucleophilic substitution by –OH, –OR, –SR

The simplest of these reactions are with alkoxides, aryloxides, and thiolates. A typical reaction is:

$$6MeO^-Na^+ + (NPCl_2)_3 \xrightarrow[0°]{benzene/methanol} [NP(OMe)_2]_3 + 6NaCl$$

Many different substituents have been incorporated in this way, with both partial and complete replacement of chlorine.[7] Trifluoroethoxy[78] and phenoxy[7] groups are introduced non-geminally, whereas ethylthio and phenylthio groups enter geminally.[7] The substitution patterns observed can to some degree be rationalized in terms of inductive, steric, and polarizability effects, but it will be unsafe to generalize about mechanics until kinetic data are available.[7]

The cyclic dichlorophosphazenes are insoluble in, and not wetted by, water, and their heterogeneous reaction with water is slow.[35] The trimer can be steam-distilled but is completely decomposed to ammonia, hydrochloric acid, and phosphoric acid on heating to 150–250° with water in a sealed tube.[35] It can be hydrolyzed, without destroying the skeleton, by means of sodium acetate in cold or warm dioxan–water.[60,61,68] The resulting well characterized sodium salt can then be converted into a free acid by, for example, ion-exchange. The acid, however, has not a phosphazene structure; it is a cyclophosphazane[35,85] resulting from a rearrangement which is described more fully in Chapter 11, Sec. III.D. The tetrameric chloride and fluoride, when hydrolyzed similarly, also rearrange in this way, giving the tetrameric cyclophosphazane acid.[35,80]

The susceptibility of the organophosphazenes towards hydrolytic substitution on phosphorus follows a pattern consistent with S_N2 attack; only electronegative organic groups can be easily hydrolyzed off, and the more electron-withdrawing they are the more easily they go. Thus, heating various phenyl and chlorophenyl derivatives of $(NPCl_2)_3$ at 150–250° for days with water or aqueous alkali breaks down the phosphazene rings without detaching the aryl groups from phosphorus.[1] The compounds $(NPMe_2)_{3,4}$ are soluble in water without decomposition.[77] Aryloxy derivatives can be substitutionally hydrolyzed with difficulty. The half-reaction time for the first substitution in $[NP(OPh)_2]_3$ is over 1000 hours in $\sim 0.01M$ NaOH in 25% aqueous diglyme at 80°; these times are lengthened by methyl substitution in the benzene rings, and shortened by nitro substitution.[6] Fluoroalkoxy derivatives hydrolyze somewhat more easily,[7] for example $[NP(OCH_2CF_3)_2]_3$ with a half-reaction time of a few minutes under similar conditions, though they are scarcely affected by hot aqueous acid or base in a heterophase system.[6] Only one aryloxy group could be hydrolytically cleaved off. Up to three fluoroalkoxy groups were lost, in the expected non-geminal sequence.[6] The tetrameric fluoroalkoxy compounds hydrolyze faster than the trimers, perhaps because (as the entropies of activation suggest) they are more flexible during the approach to the transition state.[7]

The hydrolysis of $[NP(NH_2)_2]_4$ is slow in water at room temperature and rapid at 100°. Only non-cyclic products result.[47]

(VIII)

Reaction of the chlorophosphazene trimer with dihydroxyaromatic compounds in presence of trimethylamine gives rise to spiro compounds exemplified by **VIII**, made in this way from catechol. Compound **VIII** is a very effective clathrating host compound, capable of taking up many small organic molecules as guest compounds, a property possessed in lesser degree by other similarly constituted compounds.[13]

F. Nucleophilic substitution by amino groups

The replacement of chlorine in $(NPCl_2)_{3,4}$ by amino groups has been extensively investigated by R. A. Shaw and others.[48,50,51,52,82] Simple generalizations about substitution patterns reported by Bode in the 1940s, and other workers a little later, have been shown to be unsound. In fact, the patterns are so complicated that it is difficult to generalize about them, and even difficult to identify and isolate the parameters needing further examination.

The cleanest and most fully studied reactions are those of the trimer $(NPCl_2)_3$ with secondary amines. The sterically undemanding dimethylamine and aziridine can be made to replace all the chlorines stepwise, while with the bulky dibenzylamine no more than two amino groups will go in. When two, three, or four similar amino groups are substituted, mixtures of isomers are formed (Fig. 12.3). Usually non-geminal substitution predominates, but in the particular case of trisubstitution alone, the yield of geminal isomer has been found to become large in aromatic solvents in presence of excess amine; this may be connected with a specific solvation of the phosphazene, for which there is n.m.r. evidence. The relative yields of *cis* and *trans* non-geminal isomers in di-, tri-, and tetra-substitution vary considerably with the amine and the reaction conditions, but are not easily interpreted because (as independent experiments show) *cis* and *trans* forms in some cases isomerize to each other under mild conditions. The reaction of the tetrameric chloride $(NPCl_2)_4$ with secondary amines is exceedingly complex and only at an early stage of investigation; with dimethylamine, all five possible isomers of $N_4P_4Cl_6(NMe_2)_2$ have been reported. With *wet* diethylamine, $(NPCl_2)_3$ unexpectedly gives, in 9% yield, a P–N 6-membered ring containing two unsaturated N atoms and one saturated N atom.[19,50]

The compounds $(NPCl_2)_{3,4}$ react rapidly with liquid ammonia at or below room temperature, giving mainly the fully aminated derivatives $[NP(NH_2)_2]_{3,4}$.[7,48] These are source materials for phosphams (Chapter 13, Sec. III). The diamido compound $N_3P_3Cl_4(NH_2)_2$ can be made easily by reaction of $(NPCl_2)_3$ with gaseous or aqueous ammonia.[29] Only one isomer is obtained, which suggests that substitution is geminal, since *cis* and *trans* structures are possible for a non-geminal diamido compound; and there is support from ^{31}P n.m.r. for the geminal structure (cf. Section A above), though this is not conclusive. A monoamido derivative $N_3P_3Cl_5NH_2$ can be made by pyrolysis of the diamido compound.

With primary amines and $(NPCl_2)_3$, up to six amino groups can usually be introduced, but many of the partially substituted derivatives are little known; experimental difficulties arise here because of the formation of gummy resins. Thus the geminal compounds $N_3P_3Cl_2(NHR)_4$ (R = Et, Pr^i, Bu^t, Ph) can be made, but of the tris-derivatives of these amines only a non-geminal ethyl compound and a geminal phenyl compound have been obtained, and these in low yields. In contrast to ammonia, methylamine and ethylamine substitute the first two chlorines non-geminally, but with the heavier alkylamines and with aniline, mainly geminal substitution is indicated at all stages.

The full rationalization of these substitution patterns will require much more experimental work, but some interesting tentative suggestions can be found in the publications of Shaw's group.[82]

Even tertiary amines NR_3 will react under more drastic conditions with the trimeric chloride, replacing chlorine by $-NR_2$ groups and eliminating RCl. These reactions have been little studied.[82]

Triphenylphosphine imine (triphenylmonophosphazene), Ph_3PNH, replaces one or two chlorines in $(NPCl_2)_3$ by the phosphazenyl group Ph_3PN-.[82] Hydrazides $N_3P_3F_5NHNH_2$ and $N_3P_3F_5NHNHF_5P_3N_3$ can be made similarly from $N_3P_3F_5Br$, and serve as the source of various organic derivatives.[71]

The reaction of $(NPCl_2)_4$ with ethylamine in chloroform has been found to give in good yield the compound $N_4P_4(NHEt)_6(NEt)\cdot HCl$, with an $-NEt-$ bridge across the ring, joining antipodal P atoms.[51]

Recently the deamination reactions represented by the equation

$$N_3P_3(NMe_2)_6 + 2n\,HX \xrightarrow[\text{hours}]{\text{boiling xylene}} N_3P_3X_n(NMe_2)_{6-n} + n\,Me_2NH\cdot HX$$

(X = Cl, Br; n = 2 or 3)

have been reported.[58]

G. Nucleophilic substitution by carbanions

Straightforward successive nucleophilic substitution can be effected by alkyl or aryl lithiums on the fluorocyclophosphazenes $(NPF_2)_{3,4,5}$.[7] With tetramer and pentamer, and enough methyllithium, total substitution occurs.[69] With the trimer, up to five fluorines can be thus substituted by phenyl, but only two by methyl before addition reactions begin to interfere.[69] The substitution patterns observed with methyllithium on $(NPF_2)_4$ (Fig. 12.4) cannot be explained in terms of simple σ-inductive effects. First, if only σ-inductive effects were important, the fluorines could not be substituted geminally by any reagent, yet the *gem*-dimethyl compound results in much

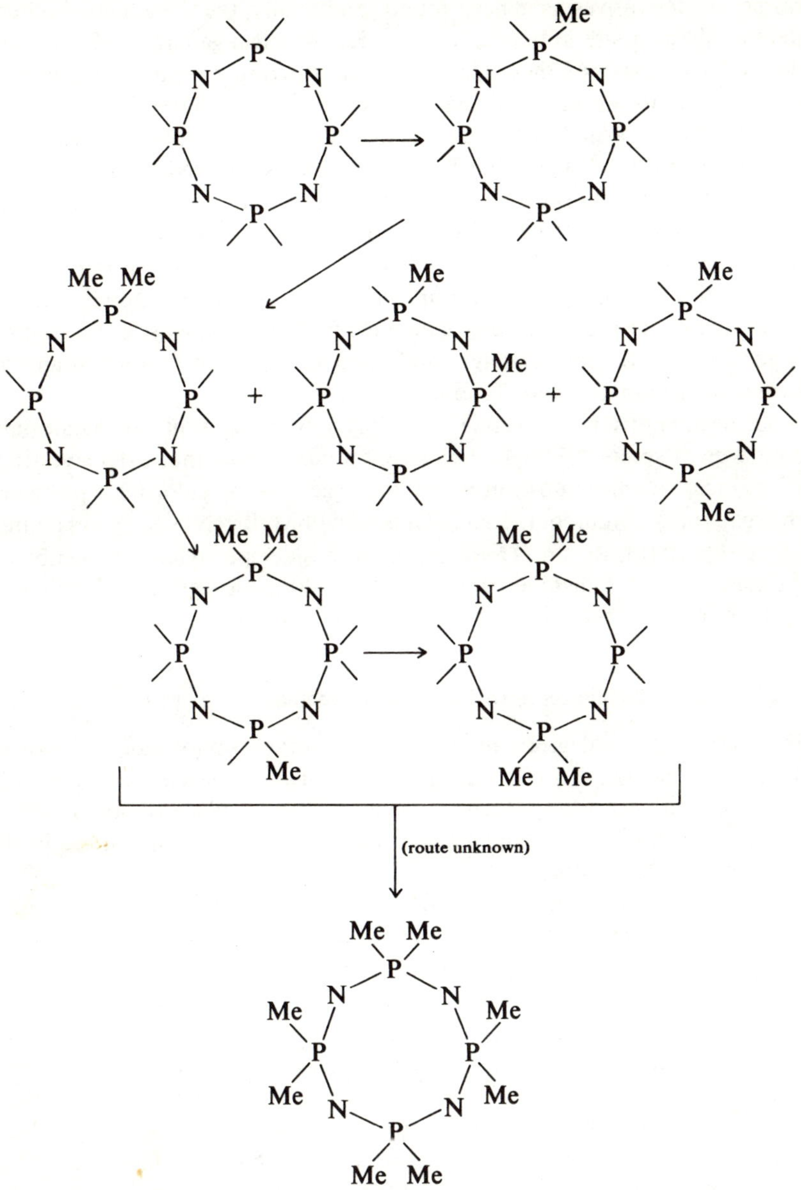

Figure 12.4. The pattern of substitutions in the reaction of $(NPF_2)_4$ with methyllithium.[69]

larger yields (40%) than any other isomer; secondly, no non-geminal tri- or tetra-methyl derivatives have been found; and finally, the third methyl enters preferentially at a site antipodal to the first two. Ranganathan, Todd, and Paddock have explained these features by considering "π-inductive effects". The bond lengths in gem-$N_4P_4F_6Me_2$ show that π electron density in the ring is predominantly located away from the methyl-substituted phosphorus atom. Thus the substitution of the first fluorine by methyl, by reducing π electron density on the phosphorus atom concerned, gives it a partial positive charge which acts counter to the σ-inductive effect, tending to favor the geminal entry of the second methyl group, and similarly to make non-geminal positioning of three or four methyl groups very unlikely. The antipodal entry of the third methyl group can also be explained in terms of π density. The theory also accounts for the even stronger preference of the trimeric and pentameric fluorides for geminal dimethylation.

It has been known for 50 years that $(NPCl_2)_3$ with excess phenylmagnesium bromide gives some $(NPPh_2)_3$, but recent studies show that extensive ring-opening accompanies substitution in the reactions of chlorophosphazenes with organomagnesium compounds. Thus with PhMgBr, $(NPCl_2)_3$ gives mainly $Ph_3P(NPPh_2)_2NMgBr$, and $(NPCl_2)_4$ gives, besides phenylated 8-membered phosphazene rings, 6-membered rings with $-N=PPh_3$ side-groups which probably arise from the recyclization of linear products.[7] Cf. Sec. J below.

H. Addition of phosphazenes to Lewis acids, including the proton

Although to varying degrees involved in π bonding to phosphorus (Sections VIII and IX below), the nitrogen lone pairs in phosphazenes are available for reaction with acceptor ions and molecules. The phosphazenes are therefore bases in both the Brønsted and Lewis senses, a character well marked in the formation of adducts and salts.

We have referred (Section V.B) to the use of salt formation with concentrated sulfuric acid for the separation of the cyclic chlorophosphazenes.[7] Most phosphazenes have enough basic strength to add one proton; in a few cases two protons may be added. pK_a values for many protonated phosphazenes have been measured by Feakins and Shaw and their coworkers, principally by means of potentiometric titration with perchloric acid in nitrobenzene. A selection of values is given in Table 12.1 As expected, the weakest bases are those phosphazenes with electron-withdrawing substituents, and the strongest those with electron-releasing substituents, but more factors than the simple inductive effect must be invoked for a detailed explanation of the relative values. The differences between trimer and tetramer are small and not always in the same direction. The changes in pK_a caused by substitution are often approximately additive and can, with some limitations, be calculated

VII. REACTIONS AND DERIVATIVES

Table 12.1. A selection of pK_a values (first protonation) for cyclic phosphazenes $(NPR_2)_{3,4}$ in nitrobenzene at 25°.[7,82]

R	Cyclic trimer	Cyclic tetramer
F	−6	−6
Cl	−6	−6
OPh	−5.8	−6.0
OMe	−1.9	−1.0
Ph	1.5	2.2
Et	6.4	7.6
NEt_2	8.5	8.3

from empirical "substituent constants".[81,82] Various phosphazene salts have been isolated in the crystalline state, such as $(NPCl_2)_3 \cdot HClO_4$ and $(NPCl_2)_2 \cdot 2HClO_4$ from the components in glacial acetic acid.[7,82] The acidic protons are believed to be bound to ring nitrogen atoms in nearly all cases, even when amino substituents are present. This view is well supported for the aminophosphazenes by n.m.r. and i.r. studies, by the relative magnitudes of first and second protonation constants, and by crystal structure determinations of $N_3P_3Cl_2(NHPr)_4 \cdot HCl$ and $N_4P_4Me_6Et_2 \cdot 2HCl$, which show lengthened skeletal P–N bonds to the protonated nitrogen atoms.[7,22] It seems reasonable to assume that the acidic protons in such compounds as $(NPCl_2)_3 \cdot HClO_4$ are also bound to nitrogen rather than phosphorus, though experimental evidence is lacking. $(NPF_2)_3$ should be an extremely weak base, and a reported[7] compound $(NPF_2)_3 \cdot 2HF \cdot 2H_2O$ may not contain protonated nitrogen atoms, but rather HF molecules hydrogen-bonded to the fluorine atoms of the phosphazene.

Stable water-soluble quaternary salts $[(NPMe_2)_{3,4}Me]^+I^-$ are readily formed by direct combination. From n.m.r. evidence the Me^+ group is attached to a ring nitrogen atom.[7] The similarly constituted $[(NPPh_2)_3Me]^+BF_4^-$ can be made from $(NPPh_2)_3$ and trimethyloxonium fluoroborate.[7] But this reagent methylates an *exo*cyclic nitrogen atom in its reaction with any of the compounds $N_3P_3Cl_{6-n}(NMe_2)_n$ ($n = 1, 2, 3, 4, 6$), in contrast to the protonation of aminophosphazenes.[82]

Various other adducts of phosphazenes with Lewis acids are known[40] but few of them have been fully characterized and few X-ray structure determinations have been reported. The adducts $(NPCl_2)_3 \cdot AlBr_3$, $(NPBr_2)_3 \cdot AlBr_3$, and $(NPBr_2)_3 \cdot 2AlBr_3$ form from the components in carbon disulfide; they are colorless crystals with i.r. spectra consistent with a covalent donor–acceptor $(N \rightarrow AlBr_3)$ formulation.[7] There is a similar adduct $(NPCl_2)_3 \cdot AlCl_3$ of unknown structure.[7] $(NPCl_2)_3$ is not a strong enough base to add to $TiCl_4$ or $SnCl_4$, but the stronger base $(NPMe_2)_3$ gives stable 1:1 crystalline adducts with both of these Lewis acids.[7] With the powerful Lewis acid SbF_5 the weakly basic fluorophosphazenes $(NPF_2)_{3-6}$ all form 1:2 crystalline adducts sublimable without decomposition *in vacuo* below 110°.[7] It seems unlikely that the nitrogen atoms in $(NPF_2)_n$ would have enough donor strength for adduct formation, and the i.r. evidence suggests that these adducts are fluorine-bridged, $-P-F \rightarrow SbF_5$ (cf. comment above on the F "salts" of fluorophosphazenes). Adducts of the trimeric chloride with $SbCl_5$ and $TaCl_5$ are reported on n.q.r. evidence to be salts with the anions $SbCl_6^-$ and $TaCl_6^-$ respectively.[49]

Two X-ray structure determinations show different ways in which a phosphazene ring may ligate to a metal, and suggest that this area of phosphazene chemistry would be worth pursuing further. First, a compound forming yellow needles, and prepared from $(NPMe_2)_4$ with $CuCl_2$ in methyl ethyl ketone, has a structure[86] with one ring nitrogen atom coordinated to a $CuCl_3^-$ group and the opposite one protonated. Secondly, reaction of $[NP(NMe)_2]_6$ with an equimolar mixture of $CuCl_2$ and $CuCl$ gives orange crystals of a compound $[N_6P_6(NMe_2)_{12}CuCl]^+[CuCl_2]^-$. In the cation of this compound the phosphazene acts as a macrocyclic ligand, using four of the six nitrogens as donor atoms while the copper is 5-coordinate.[55] The question of π-complexes of the phosphazene rings, analogous to ferrocene, now naturally arises. In 1966 D. A. Brown predicted, from consideration of the MO theory of the cyclophosphazenes, that they might form π-complexes with metal carbonyls, but experiments which might have been expected to produce such a complex have not, it seems, done so. Yellow sublimable crystals with the formula $(PNCl_2)_3Cr(CO)_3$ are formed from the phosphazene and $(MeCN)_3Cr(CO)_3$ in ether or tetrahydrofuran at 30-40°,[7] but this has not been proved to be a π-complex. $(NPMe_2)_4$ reacts almost quantitatively with molybdenum hexacarbonyl, giving a yellow solid $(NPMe_2)_4Mo(CO)_4$, but i.r. evidence points to a C_{2v} structure, with coordination to molybdenum of two opposite nitrogen atoms of the ring.[7] An X-ray structure determination on the yellow crystalline complex $[NP(NMe)_2]_4W(CO)_4$ has shown bidentate σ coordination from one ring nitrogen and one exocyclic nitrogen to the tungsten atom.[21] Putting electron-releasing amine groups into a cyclophosphazene, as in this example, may in principle facilitate π bonding of the phosphazene ring to a metal, but is not helpful if σ bonding to an exocyclic nitrogen is preferred.

When treated with excess liquid SO_3 at 40°, $(NPCl_2)_3$ adds 3 SO_3 to give a hygroscopic solid.[36] This has not recently been investigated and its structure is unknown. Yellow powders, $(NPPh_2)_3 \cdot 2ICl$ and $(NPPh_2)_3 \cdot HICl_2$, are readily prepared from the components.[88] They are probably straightforward Lewis adducts, since their infrared spectra show that they contain cyclophosphazene rings.

J. Catalytic arylation of phosphazenes

The reported reactions of this kind are of the Friedel–Crafts type, with chloro- and fluoro-phosphazenes. $(NPCl_2)_3$ can be phenylated with benzene and aluminum chloride.[7] Reaction is slow: several days' refluxing gives mainly the diphenyl derivative; for the tetraphenyl compound six weeks' refluxing is needed; and the hexaphenyl compound is only readily obtained by autoclaving at 150°. The reaction does not occur in absence of aluminum chloride. Phenylation in this and other similar cases is geminal; if, however, secondary amino groups are present they direct phenylation geminally to themselves. In explanation, it has been suggested that the first step of these reactions is the loss of a halide ion, promoted by $AlCl_3$, and that this process is favored by the presence of an electron-supplying group, such as phenyl or (especially) secondary amino, on the phosphorus atom concerned.[7]

K. Reduction of phosphazenes

The ultraviolet spectra of the cyclic phosphazenes (Section IX below) give no evidence of antibonding π levels associated with the P–N ring systems at low energies, comparable to S_4N_4 (Chapter 6, Sec. II.A.4) or organic aromatics. Such levels may not indeed exist, since phosphazene rings do not seem reducible to radical anions.[5] At potentials down to -3.0 volt in dimethylformamide, in the presence of $Bu_4N^+I^-$ as supporting electrolyte, the only phosphazenes reduced at a dropping mercury electrode have aromatic rings attached to phosphorus directly or (in a few cases) through oxygen. There is e.s.r. evidence that the site of reduction is always the aromatic ligand, and the half-wave potentials give no evidence for any broad delocalization interaction between the phosphazene ring and aromatic ligands.[5]

VIII. Molecular structure and its rationalization

The molecular structures of about fifty cyclophosphazenes have been determined by X-ray diffraction, and new results appear every year.[7,40-46] In Fig. 12.5 we show some structures chosen to illustrate particular points. We shall now outline the main structural features, and attempt at the same time to rationalize them, where appropriate, with the help of current ideas on bonding. We emphasize that the theory of bonding in phosphazenes is far from finalized, though it has often been plausibly written about, and that in comparing molecular parameters, a correlation does not necessarily imply a cause-and-effect relationship.

Some important structural features of the cyclic phosphazenes are as follows.

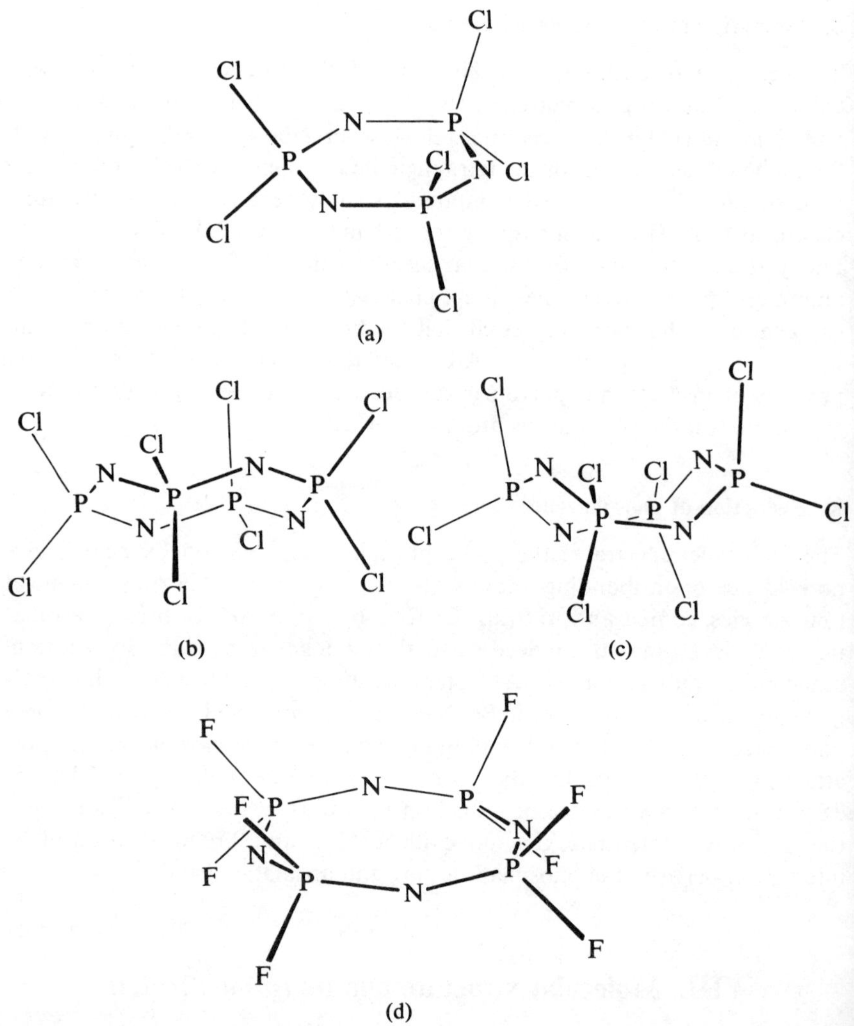

Figure 12.5. Molecular structures of cyclophosphazenes: (a) (NPCl$_2$)$_3$; (b) (NPCl$_2$)$_4$ in the T-form; (c) (NPCl$_2$)$_4$ in the K-form; (d) (NPF$_2$)$_4$ (sources cited in ref. 7).

(1) The ring bond angle at phosphorus is remarkably constant, ranging between 115° and 127° and usually close to 120°. This bond angle, at 4-coordinate P(v), differs strikingly from the angles of less than 90° found for 5-coordinate P(v) in the phosphazenes (Chapter 11), and is usually attributed to hybridization approaching sp^3 in the former as opposed to sp^{3}d in the latter. The need to comply accordingly with an angle near 120° at P seems to be a

primary constraint on the possible shapes of phosphazene rings, and may be held responsible for the non-existence of 4-membered phosphazene rings.

(2) Given $\sim 120°$ at the phosphorus atoms, reasonable bond angles of $\sim 120°$ at nitrogen are readily achieved in the cyclic trimers by planar D_{3h} or nearly planar C_{3v} configurations. So $(NPCl_2)_3$ [Fig. 12.5(a)] has an almost planar, slightly puckered ring. The same holds for the other cyclic trimers investigated; $(NPF_2)_3$ is accurately planar.

An angle of 120° at nitrogen implies sp^2 hybridization. The angle is never appreciably smaller than this in the phosphazenes. Much smaller angles would indicate sp^3 hybridization, which is scarcely possible in the phosphazenes since it would require a full negative electronic charge on each nitrogen atom; but such small angles do occur in the phosphazanes (Chapter 11).

(3) In larger phosphazene rings, where again the angle at phosphorus is always near 120°, various structures are found. Uniquely for the solid state, $(NPF_2)_4$ is planar [Fig. 12.5(d)]. $(NPCl_2)_4$ occurs as chair-form rings in the stable crystalline T-form, and tub-form rings in the metastable K-form [Fig. 12.5 (b) and (c)]. There is a recent report[20] of two conformers of a phosphazene ring in the same crystal lattice. $(NPCl_2)_5$ has two re-entrant angles at nitrogen, and $[NP(OMe)_2]_8$ has a stepped ring.[7] In all of these the angles at nitrogen are well over 120°, ranging up to near 160°. This shows varying contributions from sp hybridization, which is possible without charges on the nitrogen atoms. It has been argued[34] that these angles may be largely determined by $P\cdots P$ non-bonded interactions.

(4) The P–N bond lengths in the rings have attracted much comment. All are much shorter (147–162 pm) than the single-bond value of 178 pm. Where all N–P bonds in a ring are structurally equivalent they are equal in length, suggesting delocalized π bonding (Section IX below). However, there are significant variations between non-equivalent bonds in a ring, and between bonds in similar chemical situations in rings of different sizes. We shall discuss first the general reasons for the short bond length and next the current rationalizations of its variations.

Although the estimation of bond orders from bond lengths is quite controversial, it is generally accepted that the P–N bonds in phosphazenes are multiple;[57] the value 151 pm found in $(NPF_2)_4$, for example, could imply a π bond order of about 0.8, in addition to the σ bond. The lengths of P–N bonds in the phosphazenes, though all compatible with multiple bonding, range between 147 and 162 pm.[7] The causes of this variation are not fully understood, but two correlations have been noted.

First, Schlueter and Jacobson[74] (cf. ref.89) pointed out a fairly pronounced inverse correlation between P–N bond length and bond angle at nitrogen (Fig. 12.6). Glidewell[34] suggested that an important determinant of this relationship might be a need to maintain an almost constant distance (290.2 pm,

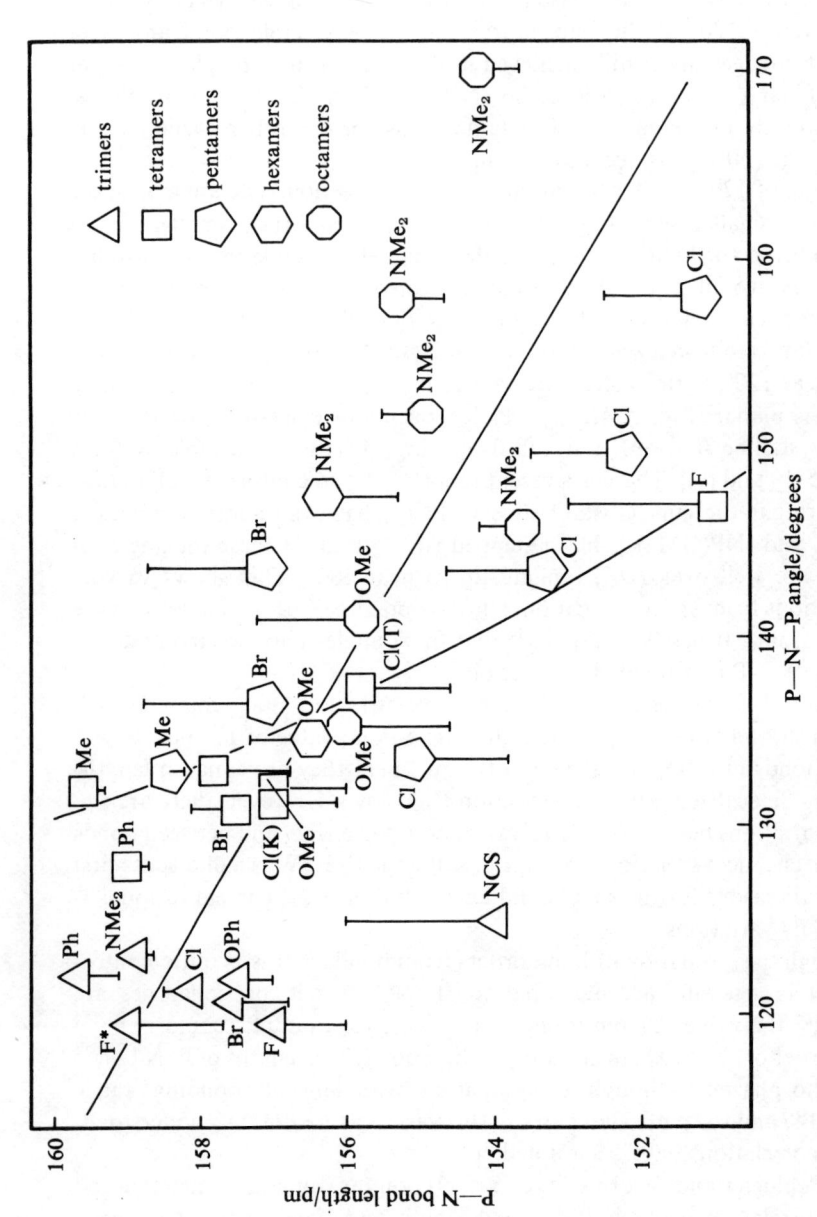

Figure 12.6. Relationship between P–N bond length and bond angle at nitrogen for homogeneously substituted phosphazenes. The markers for standard deviations should be understood as extending equally above and below each point, but for clarity they are shown in one direction only (data from refs. 7 and 41–46). The curved line is calculated from Glidewell's postulated relationship.[34] The F* point is an electron

presumably a van der Waals contact distance) between the centers of the two phosphorus atoms linked to a nitrogen atom. The bond-angle–bond-length relationship calculated using Glidewell's value for the P···P spacing is plotted on Fig. 12.6. Obviously Glidewell's idea taken by itself is too simple, but it has rightly directed attention to the importance of van der Waals interactions in determining the shapes of phosphazene rings. Another view of this bond-angle–bond-length correlation (quite compatible with Glidewell's) connects it with the type of bonding between nitrogen and phosphorus. To explain the shortening of the bond as the angle opens, it is argued on the one hand that the overlap of the in-plane nitrogen lone-pair orbital with the phosphorus $d_{x^2-y^2}$ orbitals improves, strengthening the π component of the bonds (Section IX below). Additionally or alternatively, the strength of the σ component increases because of the change of nitrogen hybridization away from sp^2 at 120° towards sp at 180°. The relative magnitude of these two effects is not known.

A *second* correlation often mentioned is that the P–N bonds shorten with increasing electronegativity of the ligands on phosphorus. Figure 12.6 does indeed show this; for example, compare the tetrameric fluoride with the tetrameric chloride. The effect is sometimes, as in this example, associated with an increase in the bond angle at nitrogen, and is less striking when the comparison is made between compounds with nearly constant bond angles at nitrogen, e.g. the trimeric halides; so it may be an oversimplification to attribute it, as is sometimes done, to shrinkage of the phosphorus bonding orbitals by the inductive effect of the ligand.

A further factor which could affect P–N bond lengths is suggested by a crystallographic study of *gem*-$N_3P_3Cl_3(NMe_2)_3$;[2] in this molecule all the substituents (Cl and N) are of equal electronegativity and the bond angles at nitrogen vary little, but the ring bond lengths range from 154.5 to 160.8 pm. In this and other similar cases there is probably conjugative transfer of electron density from the lone pairs of the exocyclic nitrogen atoms to the ring π systems, besides the usual inductive effects.

The central bonds of $N_7P_6Cl_9$ (formula **III** in Fig. 12.1) are a very interesting case. They are exceptionally long (172.3 pm), although the planar configuration around the central nitrogen atom, and the absence of any basic character associated with it, seem compatible with π bonding by its lone pair. In explanation, it has been shown that normally short central P–N bonds could only be achieved in this molecule by distortions which would weaken the π bonding.[62]

(5) Because phosphorus has available for π bonding five 3d orbitals with different orientations, it may be expected that changes in ring conformation would have relatively little effect on the energy of π bonding. This is supported by observations that P–N bond lengths are independent of dihedral angles in

(NPCl$_2$)$_4$, [NP(NMe$_2$)$_2$]$_6$, and [NP(OMe)$_2$]$_8$.[89] In consequence, minor energy factors, such as packing in the crystal, may be expected in some cases significantly to affect the ring shape. In support of this view, the vibrational spectrum of (NPCl$_2$)$_4$ in the vapor state has been interpreted to mean D_{4h}, planar, symmetry as opposed to the non-planar conformations in the crystal [Fig. 12.5 (c) and (d)]; (NPF$_2$)$_4$ becomes non-planar in a phase transition taking place at $-78°$;[38] and there are other similar cases.[7] It is unlikely too that transannular bonding could decidedly influence the conformational preferences of the larger, relatively flexible phosphazene rings.[28]

IX. Electronic structure and the question of aromaticity

The theory of bonding in the phosphazenes has been developed since the late 1950s by Craig, Paddock, Cruikshank, Dewar, and others.[7] Recently a new factor has been injected into the situation by the discovery by Perkins and Labarre that there may be a significant amount of cross-ring (transannular) bonding.[28] Although the importance of this type of bonding has not yet been fully evaluated, it seems unlikely to affect the established picture of skeletal σ and π bonding, which may be summarized as follows.[4,66] The σ framework of a cyclic phosphazene is built up from sp^3 hybrid orbitals of phosphorus, and nitrogen hybrid orbitals which approach sp^2 when the bond angle at nitrogen is near 120°, but may involve considerable sp character for the much larger bond angles often observed. We shall assume sp^2 nitrogen for the moment. The three sp^2 nitrogen orbitals are taken to be in the P–N–P plane (the xy plane, Fig. 12.7); two of them form σ bonds to neighboring phosphorus atoms and the third, projecting out of the ring, is occupied by a lone pair. This leaves nitrogen with a p$_z$ orbital, perpendicular to the local ring plane and occupied by one electron. The four σ bonds formed by phosphorus leave it too with an unpaired electron, which cannot be in any of its fully occupied sp^3 orbitals. The current hypothesis is that the odd electrons from N and P pair up in a molecular orbital of π type, involving the p$_z$ orbital of nitrogen and appropriate orbitals of phosphorus (3d, 4s, 4p, or mixtures). The π molecular orbital, like its component atomic orbitals, is antisymmetric in the P–N–P plane [Fig. 12.7 (a) and (b)] and is therefore usually designated π_a. In addition, there may be a second kind of π bonding, making use of the third sp^2 hybrid orbital of nitrogen and another phosphorus orbital (d$_{x^2-y^2}$ or d$_{xy}$), and involving the nitrogen lone pair [Fig. 12.7 (c) and (d)]; this is symmetric in the P–N–P plane and so is designated π_s. It is believed that π_a is the main contribution to π bonding,[57] while the π_s contribution is smaller and varies considerably in importance in different phosphazenes, being much affected by the bond angle at nitrogen (see preceding section) and other factors.

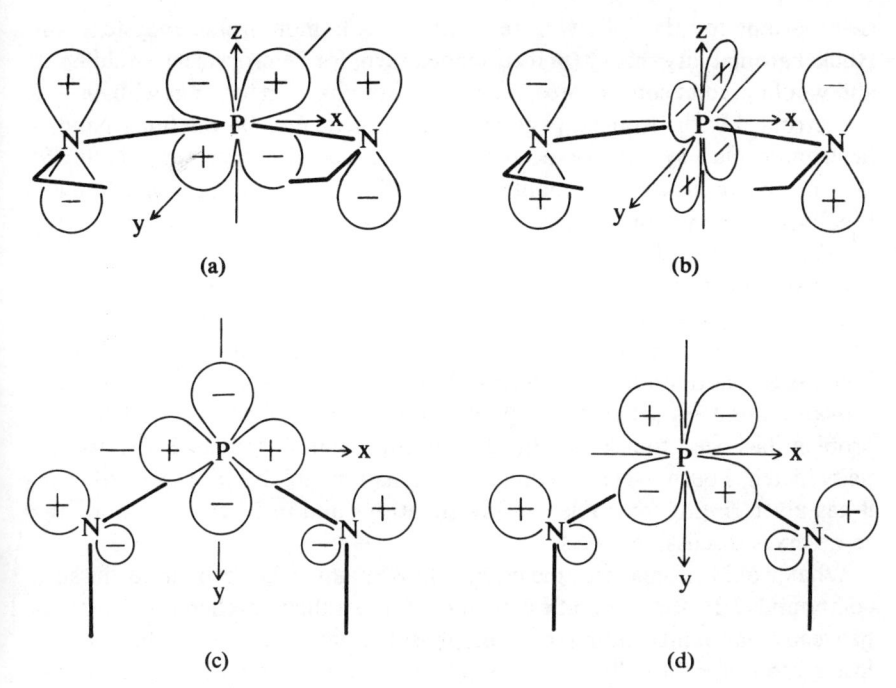

Figure. 12.7. Conceivable modes of π bonding in phosphazene rings (for trimers; diagrammatic): (a) heteromorphic π_a system; (b) homomorphic π_a system; (c) homomorphic π_s system; (d) heteromorphic π_s system.

Phosphorus 3d orbitals of suitable symmetry (Fig. 12.7) are available for both postulated types of π bonding, and the number of π electrons, exclusive of lone pairs, equals the number of ring atoms. It is necessary now to take a closer look at these different modes of π bonding. We consider planar rings, keeping in mind that the departures from planarity in actual phosphazene rings are not enormous, so that the arguments should hold at least roughly for all of them. In the π_a system it will be seen [Fig. 12.7(a)] that the use of phosphorus d_{xz} orbitals alone would give rise to delocalized bonding in which the signs of orbital interaction alternate round the ring. This is called "heteromorphic" bonding. If d_{yz} alone were used in π_a bonding [Fig. 12.7(b)] the signs of interaction would be constant round the ring, and a "homomorphic" π_a system resembling that in benzene would result. If d_{xz} and d_{yz} contributed equally, it has been shown[23,57] that the π_a system would cease to be fully delocalized and would break up into three-center P–N–P bonds, now usually called Dewar islands. Finally, it may be noted that the π_s system too could be homomorphic or heteromorphic [Fig. 12.7 (c) and (d)] depending on the phosphorus d orbitals used. The heteromorphic/homomorphic distinction is important for the following reason.[57] In a homomorphic π system, the Hückel aromaticity theory (in its extended form for heterocycles) would hold, and would predict some degree of aromatic stabilization for rings with $4n + 2$ π electrons, i.e. the trimers, pentamers, heptamers, etc. On the other hand, a heteromorphic π system is expected to show a π bond energy per P–N unit at first increasing slowly and monotonically with ring size and then levelling off. Finally, in a π system composed of ideal Dewar islands, the π bond energy per P–N unit would be independent of ring size.

Considering this variety of π bonding possibilities, together with the effects of exocyclic ligands on π bonding, the accompanying (synergistic or antagonistic) variations of σ bond strength, and the likely presence of transannular bonding and antibonding, it is obviously a formidable task to fit a unique and provable bond model to the experimental evidence. One approach to this problem has been to ask: are the cyclophosphazenes aromatic? The aromaticity idea has been overworked, and disenchantment has set in. Nevertheless, it has given rise to some relevant investigations and instructive findings, and so deserves discussion.

What would aromaticity mean, and how might it be recognized, in such compounds? It seems pointless to look for parallels in chemical reactivity between compounds so unlike as the phosphazenes and benzene, but we may instead look for the following physical evidence of electron delocalization in the phosphazene rings: equality of ring bond lengths; ring currents; aromatic stabilization; and optical transitions of electrons to low-lying empty π^* antibonding orbitals.

The P–N *bond lengths* are indeed equal within experimental error in any homogeneously substituted trimer or tetramer, and where significant variations occur in the higher oligomers they are associated with changes in the bond angle at nitrogen. This is consistent with full aromatic π delocalization but does not prove it, since the constant bond lengths are equally well explained by Dewar islands.

Ring currents in $p\pi$–$d\pi$ systems may in theory contribute to the magnetic susceptibility of a substance in either a paramagnetic or a diamagnetic sense, according to which d orbitals are involved.[4] The susceptibilities of the cyclophosphazenes show no marked effect of either sign and throw little light on any ring currents. The Faraday effect (magneto-optic rotation) has been determined for a number of phosphazenes.[28] The molecular rotations for homologous series $(NPX_2)_{3,4,5}$ are exactly or roughly in the ratio 3:4:5, and there is no sign of the quasi-exponential increase to be expected, on ascending the series, if there were ring currents. The ^{19}F n.m.r. signals of the fluoride trimer and tetramer occur at almost exactly the same chemical shift as that of the linear polymer $(NPF_2)_n$, thus showing no evidence of deshielding by magnetically induced ring currents.[9] Finally, calculations by the CNDO/2 method of

IX. ELECTRONIC STRUCTURE

Table 12.2. Dependence of enthalpy of formation of ring N–P bonds on ring size in $(NPCl_2)_n$.[39]

n	3	4	5	6	7
ΔH per P–N bond for reaction trimer → n-mer (kJ mol^{-1})	0	−6.8	−9.5	−10.5	−10.8

Pople et al. have shown charges on the ring atoms (phosphorus positive, nitrogen negative) in $(NPF_2)_{3,4}$ and $(NPCl_2)_3$ too large to make ring currents plausible.[28]

The thermal stability of phosphazene rings, though impressive (Section VI.A above), is not in itself evidence of *aromatic stabilization*. In order to look for this, one should ideally (following the precedent of benzene) compare experimental enthalpies of formation of the phosphazenes $(NPX_2)_n$ with those of hypothetical similar compounds having alternate single and double bonds in the ring, but the latter quantities cannot be satisfactorily estimated. A less ambitious but more practical idea is to compare the enthalpies of formation of the $(NPCl_2)_n$ phosphazenes with each other (Table 12.2).[39] These data show no unusual degree of stabilization associated with any particular ring size. Instead, the strength of the ring bonds increases progressively from trimer to heptamer; it seems to be levelling off for the largest rings. The increase is small—an order of magnitude smaller than the delocalization energy of benzene. It may be connected with the widening of P–N–P bond angles and associated bond shortening mentioned above. However, all arguments purporting to connect the stability of phosphazene rings with detailed features of the π bonding must be treated sceptically, in consequence of some recent results of CNDO/2 calculations, which point to transannular bonding (Fig. 12.8) of strength of the order of 10% of the strength of a P–N bond.[28]

The ultraviolet absorption spectra of the cyclophosphazene trimeric and

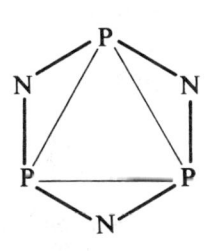

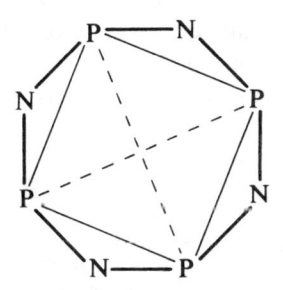

Figure 12.8. Transannular π bonding in phosphazene rings:[28] light full lines, transannular bonding interactions; light broken lines, transannular antibonding interactions.

tetrameric halides give no evidence of $\pi \to \pi^*$ transitions like those of benzene and other organic aromatics;[7] it must be remembered, however, that such transitions are forbidden in $p\pi$–$d\pi$ bonded systems and so would be expected to be weak or unobservable.[57] These phosphazenes are transparent through the near u.v. down to below 220 nm. Their longest-wavelength absorption bands, which have maxima in the region 157.5–200 nm, are attributed to $n \to \pi^*$ transitions of lone pairs (probably on halogen atoms). These bands show no delocalization-induced bathochromic shift with increasing ring size. The high polymeric chloro- and organo-phosphazenes are as transparent in the 200–700 nm region as the corresponding trimer, again supporting the view that there is no extensive π delocalization.

These lines of experimental evidence, then, give no clear evidence of circumannular delocalization of π electrons. Nevertheless, certain properties of the cyclophosphazenes do alternate with increasing ring-size in a manner suggesting a small degree of Hückel-type delocalization of the π system. Thus, the lowest ionization potentials of the fluorides and chlorides alternate in this way, being up to 0.7 volt higher for the odd-numbered rings, i.e. those with Hückel numbers, $4n + 2$, of π electrons.[17] Apart from the trimers, the odd-numbered rings are weaker bases than the even-numbered. The ^{19}F n.m.r. chemical shifts in the pentafluorophenyl derivatives alternate with ring size up to the hexamer. Rates of fluorination of the chlorides show alternation.[67]

What description of the π bonding can be given which is consistent with these facts? The best answer seems to be as follows: the main π bonding is π_a, predominantly of the three-center Dewar island type and heteromorphic in character; this accounts for the bond lengths and lack of evidence for ring currents or Hückel-type stabilization. But some contribution from a homomorphic π_s system must also be postulated to account for those properties that alternate with ring size, such as ionization potential and base strength. The ionization potentials show that the homomorphic π orbitals are at higher energy levels than the heteromorphic and so contribute less to the π bond strength. The importance of transannular bonding and antibonding has still to be fully evaluated. With regard to aromaticity, it seems fair to state that this organic concept is only marginally appropriate to the cyclophosphazenes, but that attempts to apply it have led to useful insights.

References

1. Acock, K. G., Shaw, R. A. and Wells, F. B. G., *J. Chem. Soc.* 121 (1964)
2. Ahmed, F. R. and Pollard, D. R., *Acta Cryst.* **28B**, 513 (1972).
3. Allcock, H. R., Kugle, R. L. and Valan, K. J., *Inorg. Chem.* **5**, 1709 (1966)
4. Allcock, H. R., *Heteroatom Ring Systems and Polymers*, Academic Press, New York (1967)
5. Allcock, H. R. and Birdsall, W. J., *Inorg. Chem.* **10**, 2495 (1971)

6. Allcock, H. R. and Walsh, E. J., *J. Am. Chem. Soc.* **94**, 4538 (1972)
7. Allcock, H. R., *Chem. Rev.* **72**, 315 (1972)
8. Allcock, H. R., *Phosphorus-Nitrogen Compounds*, Academic Press, New York (1972)
9. Allcock, H. R., Kugel, R. L. and Stroh, E. G., *Inorg. Chem.* **11**, 1120 (1972)
10. Allcock, H. R., *Chem. Brit.* **10**, 118 (1974)
11. Allcock, H. R., *Angew. Chem. Int. Ed. Engl.* **16**, 147 (1977)
12. Allcock, H. R., private communication (1978)
13. Allcock, H. R., *Acc. Chem. Res.* **11**, 81 (1978)
14. Bermann, M., *Adv. Inorg. Chem. Radiochem.* **14**, 1 (1972)
15. Bermann, M., *Topics Phosphorus Chem.* **7**, 311 (1972)
16. Biddlestone, M. and Shaw, R. A., *Chem. Commun.* 407 (1968)
17. Branton, G. R., Brion, C. E., Frost, D. C., Mitchell, K. A. R. and Paddock, N. L., *J. Chem. Soc. A* 151 (1970)
18. Brenner, K. S., *J. Chromatogr.* **57**, 131 (1971)
19. Bullen, G. J., Dann, P. E., Evans, M. L., Hursthouse, M. B., Shaw, R. A., Wait, K., Woods, M. and Yu, H. S., *Z. Naturforsch. B* **31b**, 995 (1976)
20. Bullen, G. J., Cameron, T. S. and Dann, P. E., *Abstracts of Second International Symposium on Inorganic Ring Systems*. Akademie der Wissenschaften, Göttingen (1978).
21. Calhoun, H. P., Paddock, N. L., Trotter, J. and Wingfield, J. N., *Chem. Commun.* 875 (1972)
22. Calhoun, H. P., Oakley, R. T., Paddock, N. L. and Trotter, J., *Can. J. Chem.* **53**, 2413 (1975)
23. Craig, D. P., in ref. 20.
24. Dalgleish, W. H. and Porte, A. L., *J. Magn. Reson.* **20**, 351, 359 (1975)
25. Emsley, J. and Paddock, N. L., *J. Chem. Soc. A* 2590 (1968)
26. Emsley, J. and Udy, P. B., *J. Chem. Soc. A* 3025 (1970)
27. Emsley, J. and Udy, P. B., *J. Chem. Soc. A* 768 (1971)
28. Faucher, J. P., Labarre, J.-F. and Shaw, R. A., *Z. Naturforsch. B* **31b**, 677 (1976)
29. Feistel, G. R. and Moeller, T., *J. Inorg. Nucl. Chem.* **29**, 2731 (1967)
30. Fluck, E., *Z. Anorg. Allg. Chem.* **315**, 181, 191 (1962)
31. Fluck, E., *Topics Phosphorus Chem.* **4**, 291 (1967)
32. Fluck, E., Schmid, E. and Haubold, W., *Z. Anorg. Allg. Chem.* **433**, 229 (1977)
33. Gieren, A., Dederer, B., Roesky, H. W. and Janssen, E., *Angew. Chem. Int. Ed. Engl.* **15**, 783 (1976)
34. Glidewell, G., *J. Inorg. Nucl. Chem.* **38**, 669 (1976)
35. *Gmelins Handbuch der Anorganischen Chemie*, Teil C, "Phosphor", Verlag Chemie, Weinheim (1965)
36. Goehring, M., Hohenschutz, H. and Appel, R., *Z. Naturforsch. B* **9b**, 678 (1954)
37. Heal, H. G., Shahid, M. S. and Garcia-Fernandez, H., *J. Chem. Soc. A* 3846 (1971)
38. Henkel, G. and Krebs, B., in ref. 20.
39. Jaques, J. K., Mole, M. F. and Paddock, N. L., *J. Chem. Soc.* 2112 (1965)
40. Keat, R. and Shaw, R. A., in *Organic Phosphorus Chemistry*, ed. G. M. Kosolapoff and L. Maier, vol. 6, p. 883, Wiley, New York (1973)
41. Keat, R., "Phosphazenes", in *Organophosphorus Chemistry* (Specialist Periodical Reports), The Chemical Society, London, **4** (1973)
42. *Idem, ibid.*, **5** (1974)

43. *Idem, ibid.*, **6** (1975)
44. *Idem, ibid.*, **7** (1976)
45. *Idem, ibid.*, **8** (1977)
46. *Idem, ibid.*, **9** (1978)
47. Kobayashi, E., *Chem. Lett.* 479 (1976)
48. Kobayashi, E., *Bull. Chem. Soc. Japan* **49**, 3524 (1976)
49. Kravchenko, E. A., Levin, B. V., Bananyarli, S. I. and Toktomatov, T. A., *Koord. Khim.* **3**, 374 (1977)
50. Krishnamurthy, S. S., Sau, A. C., Murthy, A. R. V., Keat, R. and Shaw, R. A., *J. Chem. Soc. Dalton Trans.* 1405 (1976)
51. Krishnamurthy, S. S., Sau, A. C., Murthy, A. R. V., Shaw, R. A., Woods, W. and Keat, R., *J. Chem. Res. S* 70 (1977)
52. Krishnamurthy, S. S., Sau, A. C. and Woods, M., *Adv. Inorg. Chem. Radiochem.* **21**, 41 (1978)
53. Lehr, W., *Z. Anorg. Allg. Chem.* **371**, 225 (1969)
54. Lund, L. G., Paddock, N. L., Proctor, J. E. and Searle, H. T., *J. Chem. Soc.* 2542 (1960)
55. Marsh, W. C., Paddock, N. L., Stewart, C. J. and Trotter, J., *Chem. Commun.* 1190 (1970)
56. Migachev, G. I. and Stepanov, B. I., *Russ. J. Inorg. Chem.* **11**, 929 (1966)
57. Mitchell, K. A. R., *Chem Rev.* **69**, 157 (1969)
58. Nabi, S. N., Shaw, R. A. and Stratton, C., *J. Chem. Soc. Dalton Trans.* 588 (1975)
59. Nielsen, M. L. and Cranford, G., *Inorg. Synth.* **6**, 94 (1960)
60. Nielsen, M. L. and Morrow, J. T., *Inorg. Synth.* **6**, 99 (1960)
61. Nikolaev, A. F., Dreiman, N. A. and Zyryanova, T. A., *Zh. Obshch. Khim.* **40**, 937 (1970)
62. Oakley, R. T. and Paddock, N. L., *Can. J. Chem.* **51**, 520 (1973)
63. Oakley, R. T. and Paddock, N. L., *Can. J. Chem.* **53**, 3038 (1975)
64. Paddock, N. L. and Searle, H. T., U.S.Pat. 3,008,799, appl. Dec. 4th (1958)
65. Paddock, N. L. and Searle, H. T., *Adv. Inorg. Chem. Radiochem.* **1**, 348 (1959)
66. Paddock, N. L., *Quart. Rev.* **18**, 168 (1964)
67. Paddock, N. L. and Serreqi, J., *Can. J. Chem.* **52**, 2546 (1974)
68. Radosavljević, S. D. and Šašić, J. S., *Z. Anorg. Allg. Chem.* **387**, 271 (1972)
69. Ranganathan, T. N., Todd, S. M. and Paddock, N. L., *Inorg. Chem.* **12**, 316 (1973)
70. Roesky, H. W. and Janssen, E., *Z. Naturforsch.* B **29b**, 174 (1974)
71. Roesky, H. W. and Janssen, E., *Z. Naturforsch.* B **29b**, 177 (1974)
72. Roesky, H. W. and Janssen, E., *Angew. Chem. Int. Ed. Engl.* **15**, 39 (1976)
73. Roesky, H. W. and Banek, M., *Chem.-Ztg.* **102**, 155 (1978)
74. Schlueter, A. W. and Jacobson, R. A., *J. Chem. Soc. A* 2317 (1968)
75. Schmidpeter, A. and Eiletz, H., *Chem. Ber.* **108**, 1454 (1975)
76. Schmidpeter, A. and Eiletz, H., *Phosphorus* **6**, 113 (1976)
77. Schmulbach, C. D., *Progr. Inorg. Chem.* **4**, 275 (1962)
78. Schmutz, J. L. and Allcock, H. R., *Inorg. Chem.* **14**, 2433 (1975)
79. Schmutzler, R., *Inorg. Synth.* **9**, 76 (1967)
80. Seel, F. and Langer, J., *Z. Anorg. Allg. Chem.* **295**, 316 (1958)
81. Shaw, R. A., *Pure Appl. Chem.* **44**, 317 (1974)
82. Shaw, R. A., *Z. Naturforsch.* B **31b**, 641 (1976)
83. Shaw, R. A., *Phosphorus and Sulfur* **4**, 101 (1978)

84. Sisler, H. H., Ahuja, H. S. and Smith, N. L., *Inorg. Chem.* **1**, 84 (1962)
85. Toy, A. D., "Phosphorus", in *Comprehensive Inorganic Chemistry*, ed. J. C. Bailar, H. J. Emeléus, R. S. Nyholm, and A. F. Trotman-Dickenson, Pergamon (1973)
86. Trotter, J. and Whitlow, S. H., *J. Chem. Soc. A* 455 (1970)
87. Uhlig, E. and Eckert, H., *Z. Anal. Chem.* **204**, 332 (1964)
88. Whitaker, R. D., Carleton, J. C. and Sisler, H. H., *Inorg. Chem.* **2**, 420 (1963)
89. Zoer, H. and Wagner, H. J., *Acta Cryst.* **B28**, 252 (1972)

Note added in proof

(p. 215) Several aminophosphazenes are effective against mouse leukemia and melanoma, the best being $(NPAz_2)_3$ (Az = aziridino) [Labarre, J.-F., Faucher, J. P., Levy, G., Sournies, F., Cros, S. and Francois, G., *Eur. J. Cancer* **15**, 637 (1979)].

13
Polymeric phosphorus nitrides and related compounds

I. Introduction

Phosphorus forms a range of rather refractory solid compounds with nitrogen, both alone and together with hydrogen, oxygen, or sulfur. The inertness and polymeric nature of these substances has made them difficult to characterize, and less is known about them than might have been expected for compounds between such common elements.

There potential interest for heterocyclic chemistry arises from: their structures, which are almost entirely unknown but probably include heterocyclic rings; their possible technical applications as polymers and fertilizers; and their possible use in the synthesis of heterocyclic compounds, for which there are promising indications.

For many years these compounds were assumed to be representable by simple "daltonide" formulas such as PN, P_3N_5, OPN, etc., though the appreciation that they were polymeric gradually developed during the present century. In 1958 van Wazer[26] critically reviewed the "facts" of this area of chemistry. He thought it likely that the amorphous compounds formulated in the literature as PN, OPN, SPN, and PN_2H are polymers from systems in which a wide range of composition is possible without much change of properties, and recommended their reinvestigation "from the polymer viewpoint". This is still excellent advice, though quite recently there has been progress in the preparation and characterization of crystalline phases which probably have fairly definite stoichiometry.

This subject has not been dealt with in recent texts, and some important recent results have been published in Russian in minor journals, making it difficult for English-speaking readers to get an up-to-date picture. However, a comprehensive review, originally in Russian but available in English translation,[4] has lately appeared.

There is a mild degree of technical interest in phosphorus nitrides; their synthesis from the elements, followed by hydrolysis, would be a more direct route from raw materials to ammonium phosphate for fertilizer than the conventional procedure via Haber ammonia.[4]

II. History

The conditions needed for direct combination of phosphorus with nitrogen are rather exacting and are only now becoming properly understood. Hence early approaches to the phosphorus nitrides and related compounds were indirect.

The longest-known of this group of compounds is "phospham", $(PN_2H)_x$. It was reported in 1810 by H. Davy, who made it by igniting the products of the PCl_5–NH_3 reaction, and was more fully examined between 1832 and 1846 by Wöhler, Liebig, Gerhardt, and Rose.[13] The first authentic report of a phosphorus nitride came in 1862 from Briegleb and Geuther, who got an inert material analyzing approximately as $(P_3N_5)_x$ by heating magnesium nitride in the vapor of PCl_5.[13] The direct combination of phosphorus vapor with nitrogen was first seriously studied by Moldenhauer and Dörsam (1926)[15] who made $(PN)_x$ in this way under the action of an electric discharge, basing their work on earlier incomplete observations, including the long-known fact that phosphorus will remove the last trace of gaseous nitrogen from electric light bulbs. At that time, the possibility of purely thermal combination of the elements was not recognized and not clearly distinguished from the action of discharges; thermal combination, however, was soon afterwards accomplished in the important work of Moureu and his group in the 1930s.[13,16,17,18] In recent times, the most important development in the chemistry of phosphorus nitrides has been the preparation and study of crystalline forms of P_3N_5 by Huffman and Tarbutton in America,[10] by T. Miller and his group in Riga,[1,24] and by Boden, Sadowski, and Lehmann in Dresden.[3]

The oxonitride $(OPN)_x$, first described by Gerhardt in 1846, and the thionitride $(SPN)_x$, first made by Glatzel in 1893, have received only a little attention in recent times.[4,13,19]

III. The phosphorus nitrides and phospham

A. The gas-phase molecule PN

Monomeric PN is the only phosphorus–nitrogen compound known in the gas phase. It was first recognized by its emission spectrum from electric discharges through gaseous mixtures of the elements.[13]

There have been several studies of the equilibrium $P_2(g) + N_2(g) \rightleftharpoons 2PN(g)$. Perhaps the best results are those of Gingerich,[6] who studied the equilibrium at 2000 K mass-spectrometrically in a Knudsen cell. At this temperature equilibrium was probably reached virtually instantaneously. Gingerich found $\Delta H°_{298} = +199$ kJ mol^{-1} and $-R \ln K_p = 20.910$ at 1985 K. Since the forward reaction starting from P_2 or P_4 is endothermic, the equilibrium concentration of PN rises with temperature in the region (roughly up to 4000 K) where P_4 and then P_2 are predominant molecular forms of phosphorus. At higher temperatures atomic phosphorus predominates, the formation of PN consequently becomes exothermic, and so the equilibrium concentration of PN falls off. Graphs showing the equilibrium composition of the system have been published.[4] At 1 atmosphere total pressure and equal atom percentages of N and P, the yield of PN is a maximum, 28 mol per cent, at about 3400 K.[4,5]

Monomeric gas-phase PN is present in high concentrations, together with P_2 and N_2, in the vapors emitted by the solid nitrides $(PN)_x$ or $(P_3N_5)_x$ in the 700–1000°C region,[9,25] but is not then in thermodynamic equilibrium; the equilibrium concentration at 1200 K is only about 2%.[4,9]

B. Amorphous phosphorus nitrides, including "phosphorus mononitride", $(PN)_x$

Polymers amorphous to X-rays and with compositions in the range $PN_{0.9}$ to $PN_{1.7}$ have been prepared. They are insoluble, chemically inert powders.

The first good preparation of the compound of this series with composition $(PN)_x$, usually called phosphorus mononitride, was devised by Moureu.[18] A stoichiometric mixture of white phosphorus and nitrogen, in amount to give about 1 atmosphere pressure, is heated at 500° in a sealed bulb so that the phosphorus is entirely vaporized. At this temperature the rate of combination is negligible, and anyway (see preceding section) the equilibrium conversion into PN monomer is probably slight. Combination is made to occur by heating a tungsten filament in the gas mixture electrically to 1800°. A deposit of $(PN)_x$ then starts to form on the walls (still at 500°) and combination is complete after 2 hours. Moureu's explanation, which still seems acceptable, is that PN monomer is produced fairly fast in the hot zone near the filament, and polymerizes on to the cooler glass walls before it has time to decompose into the elements.

A nitride of similar composition, $(PN)_x$, can be made chemically by dropping PCl_3 (in relatively large amount) into liquid ammonia. The yield is high and the product is easily separated from NH_4Cl formed at the same time.[2]

The $(PN)_x$ polymers are red to yellow powders. At 450°, Moureu's product evolves spectroscopically detectable amounts of PN vapor;[13] between 600° and 800° it can be slowly sublimed *in vacuo*;[17] and at 900° it volatilizes quickly, breaking down very largely to the elements.[10] Its enthalpy of depolymerization has been estimated at 172 kJ mol^{-1}, corresponding to an average P–N bond-energy term of 286 kJ mol^{-1}.[13] This suggests a framework structure of essentially single bonds. Such a framework would be, in a broad sense, isoelectronic with polymeric forms of phosphorus itself. As shown by red and black phosphorus, there is a wide range of possible structures; but nothing can be deduced from the amorphous X-ray diffraction pattern of (PN)$_x$ other than the lack of long-range order. It seems likely, by analogy with phosphorus, that (PN)$_x$ may contain 6- and/or 5-membered rings in its structure.

From the standpoint of heterocyclic chemistry, the most interesting reaction of (PN)$_x$ is that with chlorine, which gives (NPCl$_2$)$_3$ (Chapter 12) and some PCl$_5$.[13,17] Reaction only begins at 500° and for completeness 800° is required; thus the cyclic phosphazene trimer is probably formed via PN monomer and not directly from rings present in the (PN)$_x$ structure.

(PN)$_x$ is very inert towards hydrolysis, which requires 3 days with water at 215° in a sealed tube.[13]

Amorphous P–N polymers with N : P ratios up to 1.7 (i.e. up to the composition of P$_3$N$_5$; Section III.C below) can be made either by the discharge synthesis mentioned above in Section II, or by heating (PN)$_x$ in nitrogen or ammonia at 800–900°.[10] Crystalline P$_3$N$_5$ is usually formed simultaneously.[10]

C. Triphosphorus pentanitride

This compound apparently exists in three crystalline forms, making it "better defined" than (PN)$_x$.

Most preparative routes to (P$_3$N$_5$)$_x$ involve the pyrolysis of compounds containing nitrogen and phosphorus(v). In a recent method,[21] (NPCl$_2$)$_3$ (Chapter 12) or PCl$_5$, in a suitable solvent, is added to excess liquid ammonia; after a longish reaction time the product consists mainly of cyclo-[NP(NH$_2$)$_2$]$_3$, and of course NH$_4$Cl. The [NP(NH$_2$)$_2$]$_3$ is heated for 24 hours *in vacuo* at 300° to give phospham, (PN$_2$H)$_x$ (Section D below)[7]. With many hours' heating *in vacuo* at 728°, very pure P$_3$N$_5$ results, in yields of 88%.[16] The temperature for this stage is rather critical, since at 710° there is also some volatilization of PN monomer and N$_2$ from the P$_3$N$_5$.[16] (P$_3$N$_5$)$_x$ may also be prepared by Stock and Hoffman's method of pyrolyzing P$_2$S$_5 \cdot$7NH$_3$, first in ammonia at 230° and then at a bright red heat under N$_2$ or H$_2$.[12]

(P$_3$N$_5$)$_x$ is obtained in these ways as a white amorphous powder, odorless and tasteless, insoluble in all the usual solvents in the cold, but slowly hydrolyzed by water at 180° to ammonium phosphate.[13]

The hexagonal crystalline α-form of P$_3$N$_5$ results from long heating of the amorphous material at 800° under N$_2$ or NH$_3$.[8,16] A tetragonal β-form was first reported by Miller's group in 1967. Details of its preparation were not given at the time, but this form was later said[24] to arise from the action of ammonia on $(OPN)_x$ (Section IV.A below) for 130 hours at 850°; whether this was the original method of preparation is not clear. Recently a third (γ) crystalline form has been reported; the α-form is converted completely into it when heated at 500° for several days under 40 atm- pressure of ammonia.[3]

Complete X-ray structure determinations have not been performed for any of the forms of P_3N_5, though unit-cell dimensions for the α- and the β-form have been reported.[1] I.r. spectra give some indication of the structures.[1,3,22] Amorphous and α- and γ-crystalline P_3N_5 give strong bands in the 1250–1345 cm^{-1} region, resembling similar bands in the spectra of phospham and high-polymer $(NPCl_2)_x$, and possibly representing ν_{as} as for the configuration –P–N=P–. The only detailed i.r. spectrum of α-P_3N_5 so far published[3] contains a strong band in the N–H stretching region, which shows that the sample was not quite pure and raises the question whether impurities may be essential for formation of some of the crystalline modifications. Nitrogen 1s and phosphorus 2p X-ray photoelectron spectra for P_3N_5 have been published; they give no evidence for more than one type of N or P environment; unfortunately it was not stated which modification was examined.[8,20]

Amorphous or α-crystalline P_3N_5 slowly vaporizes to a mixture of monomeric PN and nitrogen when heated above 800° *in vacuo*.[16,25] A deposit of red-brown $(PN)_x$ has been observed to form on the cooler parts of the vessel.[16]

At 700°, P_3N_5 burns slowly in chlorine, giving a sublimate of $(NPCl_2)_3$, with a little PCl_5.[13] $(NPF_2)_3$ and $(NPF_2)_4$ likewise result in moderate yields when P_3N_5 is heated in CF_3SF_5 or NF_3;[13] here, too, temperatures of around 700° are required. These high reaction temperatures suggest that PN monomer is the reactive entity.

α-Crystalline P_3N_5 is now obtainable from specialist suppliers of research chemicals. It is expensive.

D. Phosphams, $(PN_2H)_x$

The preparation of phospham by heating $[NP(NH_2)_2]_3$ has been mentioned in the previous section. A recent investigation has confirmed this procedure without introducing any major improvement.[7] Phospham is also formed in an equilibrium reaction from red phosphorus and ammonia at 500–600°/17–37 atm.[23]

Phospham is a white, very light powder which does not melt but above 500° (see preceding section) decomposes to P_3N_5. It is insoluble in ordinary solvents and unaffected by many acids and bases.[7]

Phospham is amorphous to X-rays and has never been prepared in a crystalline form. It is obviously a high polymer of fairly disordered structure. Its i.r. spectrum shows a broad N–H stretch (2400–3400 cm^{-1}), a strong band at 1250 cm^{-1} attributed to ν_{as}(P–N=P), and an equally strong band at 950 cm^{-1} attributed to ν_{as}(P–NH–P).[7] It is interesting that phosphams derived respectively from $N_3P_3(NH_2)_6$ and $N_4P_4(NH_2)_8$ differ in density, in line with differences observed between homologous cyclophosphazenes $N_3P_3X_6$ and $N_4P_4X_8$, suggesting that 6- and 8-membered phosphazene rings respectively remain as structural units in the phosphams.[14]

With sodium or potassium amide at 230° and then 370°, phospham gives the salts $(PN_2Na)_x$ and $(PN_2K)_x$.[7] These too are amorphous to X-rays. In their i.r. spectra the N–H stretch is lacking and the two ν_{as} bands just mentioned become one band at about 1085 cm^{-1}, presumably because of resonance in the polymeric ion $(PN_2^-)_x$. $(PN_2^-)_x$ is isoelectronic with $(SiO_2)_x$; and the i.r. spectra of $(PN_2K)_x$ and α-quartz are similar.[7]

IV. Polymeric phosphorus oxonitride and thionitride

A. Phosphorus oxonitride, (OPN)$_x$

This compound, discovered in 1846, has been relatively little investigated. As van Wazer recognized, it is a high polymer probably structurally related to the phosphorus nitrides and phospham.[26]

The usual preparation is by pyrolysis of phosphoryl triamide, $OP(NH_2)_3$ (Chapter 11, Sec. III.B).[4,11] Phosphoryl amide resembles sulfuryl amide (Chapter 4, Sec. III.B) in undergoing condensation polymerization with the elimination of ammonia. On mild heating or even by treatment with HCl in ether below room temperature, linear condensation products $H_2N[P(O)(NH_2)NH]_nH$ ($n = 2$ to 5) are formed; these are soluble in water and have been separated by paper chromatography.[13] In contrast to the behavior of sulfuryl amide, no small rings appear to be formed. Towards 300°, however, crosslinking sets in, with elimination of more ammonia and formation of a white, amorphous, water-insoluble powder with the empirical formula $[OP(NH)_{1.5}]_x$.[13] This may be based on the repeating unit

$$\begin{array}{cc} O & O \\ \| & \| \\ -P-NH-P-NH- \\ | & | \\ NH & NH \\ | & | \\ -P-NH-P-NH- \\ \| & \| \\ O & O \end{array}$$

At 600° *in vacuo* it decomposes giving $(OPN)_x$, phosphorus oxonitride, in a form amorphous to X-rays.[11,24] The amorphous material is reported to crystallize when heated at 650–700° in ammonia,[24] but Boden, Sadowski, and Lehmann[3] have pointed out that the X-ray diffraction pattern of the presumed crystalline $(OPN)_x$ has features in common with that of their γ-P_3N_5. Anyway, prolonged action of ammonia on $(OPN)_x$ at 850° gives rise to P_3N_5.[24] In the course of conversion of $(OPN)_x$ into P_3N_5 by heating in ammonia, other X-ray diffraction patterns arise which have been attributed to $OPN \cdot P_3N_5$ and $OPN \cdot 2P_3N_5$.[24]

New high-temperature syntheses of $(OPN)_x$ from phosphoric acids or phosphates and urea or other amino compounds have been patented.[4]

Phosphorus oxonitride, like the nitrides, is resistant to hydrolysis, but water acts on it slowly at 175°. At 800°, $(OPN)_x$ with chlorine gives $POCl_3$ and, in contrast to the nitrides, only a trace of chlorophosphazenes.[13]

B. Phosphorus thionitride, $(SPN)_x$

Glatzel's original synthesis of this compound (1893) is still essentially the only one. An intimate mixture of ammonium chloride and P_4S_{10} is heated to 300°:[19]

$$nP_4S_{10} + 4nNH_4Cl \xrightarrow[300°]{2\,h} 4(SPN)_n + 6nH_2S + 4nHCl$$

Phosphorus thionitride is an insoluble grey-white solid. The X-ray diffraction pattern has been reported, but the compound is probably amorphous. Its i.r. spectrum contains a diffuse band at 1150 cm^{-1} in the general region of ν_{as} for $-P-N=P-$.[19]

The thionitride does not melt, but decomposes at about 600°. It is more reactive than the nitrides and oxonitride. Thus hydrogen at 450° gives H_2S, NH_3, and PH_3 among other products.[19] Chlorine at 250° gives, in good yield, a mixture of $(NPCl_2)_3$ and higher $NPCl_2$ polymers, as well as $SPCl_3$.[19] An interesting reaction with $AlCl_3$ at 350°[19] gives an adduct $SPN \cdot AlCl_3$ which is stable up to at least 700° and unaffected by water; it is cleaved by dry HCl to $SPCl_3$ and $AlCl_3 \cdot NH_3$, suggesting that the donor atoms in the SPN adduct are nitrogens. This adduct recalls complexes of $(NPCl_2)_3$ with $AlCl_3$ (Chapter 12, Sec. VII.H).

V. Desirable lines of work in this area

The different crystalline phases obtained by heating amorphous $(P_3N_5)_x$ and $(OPN)_x$ in ammonia may not all have been correctly identified, and obviously need better characterization. Complete structure determinations are badly needed, especially one for hexagonal P_3N_5. Better infrared spectra and Raman spectra, of all the compounds mentioned in this chapter, on purer samples than hitherto, might well give good information about structures.

It may be possible to develop a reaction chemistry and derivative chemistry of these compounds, with points of contact with more conventional heterocyclic chemistry. The inertness of the nitrides and the oxonitride is, of course, a difficulty, but their reaction with water at 200° suggests that, by autoclaving with other suitable liquid reagents, heterocyclic fragments might be extracted from them. The thionitride might react similarly but more easily. The sodium and potassium salts of phospham should be capable of exploitation in many syntheses, and no doubt similar metal salts with similar possible applications could be made from the crosslinked condensation products of phosphoryl triamide.

References

1. Baltkaula, A. and Millers, T., *Latv. PSR Zinat. Akad. Vestis, Kim. Ser.* 389 (1971)
2. Becke-Goehring, M. and Schulze, J., *Chem. Ber.* **91**, 1188 (1958)
3. Boden, G., Sadowski, G. and Lehmann, H.-A., *Z. Chem.* **11**, 114 (1971)
4. Borisov, E. V. and Nifant'ev, E. E., *Russ. Chem. Rev.* **46**, 842 (1977)
5. Fedorov, V. I., Krasnokumski, Yu I., Gans, S. N. and Parkomenko, V. D., *Zh. Priklad. Khim.* (Leningrad) **48**, 1656 (1975)
6. Gingerich, K. A., *J. Phys. Chem.* **73**, 2734 (1969)
7. Goubeau, J. and Pantzer, R., *Z. Anorg. Allg. Chem.* **390**, 25 (1972)
8. Hendrickson, D. N., Hollander, J. M. and Jolly, W. L., *Inorg. Chem.* **8**, 2642 (1969)
9. Huffman, E. O., Tarbutton, G., Elmore, K. L., Cate, W. E., Walters, H. K., Jr. and Elmore, G. V., *J. Am. Chem. Soc.* **76**, 6239 (1954)
10. Huffman, E. O., Tarbutton, G., Elmore, G. V., Smith, A. J. and Rountree, M. G., *J. Am. Chem. Soc.* **79**, 1765 (1957)
11. Klement, R. and Koch, O., *Chem. Ber.* **87**, 333 (1954)
12. Klement, R., in Brauer's *Handbook of Preparative Inorganic Chemistry*, 2nd ed., vol. 1, p. 574, Academic Press, New York and London (1963)
13. Mellor, J. W., *A Comprehensive Treatise on Inorganic and Theoretical Chemistry*, vol. VIII, Longmans, London (1928), and Supplement III to vol. VIII, Longmans, London (1971)
14. Miller, M. C. and Shaw, R. A., *J. Chem. Soc.* 3233 (1963)
15. Moldenhauer, W. and Dörsam, H., *Ber.* **59b**, 926 (1926)
16. Moureu, H. and Rocquet, P., *Bull. Soc. Chim. France* 1801 (1936)
17. Moureu, H. and Wetroff, G., *Compt. rend.* **204**, 51 (1937)
18. Moureu, H. and Wetroff, G., *Compt. rend.* **207**, 915 (1938)
19. Nabi, S. N., Nabi, Mrs. S. N. and Das, N. K., *J. Chem. Soc.* 3857 (1965)
20. Pelavin, M., Hendrickson, D. N., Hollander, J. M. and Jolly, W. L., *J. Phys. Chem.* **74**, 1116 (1970)
21. Sowerby, D. B. and Audrieth, L. F., *Chem. Ber.* **94**, 2670 (1961)
22. Steger, E., *Chem. Ber.* **94**, 266 (1961)

23. Sullivan, J. M., *Inorg. Chem.* **15**, 1055 (1976)
24. Sveic, D., Miller, T., Vitola, A. and Ozolins, G., *Latv. PSR Zinat. Akad. Vestis, Kim. Ser.* 406 (1974)
25. Uy, O. M., Kohl, F. J. and Carlson, K. D., *J. Phys. Chem.* **72**, 1611 (1968)
26. van Wazer, J. R., *Phosphorus and its Compounds*, vol. I, Interscience, New York (1958)

Appendix

Nomenclature proposals for inorganic heterocyclics

Nomenclature was discussed at the second meeting of the First International Symposium on Inorganic Ring Systems (Madrid, 1977) and at the Second International Symposium (Göttingen, 1978), where certain suggestions were made to the Inorganic Nomenclature Commission of IUPAC. No definitive scheme has yet emerged, but a consensus seems to be developing along the following lines:

(1) The Hantzsch–Widman system will be dropped, and inorganic rings clearly designated by a prefix (e.g. "cyclo") or suffix (e.g. "cycle" or "ring") which will be the same for rings of all sizes.

(2) Names will be *additive* as opposed to *substitutive*, simply listing the ring atoms and ligands with appropriate locant numbers, and not treating ring compounds as derived from real or hypothetical parent substances.

(3) In consequence of (2), it will be possible to dispense with any indication of oxidation states, ligancy numbers (λ), or unsaturation in the names.

(4) Ionic charges will be given by means of Ewens–Bassett numbers, and special ion terminations such as "onium" will probably be scrapped as superfluous.

Such a new scheme would represent a great and logical simplification compared with the present *Chemical Abstracts* system, but would be completely out of line with conventional organic nomenclature. There is some hope of bringing the inorganic and organic nomenclatures into line with each other through the "nodal nomenclature" system recently proposed by Lozac'h, Goodson, and Powell [*Angew. Chem. Int. Ed. Engl.* **18**, 887 (1979)], which, like the scheme just outlined, is essentially additive as far as molecular skeletons are concerned.

General index

Amidosulfinic acid, *see* Amidosulfurous acid
Amidosulfurous acid, 52
  acyl halides, 55
  organic derivatives, 55
Ammonia, reactions with
  P(III) halides, 188
  P(V) halides, oxohalides, and thiohalides, 197–199
  phosphorus nitrides, 253
  phosphorus sulfides, 174, 253
  sulfur chlorides, 18–20
  sulfur dioxide, 53–54
  sulfur nitrides, 107–108, 124
  sulfur trioxide, 68–69
  sulfuryl halides, 63–64
  thionyl chloride, 43–44, 150
  trithiazyl trichloride, 124
Aromaticity
  in phosphazene rings, 244–246
  in sulfur–nitrogen rings, 159–161

Birdcage compound, *see* Phosph(III)azanes
Bis(heptasulfurimido) sulfoxide $(S_7N)_2$SO, *see* Sulfur imides, cyclic
Bis(thionylimino) sulfur, *see* Thionyl imide, S(II) derivative
Bis(trifluoromethyl)sulfimide and Li derivative, 57
Bis(trimethylsilyl) sulfur diimide, *see* Sulfur diimide
Bis(trimethylstannyl) sulfur diimide, *see* Sulfur diimide

Bonds, P–N, in
  phosph(III)azanes, 187, 191
  phosph(V)azanes, 186–187, 199–200, 203, 207
  phosphazenes, 237–246
Bonds, P–S, 166
Bonds, S–N, 152–165
  aromaticity in S–N rings, 159–161
  cluster treatment of S–N cages, 161–164
  energy–order correlation, 154
  length-force constant correlation, 153
  length–order correlation, 153
  lengths and angles in saturated rings, 156
  lengths and angles in unsaturated rings, 157–159
  transannular bonding in S–N rings, 164
Bonds, transannular, 164, 245

Chemical Abstracts system, *see* Nomenclature
Cluster treatment of S–N cages, 161–164
Conformation, *see* Phosphazenes, cyclic, molecular structure; Sulfur imides; Sulfanuric derivatives, isomerism
Cyclopentathiazenium cation $S_5N_5^+$ and salts, 120, 121
Cyclophosphazane-sulfimide, *see* Cyclosulfimide-phosphazane
Cyclosulfimide-phosphazane mixed saturated rings
  4-membered, with S(IV) and P(V), 58

GENERAL INDEX

4-membered, with S(VI) and P(V), 76
  spiro, with S(VI) and P(V), 76, 77
Cyclosulfimide-phosphazene mixed unsaturated rings with S(VI), 74
Cyclosulfimides
  mixed S(VI) and S(IV), 74
  regular alternant, see Sulfamide, cyclic and linear polymers
Cyclotetrathiazenium cation, see $S_4N_4^{2+}$
Cyclotrithiazyl anion, see $S_3N_3^-$

Diazadiphosphetidines, see Phosph(III)azanes and Phosph(V)azanes, cyclic dimers
Dilution technique, 10, 31–33, 36
Dimethylsulfimide, N-trimethylsilyl-, synthesis agent, 57
Dinitrogen disulfide, see Disulfur dinitride
Dinitrogen tetrasulfide, see Tetrasulfur dinitride
Disulfur dinitride
  addition to Lewis acids, 111, 112
  polymerization, 111
  preparation and properties, 111
  reaction with ammonia, 112
  structure, 111
Dithiadiazete, see Disulfur dinitride
Dithiadiphosphetanes, see P–S rings with organo ligands, P_2S_2 rings
Dithiatetrazine, dithiatetrazetidine, see Sulfur hydrazides, cyclic
Dithiatriphospholanes, see P–S rings with organo ligands, P_3S_2 rings

Heptasulfur imide, see Sulfur imides, cyclic
Heptathiazocine, see Sulfur imides, cyclic
Heptathioimide, see Sulfur imides, cyclic
Hexanitrogen pentasulfide, see Pentasulfur hexanitride
Hexasulfur diimides, see Sulfur imides, cyclic saturated, with 8-membered rings
Hexathiadiazocine, see Sulfur imides, cyclic saturated, with 8-membered rings
Hexathiadiphosphocane, see Phosphorus sulfide bromides

Hexathiodiimides, see Sulfur imides, cyclic saturated, with 8-membered rings
Hydrazido compounds, cyclic, see Phosph(III)azanes; Phosph(V)azanes; Sulfur hydrazides

Imidodisulfinamide, 56
Imidodisulfinic acid, see Imidodisulfurous acid
Imidodisulfurous acid, salts, 54
Inorganic rubber, see Phosphazenes, elastomers
International Union of Pure and Applied Chemistry, nomenclature discussions, 259

Kinetic control, see Synthesis
Kirsanov reactions, 72, 75, 197–198

Mercury–nitrogen–sulfur compounds
  of S(IV), see Thionyl imide, Hg derivative
  of S(VI), 79

Nitrilohexaphosphonitrilic chloride, see Phosphazenes, fused-ring compound $P_6N_7Cl_9$
Nitrogen system (Franklin), 41–43, 60–61, 185–186, 193–194
Nomenclature, 4–8, 258

Pentasulfur hexanitride, 125
Pentasulfur triimides, see Sulfur imides, cyclic saturated, with 8-membered rings
Pentathiatriazocine, see Sulfur imides, cyclic saturated, with 8-membered rings
Pentathiazyl cation $S_5N_5^+$, see Cyclopentathiazenium
Pentathiotriimides, see Sulfur imides, cyclic saturated, with 8-membered rings
Perthionitrate anion NS_4^-, 125
Phosphams, 254–255
Phosph(III)azanes, 187–196
  birdcage compound $P_4(NMe)_6$, 192–195

cyclic dimers, preparation, structure, and properties, 189–192
cyclic trimers and tetramers, 192
dimeric, stability, 187, 191
high polymer, 196
hydrazido compounds, 195–196
monomeric, 187, 191
Phosph(v)azanes, 197–211
 cage compounds, 208–209
 cage phosphazane-phosphazene anion $P_{12}N_{14}S_{12}{}^{6-}$, 210–211
 cyclic dimers, preparation, structure, and properties, 199–205
 cyclic trimers and tetramers, 205–207
 dimeric, stability, 186, 204
 hydrazido compounds, 209–210
 spiro compounds, 76–77, 206–209
Phosphazenes, 214–249
 cancer therapy with, 249
 catechol and similar derivatives as clathrating agents, 231
 coupled-ring compound, 225
 cyclic, alkyl derivatives, 217, 232–235
 cyclic, amino derivatives, 231–232
 cyclic, aryl derivatives, 234, 237
 cyclic, bonding, 242–246
 cyclic, as Brønsted bases, 218–219, 234–235
 cyclic, complexes with Lewis acids, 234–236
 cyclic, complexes with Lewis bases, 227–228
 cyclic, electrolytic reduction, 237
 cyclic, fluorination, 228–229
 cyclic, general properties, 217
 cyclic, hydrolysis, 230
 cyclic, isomerism, 226–227
 cyclic, isotopic halogen exchange in, 228
 cyclic, molecular structure, 237–242
 cyclic, OH, OR, and SR derivatives, 229–231
 cyclic, preparation, 218–221
 elastomers, 222–223
 enthalpy of formation, 244–245
 fused-ring compound $P_6N_7Cl_9$, 224, 229, 241
 high polymers, synthesis and applications, 221–223
 history, 216
 ionization potentials, 246
 linear oligomers, 223–224
 literature, 216–217
 spiro compounds, 224–225
 structural types, 214–215
 u.v. absorption spectra, 245–246
Phosphazene-sulfanuric mixed rings, see Sulfanuric-phosphazene
Phosphonitrilic halides and derivatives, see Phosphazenes
Phosphorus nitrides, 251–254
 cyclic phosphazenes from, 253, 254
 monomeric PN, 251–252
 mononitride, polymeric solid, 252–253
 triphosphorus pentanitride, 253–254
Phosphorus oxides
 analogy to P–N cages, 185, 194
 analogy to P–S cages, 168–170
Phosphorus oxonitride, 255–256
Phosphorus oxosulfide, 175
Phosphorus sulfide bromides, 177
Phosphorus sulfide halides and pseudo-halides, 176–177
Phosphorus sulfide iodides
 preparation, 176
 α-$P_4S_3I_2$ as ligand, 176–177
 reactions, 176–177
 structure, 176
Phosphorus sulfides, 166–177
 phase diagram 167
 polymeric, 181–182
 preparation and properties, 168–172
 P_4S_3 as ligand, 172–173
 reactions, 172–175
 structure, 170–171
 uses, 175
Phosphorus thionitride, 256
Phosphorus tri-N-methylimide, see Phosph(III)azanes, birdcage compound
P–N rings, saturated, see Phosph(III)-azanes; Phosph(v)azanes
P–N rings, unsaturated, see Phosphazenes
P–S rings with organo ligands
 P_2S_2 rings, 180–181
 P_3S_2 rings, 178
 P_3S_3 rings, 178
 P_4S rings, 179

GENERAL INDEX

Ruggli–Ziegler dilution technique, see Dilution technique

S–N chains, unsaturated, 108–110
$S_3N_2^{2+}$ cation, 123
$S_3N_2^+$ cation and salts, 122
S_3N_2 ring, covalent derivatives, 141–143
$S_3N_2Cl^+$ cation and salts, 139–140
$S_3N_3^-$ anion and salts, 123
$S_4N_3^+$ cation, see Thiotrithiazyl
$S_4N_4^{2+}$ cation and salts, 122
$S_4N_5^-$ anion, 124
$S_4N_5^+$ cation, 124
$S_5N_5^+$ cation, see Cyclopentathiazenium
$S_6N_4^{2+}$ cation and salts, 121, 122
Sulfamide
  cyclic and linear polymers, 67–73
  metal derivatives, 64
  N-halogeno derivatives, 65
  organic derivatives, 65
  preparation and properties, 63
  trimethylsilyl derivatives, 65
  urea, analogy with, 61–62
Sulfanuric anion $N_3S_3O_4F_2^-$, 89
Sulfanuric chloride, see also Sulfanuric halides
  adducts with Lewis bases, 87
  isomers, 84, 85
  preparation and properties, 83, 84
  structure, 85
Sulfanuric chlorofluorides, 87
Sulfanuric derivatives, isomerism, 89
Sulfanuric fluoride
  high polymer, 86
  tetrameric, 86
Sulfanuric fluorides, trimeric, see also Sulfanuric halides
  preparation and properties, 85
Sulfanuric halides
  alkyl and aryl derivatives, 89
  amino derivatives, 88, 89
  hydrolysis, 88
  nucleophilic substitution, 87–89
  reactions, 87
  structural types, 82
Sulfanuric-phosphazene mixed rings
  derivatives, 95
  nucleophilic substitution, 95
  preparation and properties, 94, 95
  8-membered, 96

Sulfanuric ring compounds
  containing also S(IV), 92, 93, 148
  with saturated ring segment, 94
Sulfimide, monomeric, sulfur(VI), 62
Sulfimides, cyclic oligomers, see Sulfamide, cyclic and linear polymers
Sulfodiimides, S,S-dialkyl-, 77
  halogeno derivatives, 79
  preparation and properties, 78
  as synthesis agents, 78, 79
  trimethyltin derivative, 79
Sulfur(II) amides, 18
Sulfur diimide, 49
  bis(trifluoromethylsulfenyl), 52
  bis(trimethylsilyl), synthesis intermediate, 50
  bis(trimethylstannyl), synthesis intermediate, 52
  halogeno derivatives, 52
  monosilyl derivative, 50
  organo derivatives, 49
Sulfur dioxide
  adducts with ammonia and amines, 53, 55
  nitrogen-system derivatives, 42
Sulfur hydrazides, cyclic, 37
  derivatives with 6-membered rings, 38
  derivatives with 8-membered rings, 38
Sulfur imides, cyclic saturated, with 8-membered rings
  anions and use as intermediates, 24
  boron derivatives, 24, 25
  Brønsted acid properties, 23
  Brønsted or Lewis base properties, 24
  conformational stability, 35
  general properties, 21
  history, 16
  metal derivatives, 25, 27
  molecular structure, 22
  nitride-imides from, oligomeric and polymeric, 34
  of S(IV), see Thionyl imide
  of S(VI), see Sulfamide, cyclic and linear polymers
  organic derivatives, 28
  oxidation, 23
  phenylmercury derivatives and use as intermediates, 28
  phosphorus derivatives, 24
  physical properties, 21

preparation, 18–21
reactions with halogens and halogen compounds, 23
reduction, 23
silicon, tin, and lead derivatives, 25
stability and decomposition, 23
sulfur nitride oxide $(S_7N)_2SO$ from, 33
Sulfur imides, cyclic saturated, with 6-membered rings
 organic derivatives of S_2N_4 ring, 38
 organic derivatives of S_4N_2 ring, 36
Sulfur imides, cyclic saturated, with 12-membered rings, 36, 37
Sulfur(IV) imides and amides, classification and relationships, 42, 43
Sulfur(VI) imides and amides, classification and relationships, 60, 81
Sulfur nitride halides, see Thiazyl halides; $S_3N_2Cl^+$; $S_6N_4^{2+}$; Thiotrithiazyl
Sulfur nitride, polymeric, see Thiazyl polymer
Sulfur nitrides, saturated
 coupled-ring nitrides of the series $(S_7N)_2S_x$, 29
 fused-ring nitride $S_{11}N_2$, 31
 fused-ring nitrides other than $S_{11}N_2$, 33
Sulfur nitrides, unsaturated, see Disulfur dinitride; Pentasulfur hexanitride; Tetrasulfur dinitride; Tetrasulfur tetranitride; Thiazyl monomer
Sulfur nitride-imides, cyclic saturated, see Sulfur imides, cyclic saturated, with 8-membered rings
Sulfur nitride S-oxides, 145–151
 S_3N_2 ring, S_3N_2O, 148–149
 $S(NSO)_2$, see Thionyl imide
 S_3N_3 ring, $S_3N_3O_2Cl$, 93, 148
 S_3N_3 ring, ion $S_3N_3O_4^-$, 148
 S_4N_4 ring, $S_4N_4O_2$, 146
 S_4N_4 ring, $S_4N_4O_4$, 146
 S_4N_4 cage, $S_4N_5O^-$ ion, 150
 $(S_7N)_2SO$, see Sulfur imides, cyclic saturated, with 8-membered rings
 unknown structure, $S_3N_2O_5$, 150
Sulfur–nitrogen cages, 123, 150, 161–164
Sulfurous acid, nitrogen-system derivatives, 43

Sulfuryl diamide, see Sulfamide
Synthesis methods, 8–14
 addition polymerization, 9
 direct combination, 9
 heterofunctional condensation, 9
 homofunctional condensation, 12–13
 metal–nitrogen and metal–sulfur compounds in, 12
 ring expansions and contractions, 13
 silylamines and stannylamines in, 10–11
 thermodynamic vs. kinetic control, 9–10

Tetranitrogen tetrasulfide, see Tetrasulfur tetranitride
Tetrasulfur dinitride
 preparation, 115
 reactions, 117
 structure, 115–117
Tetrasulfur tetraimide, see Sulfur imides, cyclic saturated, with 8-membered rings
Tetrasulfur tetranitride
 adducts with Lewis acids, 105
 adducts with olefins, 103
 anion, 104
 cation, see $S_4N_4^{2+}$
 hydrolysis, 107
 metal complexes from, 105, 107
 oxidation, 103, 104
 preparation, 99–101
 properties, 101
 reactions with ammonia and amines, 107, 108
 reactions with halogens, 103
 reduction, 104
 ring-opening by organic nucleophiles, 108–110
 structure, 102
 thermal dissociation, 103
Tetrathiadiazine, see Tetrasulfur dinitride
Tetrathiatetrazocine, see Sulfur imides, cyclic saturated, with 8-membered rings; Tetrasulfur tetranitride
Tetrathiazyl dichloride, see Thiazyl halides
Tetrathiazyl tetrachloride, see Thiazyl halides

Tetrathiazyl tetrafluoride, *see* Thiazyl halides
Tetrathiotetraimide, *see* Sulfur imides, cyclic
Tetrazadiphosphorine, *see* Phosph(III)-azanes *and* Phosph(V)azanes, hydrazido compounds
Tetrazatetraphosphocine, *see* Phosphazenes
Thiazenium cation SN^+, 125
Thiazyl cation, *see* Thiazenium
Thiazyl halides, 131–138
  cyclic tetramers $(NSF)_4$ and $(NSCl)_4$, 131–133
  cyclic trimers $(NSF)_3$ and $(NSCl)_3$, 133–136
  derivatives, 137–138
  monomers NSF, NSCl, NSBr, 136–137
  tetrathiazyl dihalides $N_4S_4X_2$, 133, 144
  trithiazyl monohalides $(NS)_3X$, 136
Thiazyl monomer, 114
Thiazyl polymer $(SN)_x$, 112–114
  electrical properties, 113
  preparation, 112
  reactions, 114
  structure, 113
Thiodithiazyl cation, *see* $S_3N_2^{2+}$
Thiodithiazyl chloride, *see* $S_6N_4^{2+}$
Thiodithiazyl dichloride, *see* $S_3N_2Cl^+$
Thiodithiazyl difluoride $S_3N_2F_2$, 140–141
Thiodithiazyl dioxide, *see* Thionyl imide, S(II) derivative
Thionophosphine sulfides, *see* P–S rings with organo ligands, P_2S_2 rings
Thionyl diamide, 56
  organic derivatives, 57
Thionyl imide, 43
  CF_3S derivative, 48
  $CF_3S(O)$ derivative, 48
  $CF_3S(O_2)$ derivative, 48
  $FS(O_2)$ derivative, 48
  halogeno derivatives, 47
  isomer, red, 46
  mercury derivative, 47
  molecular structure, 44
  organic derivatives, 46
  polymers, cyclic, 44
  polymers, linear, 44
  preparation, 43
  properties, 44
  S(II) derivative $S(NSO)_2$, bis(thionylimino)sulfur, 48
  trimethylsilyl derivative (synthesis intermediate), 46
Thiophosphonic acid anhydrides, *see* P–S rings with organo ligands, P_2S_2 rings
Thiotrithiazyl cation $S_4N_3^+$ and salts
  preparation, 117–119
  properties, 119
  reactions, 119, 120
  structure, 119
Thiotrithiazyl chloride, *see* Thiotrithiazyl cation and salts
Triazatriphosphorine, *see* Phosphazenes
Trimethylsilyl sulfinylamine, *see* Thionyl imide, trimethylsilyl derivative
Trithiadiazole cation, *see* $S_3N_2^+$; $S_3N_2^{2+}$
Trithiadiazole ring, covalent derivatives, *see* S_3N_2 ring
Trithiadiphospholane, *see* P–S rings with organo ligands, P_3S_2 rings
Trithiatriazine, saturated ring, *see* Sulfamide, cyclic and linear polymers
Trithiatriazine, unsaturated ring, *see* Sulfanuric
Trithiazyl trichloride, *see* Thiazyl halides
Trithiazyl trifluoride, *see* Thiazyl halides

Ring index

Principles of arrangement: general

This index is intended as a guide to all the *inorganic* ring systems in the book. For the purposes of counting rings and specifying their composition, any organic rings or organic chelating groups present in a structure are disregarded. For example, phenyl ligands attached to an inorganic ring are not counted as rings, though they are included in the count of ligands.

The *ring formulas* below give the atoms of the inorganic ring system only, exclusive of ligands.

To a large extent, the present index follows the organization of the Ring Indexes of *Chemical Abstracts*. However, when (as often happens) several ring systems have the same ring formula, it seems undesirable to follow the *Chemical Abstracts* practice of listing them in alphabetical order of name, because of the current confusion over nomenclature in this field. Instead, they have been classified by a simple method explained below.

How to find a ring system or compound

(1) Count the rings in the molecular formula, and search for the appropriate ring size under "one-ring systems", "two-ring systems", etc., as the case may be. Ring sizes are shown in heavy type.

Where two or more inorganic rings are coupled together with no atoms in common, *each* ring will be found listed under "one-ring systems", the other ring(s) being counted as ligands.

In fused-ring, bridged-ring, or spiro systems, where rings can be counted in alternative ways, count the *smallest* possible number of *smallest* rings.

(2) Within the section of the index just located, find the formula of your ring. Formulas are listed as in *Chemical Abstracts*, in alphabetical order and in order of increasing number of atoms of each kind (the Hill system). All the ring atoms of a fused-ring, bridged-ring, or spiro system are lumped together for the entry under "ring formula".

RING INDEX

(3) When, as often happens, several ring systems have the same ring formula, they are listed in order of *increasing total number of exocyclic ligands* on the ring system. A doubly bonded ligand (=O, =S) counts as one ligand. In fused-ring, bridged-ring, or spiro systems, an atom attached to a ring is not counted as a ligand if it forms part of another ring.

(4) When two or more ring systems have both the same ring formula and the same number of ligands, a note of "further characterization" fully identifying each ring system will be found in the last column of the index.

One-ring systems

4

Ring formula	Number of ligands	Page	Further characterization
NP_2S	5	180	
N_2PS	5	58	
N_2PS	6	58	
N_2PS	7	76	
N_2P_2	4	189	
N_2P_2	5	190	
N_2P_2	6	190, 201, 202	
N_2P_2	8	199	
N_2S_2	0	111	
N_2S_2	2	112	N_2S_2 adducts
N_2S_2	2	138	$(NSCF_3)_2$
N_2S_2	6	74	
P_2S_2	4	180	

5

Ring formula	Number of ligands	Page	Further characterization
AsN_2S_2	1	51	
N_2PbS_2	1	108	
N_2S_3	0	122	
N_2S_3	1	139, 140, 148	$S_3N_2Cl^+$ and derivatives
N_2S_3	1	115, 142	hypothetical S_4N_2 structure
P_3S_2	4	178	
P_3S_2	6	179	
P_4S	4	179	

6

Ring formula	Number of ligands	Page	Further characterization
N_2S_3O	4	133, 150	
N_2S_4	0	115	
N_2S_4	2	36	
N_3PS_2	6	94	
N_3P_2S	6	94	
N_3P_2S	7	75	two structural isomers

Ring formula	Number of ligands	Page	Further characterization
6 (*cont.*)			
N_3P_3	6	205	saturated phosph(III)azane
N_3P_3	6	142, 214ff	unsaturated phosphazene
N_3P_3	8	206	
N_3S_3	1	110	Ph_3PN ligand
N_3S_3	1	136	possible compound with Cl ligand
N_3S_3	3	104, 133	thiazyl halide trimers and esters
N_3S_3	3	147	sulfur nitride oxide
N_3S_3	4	148	sulfur nitride oxide anion
N_3S_3	4	93, 148	sulfur nitride chloride oxide
N_3S_3	4	93	sulfanuric with S(IV) segments
N_3S_3	5	74	sulfimide with unsaturated segment
N_3S_3	5	92	sulfanuric with S(IV) segment
N_3S_3	6	83ff	
N_3S_3	9	67ff	
N_4P_2	6	196	
N_4P_2	8	210	
N_4S_2	4	38	
P_2S_4	4	177	
P_3S_3	6	178	
7			
N_3S_4	0	117	
8			
$As_2N_4S_2$	2	51	
NS_7	1	18ff	
N_2S_6	2	18ff	three sulfur imide isomers and derivatives
N_2S_6	2	38	hydrazido ring
N_3S_5	3	18ff	two sulfur imide isomers and derivatives
$N_4P_2S_2$	6	138	
$N_4P_2S_2$	8	78, 138	
N_4P_4	8	192	saturated phosph(III)azane
N_4P_4	8	142, 214ff	unsaturated phosphazenes
N_4P_4	12	206	
N_4S_4	0	99ff	sulfur nitride
N_4S_4	0	122	sulfur nitride cation
N_4S_4	2	133, 144	sulfur nitride chloride and fluoride
N_4S_4	2	146	sulfur nitride oxide

RING INDEX

Ring formula	Number of ligands	Page	Further characterization
8 (*cont.*)			
N_4S_4	4	18ff	saturated sulfur imide and derivatives
N_4S_4	4	36	hydrazido ring
N_4S_4	4	131ff	$(NSF)_4$ and derivative
N_4S_4	4	146	sulfur nitride oxide
N_4S_4	4	146	two O and two $SnMe_3$ ligands
N_4S_4	6	78, 138	
N_4S_4	8	44	thionyl imide tetramer
N_4S_4	8	86	sulfanuric tetramer
N_4S_4	12	67ff	
10			
N_5P_5	10	142, 214ff	
N_5P_5	15	207	
N_5S_5	0	120	
12			
N_4S_8	4	37	
N_6P_6	12	214ff	
N_6P_6	18	207	
N_6S_6	18	67ff	
14			
N_7P_7	14	214ff	
16			
N_8P_8	16	214ff	

Two-ring systems

4,4

Ring formula	Number of ligands	Page	Further characterization
N_4PS_2	9	76	
N_4P_3	11	209	

Ring formula	Number of ligands	Page	Further characterization
5,5			
N_5P_2	5	196	
P_4S_3	2	176	
6,6			
N_5PS_3	2	51	
N_5S_4	0	124	
N_5S_4	1	150	
N_6P_2	6	195	
N_6P_2	8	210	
N_6P_5	9	225	
6,8			
N_6P_4S	6	225	
8,8			
N_2S_{11}	0	31	
N_6S_5	0	125	

Three-ring systems

Ring formula	Number of ligands	Page	Further characterization
3,5,5			
P_4S_3	0	167ff	
4,4,4			
N_4P_3	10	209	
N_4P_3	11	209	
$N_6P_2S_2$	12	77	
N_6P_4	14	207	
4,5,5			
N_4S_6	0	121	
P_4S_4	0	167ff	
P_4S_4	1	167ff	

RING INDEX

Ring formula	Number of ligands	Page	Further characterization
5,5,5			
P_4S_4	0	167ff	
5,5,6			
P_4S_5	0	167ff	
P_4S_5	2	167ff	
6,6,6			
N_6P_4	6	192ff	
N_6P_4	7	193	
N_6P_4	8	193	
N_6P_4	9	193	
N_6P_4	10	193	
N_7P_6	9	224	
P_4S_6	3	167ff	
P_4S_6	4	167ff, 175	

Four-ring systems

4,4,4,4

Ring formula	Number of ligands	Page	Further characterization
N_4P_4	16	209	
N_6P_4	14	209	

Twelve-ring system*

6,6,6,6,6,6,6,6,6,6,6,6

Ring formula	Number of ligands	Page	Further characterization
$N_{14}P_{12}$	12	210	

* Most easily visualized as a "sphere" made up of twelve fused rings, though fewer than twelve *ring-closures* would be required to construct it (cf. preamble to this index).

THIS BOOK IS DUE ON THE LAST D
STAMPED BELOW
APR 2 4 1981

KS REQUESTED BY ANOTHER BO
SUBJECT TO RECALL AFTER O
RENEWED BOOKS ARE SUBJEC
IMME

AUG
APR 26
SEP 11
APR 23 '85
NOV 7 '85